Assignment Methods in
Combinatorial Data Analysis

STATISTICS: Textbooks and Monographs

A SERIES EDITED BY

D. B. OWEN, Coordinating Editor

Department of Statistics
Southern Methodist University
Dallas, Texas

OTHER VOLUMES IN PREPARATION

Assignment Methods in Combinatorial Data Analysis

LAWRENCE J. HUBERT

University of California, Santa Barbara
Santa Barbara, California

MARCEL DEKKER, INC.   New York and Basel

Library of Congress Cataloging-in-Publication Data

Hubert, Lawrence
   Assignment methods in combinatorial data analysis.

   (Statistics, textbooks and monographs ; v. 73)
   Includes bibliographies and index.
   1. Combinatorial analysis. 2. Linear models
(Statistics) I. Title. II. Series.
QA164.H83   1986   511'.6   86-19608
ISBN 0-8247-7617-8

MARCEL DEKKER, INC.
270 Madison Avenue, New York, New York 10016

Current printing (last digit):
10 9 8 7 6 5 4 3 2 1

PRINTED IN THE UNITED STATES OF AMERICA

Preface

Statisticians who consult on problems of data analysis regularly face some variant of the following situation. A client arrives with a data set already collected and requests advice regarding the most appropriate method of analysis. For convenience, we assume the data have a simple two-independent group structure and that the client is concerned with a difference between means. Unfortunately, the data were generated without an initial random assignment to the groups or a random sampling from two independent populations. Thus, the two standard models that would justify the use of a typical statistical method, such as the t-test, are missing. At this point, the statistician is placed in somewhat of a dilemma. Because a recommendation for analysis must be made, the easiest alternative would be to suggest the usual t-test or possibly some comparable nonparametric alternative. For us, however, the real question of interest concerns the meaning, if any, such a test actually has in the absence of a context that could provide one of the usual formal justifications.

There is an interpretation for a significance test in a nonexperimental context that may help alleviate a few of the misgivings a statistician might have after suggesting the use of a standard inference strategy. A t-test, for example, provides a reasonable approximation to the results obtained from a randomization test, i.e., a procedure that would relate the observed mean difference between the two groups to all possible two-group splits of the same size formed from the given data. A reference distribution is constructed by considering all possible mean differences to be equally likely. The observed difference for the way the data were actually subdivided can be compared to this reference distribution and a significance level constructed. The latter is defined as the proportion of mean differences from the reference distribution that are as extreme or more so than the observed difference.

Stated another way, the significance level has meaning for the given process of randomization irrespective of how the data actually arose. It is simply the probability of obtaining such an extreme result if the two groups were reconstructed by chance from the available data. If the significance level is fairly large, the researcher may be willing to argue that the observed difference is relatively unimportant because similar differences could often arise merely by randomly separating the available data into two groups. When the significance level is small, however, the researcher may be comfortable in assuming that the observed difference is real in the very explicit sense that it cannot be reconstructed very easily by simply subdividing the data at random. Whether the observed difference actually results in some causal manner from the given group structure cannot be answered unambiguously. All we know is that a real difference probably exists with regard to this process of randomization; the reason for the difference may be due to the group structure, or possibly, to some other factors that are related to it. This same point of view is discussed explicitly by Winch and Campbell (1969) in a general discussion of evaluating quasi-experiments by the process of successively eliminating rival hypotheses that could account for an observed mean difference.

This book carries the same logic over to a variety of different data analysis situations. In all cases, the concern is with evaluating some observed index in relation to a specific randomness conjecture that is defined only in terms of the obtained data, but which hopefully would be of value by itself irrespective of the particular data generation process. We make no pretense of doing much beyond this. However, for a few of the problems we address, particularly in the later chapters, this may be about the best that can be done given the complexity of the models that would have to be defined in a more classical framework.

At a more specific level, two comprehensive statistical paradigms are reviewed that, in addition to encompassing a wide range of existing nonparametric methods, suggest ways of analyzing data sets based on generic notions of proximity or "closeness" between objects. These related paradigms, referred to as the linear and quadratic assignment models, depend on simple indices of correspondence defined between object sets that have been matched in some a priori manner. Both linear and quadratic assignment, in this broad sense, can be interpreted as very general correlational strategies. Along with the attendant problems of significance testing through randomization and the construction of normalized descriptive indices that measure the degree of object set correspondence, a variety of extensions are discussed to multiple object sets and to higher-order assignment indices.

We start with a discussion of the linear assignment model in Chapter 1 and how it may be applied in several common data analysis contexts. This introduction then serves as a basis for the extensions in Chapters 2 and 3 to more complicated data analysis situations of comparing rectangular data matrices and multiple object sets. Chapter 4 introduces a second notion of assignment or matching based on pairs of objects. In turn, this idea provides the motivation for the generalizations in Chapters 5 to 7 that discuss, respectively, alternative indices of object set correspondence and higher-order assignment measures, multiple proximity matrices, and restrictions that may be imposed on the randomization scheme.

As far as background expected of the reader, we will assume, as a minimum, a thorough understanding of the material presented in a typical year-long sequence in applied statistics at the level of Hays (1981). This would include, in particular, the material on nonparametric statistics presented in Chapters 15 and 16. Some broad understanding of the techniques of multivariate analysis would also be helpful, especially in the later sections of the book.

For the reader's convenience, the sections of the book are marked by the letters (A), (B), and (C) to indicate the priority they might be given in a first reading. Sections marked (A) develop the major ideas and should be read to maintain reasonable continuity. Those marked (B) typically present interpretations or applications of these more general notions and should at least be skimmed. Finally, sections marked (C) could be considered optional, particularly in a first reading.

Lawrence Hubert

REFERENCES

Hays, W. L. Statistics, Holt, Rinehart & Winston, New York, 1981.
Winch, R. F., and Campbell, D. T. Proof? No. Evidence? Yes. The significance of tests of significance, The American Sociologist, 4 (1969), 140-143.

Contents

Assignment Methods in
Combinatorial Data Analysis

PART I
APPLICATIONS AND EXTENSIONS OF THE LINEAR ASSIGNMENT MODEL

1

The Linear Assignment Model

1.1 INTRODUCTION (A)[*]

Students who take elementary courses in applied statistics typically comment about their difficulty in determining when particular procedures
should be used and how various data analysis strategies relate to one
another. This latter task of integration can be solved, at least to some
extent, through the study of comprehensive statistical paradigms which
would include various special cases appropriate for different situations.
The best-known prototype is probably the general linear model (GLM), encompassing multiple regression, analysis of variance, and analysis of
covariance. When a student proceeds this far into the curriculum, the uni-
fication offered by the GLM can be extremely helpful. In fact, some may
even consider the late introduction of the GLM a serious pedagogical mis-
take because only at this point does the field become truly understandable
for them. Still, it is not at all obvious whether a top-down approach that
would start a novice out with the GLM is the most appropriate teaching
strategy to follow. There may be a need to have the special cases fairly
well understood before the real purpose of defining a more comprehensive
framework becomes apparent. In any event, these more inclusive methods
should be introduced at some point both to help a student organize his or
her knowledge of applied statistics and to suggest ways of approaching
future data analysis tasks that might not fit one of the standard textbook
examples.

A number of integrative structures similar to the GLM can be found
in the statistical literature. This chapter introduces one such paradigm
that includes a variety of so-called nonparametric strategies as special
cases. We do not begin with the most general scheme possible, however,

---

[*]The reader is referred to the Preface for an explanation of the use of (A),
(B), and (C) in text headings.

but at a somewhat more intermediate level by presenting what will be called the "linear assignment model." Once an understanding of the basic strategy is given, it will be extended and used as a building block, both in this chapter and later on. In the course of the discussion, various special techniques for particular data analysis situations become clear that might go unrecognized without the benefit of such an integrative framework.

1.2 BACKGROUND (A)

As a topic in optimization, the idea of linear assignment defines a classical subfield of operations research. Still, even though someone might be very familiar with this particular usage, it may not be at all obvious how the same linear assignment concept is also relevant to statistical inference. As a short summary that will be clarified and expanded on in the ensuing sections, the linear assignment function that is optimized in the operations research context is of exactly the same form that can be considered in a distributional framework relevant to a variety of data analysis situations. Thus, we try to identify the extremes of a distribution when our concern is optimization, whereas the actual distribution of function values is of interest when problems of statistical inference are being considered.

To provide a motivating historical context for the class of statistical methods discussed in this chapter, the notion of linear assignment (LA) is introduced in Section 1.2.1, not as a statistical problem but as one in optimization. Once some notation is established and this background developed sufficiently, the statistical implications of LA will be discussed in detail starting with Section 1.2.2 and continuing throughout the first chapter. We shall not return to the optimization aspects of LA after this initial introduction of Section 1.2.1, but instead concentrate solely on problems of data analysis that can be based on distributions associated with the linear assignment function.

1.2.1 The LA Model in Optimization (A)

In an optimization context, the LA task has a very long history both in psychometrics (Thorndike, 1950; Ward, 1958) and in the operations research literature (Garfinkel and Nemhauser, 1972). In its popularized form, we are given n people and n jobs and a numerical index of value or an assignment score, say c_{ij}, that results when the ith person is assigned or matched to the jth job. Presumably, these assignment scores are generated from aptitude tests or other ancillary information available on each person. If maximization of total value is the desired goal, then each person must be paired with or assigned a given job in such a way that the sum of all assignment scores for the n pairs is maximized (depending on how the quantity c_{ij} is keyed, it may be appropriate to phrase the optimization task as a minimization, but for convenience, a maximization task is assumed throughout this section).

Formally, the LA problem can be phrased in the format of (integer) linear programming. We wish to maximize the objective function or index of merit

$$\sum_{i,j} c_{ij} x_{ij}$$

subject to three constraints:

1. The unknown indicator, x_{ij}, is 1 or 0 depending on whether person 1 is assigned job j or not, respectively, $1 \leqslant i, j \leqslant n$.
2. $\Sigma_j x_{ij} = 1$, implying that each person is assigned to one and only one job, $1 \leqslant i \leqslant n$.
3. $\Sigma_i x_{ij} = 1$, implying that each job has one and only one person assigned to it, $1 \leqslant j \leqslant n$.

The optimization task of finding the appropriate set of dichotomous indicator functions can be approached by an application of the well-known simplex method for solving linear programs. Even though this technique ignores integer constraints on the indicator functions, the LA problem is so well structured that optimal solutions obtained without this restriction will still turn out to be integers. We note that because the total index of merit is a linear function of the unknown indicators, this assignment task is referred to as linear. For now, the qualifier could be omitted, but the distinction will become important again in Chapter 4, where a notion of quadratic assignment is introduced.

Example: As a simple example, suppose n is 4 and assume the following 4×4 matrix, $\underset{\sim}{C} = \{ c_{ij} \}$, represents the assignment scores:

		Job			
		1	2	3	4
	1	5	4	2	10
	2	14	9	13	7
Person	3	3	8	1	12
	4	15	6	11	16

The optimal solution is represented by a second 4×4 matrix, $\{ x_{ij} \} = \underset{\sim}{X}$, which picks out one and only one entry in each row and column of $\underset{\sim}{C}$ by the placement of its zeros and ones:

		Job 1	Job 2	Job 3	Job 4
	1	0	0	0	1
	2	0	0	1	0
Person	3	0	1	0	0
	4	1	0	0	0

Specifically, persons 1, 2, 3, and 4 are assigned to jobs 4, 3, 2, and 1, respectively, and the total value for the assignment is $10 + 13 + 8 + 15 = 46$.

Because this 4×4 problem is so small, there is a fairly easy way of convincing someone that the assignment represented by $\underset{\sim}{X}$ is optimal without resorting to a linear programming algorithm. Each assignment can be represented by a permutation of the first n integers, and thus, the optimization task can be solved merely by trying out all possibilities. For example, the optimal assignment or matching defines the permutation 4321, or more explicitly, a function $\rho(\cdot)$ given as $\rho(1) = 4$, $\rho(2) = 3$, $\rho(3) = 2$, $\rho(4) = 1$, which assigns persons (rows) to jobs (columns). Any of the n! possible one-to-one functions on the first n integers back to this same set provides one possible assignment, and in turn, a total index of merit. In our simple illustration, for instance, there are $4! = 24$ possibilities that are given in Table 1.2.1a along with the corresponding values of the objective function. By checking the listing of all 24 possible assignments in Table 1.2.1a, it is easy to verify that the permutation indicated earlier is optimal because it provides the largest of all entries in the merit index column.

Given this interpretation through a function $\rho(\cdot)$, which is denoted more simply as ρ from now on, an alternative strategy for specifying the LA optimization task is by the search for an optimal permutation. If

$$\Gamma(\rho) = \sum_i c_{i\rho(i)}$$

defines the value of the objective function or index of merit for a given permutation ρ, then the optimization problem can be rephrased as

Table 1.2.1a The Distribution of Merit Indices over the
24 Possible Assignments for the Simple
Example Discussed in the Text

Permutation	Index of merit
3412	18
1432	19
2413	25
2431	27
3214	30
1423	31
1234	31
4132	31
3421	32
4312	32
4213	33
3142	34
4231	35
2134	35
1342	36
2314	36
1243	37
3241	38
3124	40
2143	41
1324	42
4123	43
2341	44
4321	46

$$\max_{\rho} \sum_{i} c_{i\rho(i)}$$

where the maximum is taken over all n! possible permutations. From a statistical perspective, however, our interest will center on the index $\Gamma(\rho)$ itself and on its distribution over all possible permutations. It is this specific measure that will provide the basis for a framework that integrates a variety of different statistical topics.

1.2.2 The LA Model as a Basis for Inference (A)

The LA problem has a variety of interesting connections to several elementary statistical techniques, but not directly in terms of the optimization task. In fact, in many of the applications of interest, the optimal permutation could be found easily by simple inspection, e.g., if the largest entry in each row is also the largest entry in its respective column, then the n row maxima define an optimal assignment. For us, the important connections exist through the distribution of the objective function $\Gamma(\rho)$ over all permutations. Or in other words, the criterion of overall merit for a particular assignment is of major interest irrespective of that specific assignment's optimality.

The usual permutation inference strategy of nonparametric statistics will be assumed throughout (see Lehmann, 1975). Specifically, some permutation, say ρ_0, is identified as providing an a priori conjecture and we wish to evaluate the relative size of the associated (or observed) statistic $\Gamma(\rho_0)$ against what could be expected if ρ_0 had been randomly selected. Thus, given the availability of a complete enumeration, $\Gamma(\rho_0)$ could be compared to this distribution, and if extreme enough, declared significantly different from what would be expected if ρ_0 represented a random matching or assignment. When large (small) values of $\Gamma(\rho_0)$ are of concern, the exact significance level reported would be the proportion of permutations giving indices as large (small) or larger (smaller).[†]

[†] We note the ambiguity in specifying a two-tailed significance level if the reference distribution is not symmetric around its mean value. Two options are available depending on how the term "as extreme" is interpreted. One strategy would merely find the one-tailed probability and double it to report a two-tailed value; a second approach would try to construct a counterpart of the observed mean difference on the other "side" of the reference distribution and add both tail probabilities together. These comments will be germane throughout the book. We will implicitly assume that one-tailed tests are of interest, but in those cases where an option actually exists, one of these two alternatives could be followed.

This short paragraph gives a succinct summary of our use of LA; however, we should also reiterate at the outset the basic conceptual difference between this approach and what is typically done in nonparametric statistics from a distributional perspective. In the latter, it is common to assume that some particular random mechanism is responsible for generating the given data, i.e., we suppose a set of distributional assumptions. The "null" hypothesis of interest is then defined in terms of some restrictive characteristic of this mechanism, e.g., equality of distributions, distributional symmetry, factorization of a joint distribution because of independence, and so on. Although we may be led eventually to an LA problem because of the restrictive structure under the "null" hypothesis, this use of LA is more or less indirect because the basic aim is with testing the restrictions assumed for the data generation mechanism. The emphasis we take throughout this book is different in that the data are assumed fixed and given (say, as a matrix $\underset{\sim}{C}$) and our interests are primarily with assessing whether some assignment, specified a priori, could be considered a random draw from all possible assignments. Thus, we will not be explicitly concerned with the particular data generation mechanism that might have led to an LA task (the numerical example presented in the next section illustrates this point very well).

As the small example in Section 1.2.1 illustrated, it is possible, at least theoretically, to obtain the distribution of $\Gamma(\rho)$ for all n! permutations by complete enumeration. Moreover, for some special forms of the matrix $\underset{\sim}{C} = \{c_{ij}\}$, the tabling of the permutation distribution may be made feasible even for moderately large n by deriving an appropriate recurrence relation, i.e., some formal way of cutting down computation by relating one particular permutation and its index value to another. In the general case, however, no simple alternative to a complete listing appears possible, and thus, it is of some importance to have at least the mean and variance of $\Gamma(\rho)$ by simple formula.

Assuming that all n! possibly nondistinct values of $\Gamma(\rho)$ are equally likely, the moment expressions given below can be obtained through the indicator strategy to be discussed in Section 1.4.1. Let

$$A_1 = \left(\sum_{i,j} c_{ij}\right)^2 \qquad A_2 = \sum_j \left(\sum_i c_{ij}\right)^2$$

$$A_3 = \sum_i \left(\sum_j c_{ij}\right)^2 \qquad A_4 = \sum_{i,j} c_{ij}^2$$

then

$$E(\Gamma(\rho)) = \frac{1}{n} \sum_{i\ j} c_{i,j}$$

$$V(\Gamma(\rho)) = \left[\frac{1}{n(n-1)}\right]\left[\frac{1}{n} A_1 - (A_2 + A_3) + nA_4\right]$$

$$(1.2.2a)$$

These expressions may be verified numerically with the simple 4×4 example of Section 1.2.1. The exact mean and variance can be found either from the complete enumeration or from the formulas given above. Here,

$$A_1 = 18,496 \qquad A_2 = 5,170$$

$$A_3 = 4,852 \qquad A_4 = 1,496$$

$$\sum_{i,j} c_{ij} = 136$$

Thus,

$$E(\Gamma(\rho)) = \frac{1}{4} \cdot 136 = 34$$

$$V(\Gamma(\rho)) = \left[\frac{1}{4(4-1)}\right]\left[\frac{1}{4} \cdot 18,496 - (5,170 + 4,852) + 4(1,496)\right] = 48.8333$$

To provide alternative expressions that are somewhat simpler to use, suppose $\bar{c}_{i\cdot}$ and $\bar{c}_{\cdot j}$ denote, respectively, the average of the ith row and the jth column of $\underset{\sim}{C}$, and let $\bar{c}_{\cdot\cdot}$ be the average of all entries in $\underset{\sim}{C}$. If a new matrix, $\underset{\sim}{D} = \{d_{ij}\}$, is defined, where [†]

$$d_{ij} = c_{ij} - \bar{c}_{i\cdot} - \bar{c}_{\cdot j} + \bar{c}_{\cdot\cdot}$$

$$(1.2.2b)$$

then for any permutation ρ, an assignment index based on $\{d_{ij}\}$ is equal to that based on $\{c_{ij}\}$ minus the latter's expectation, $n\bar{c}_{\cdot\cdot}$. Thus, the two matrices provide indices that are equivalent up to an additive constant, where the constant is not affected by the permutation ρ.

[†]Throughout the book, a dot replacing a subscript implies that a summation was taken over that particular subscript.

The moment formulas reduce considerably using $\underset{\sim}{D}$. Based on this matrix,

$$E(\Gamma(\rho)) = 0$$

$$V(\Gamma(\rho)) = \frac{1}{n-1} \sum_{i,j} d_{ij}^2 \qquad\qquad (1.2.2c)$$

In an analysis of variance context with one observation per cell, expression (1.2.2b) removes the row, column, and grand means; thus, d_{ij} represents an interaction in the cell defined by row i and column j. The variation in $\Gamma(\rho)$ over different permutations is due solely to the interaction that the row and column objects produce. For convenience in our discussions later on, the matrix D constructed by equation (1.2.2b) will be referred to as the "interaction matrix."

Obviously, when complete enumeration is prohibitive, the availability of mean and variance formulas suggests a natural significance testing strategy through the Z statistic:

$$Z = \frac{\Gamma(\rho_0) - E(\Gamma(\rho))}{\sqrt{V(\Gamma(\rho))}}$$

Assuming the adequacy of the usual approximation when n is reasonably large, Z could be compared to the standard normal distribution. For now, asymptotic normality will be the significance testing strategy of choice, but the topic will be taken up in more detail in Section 1.4.2 where several other alternatives will be suggested. We might mention here, however, that one of these options, based on random sampling from the complete distribution of $\Gamma(\rho)$, will be the significance testing strategy suggested for routine use, possibly in conjunction with a tentative comparison of Z to the standard normal or other approximating distribution.

The index $\Gamma(\rho)$ and the inference model based on random permutations has been discussed before in the statistical literature but under the name of a bilinear permutation statistic (see Puri and Sen, 1971, pp. 75-76). The topic of asymptotic normality, in particular, has been of concern, and several sufficient regularity conditions have been stated. The one that is probably the easiest to implement, due to Hoeffding (1951), requires the expression

$$\frac{\underset{i,j}{\max}\ d_{ij}^2}{V(\Gamma(\rho))}$$

to converge to zero as $n \longrightarrow \infty$. In words, the terms in the matrix $\underset{\sim}{C}$ must be reasonable in size compared to the variance. This is a fairly innocuous condition and will be satisfied routinely in many of the applications considered.

There are instances, however, in which the Hoeffding condition is not satisfied and different limiting distributions are necessary. As one very simple example, suppose $c_{ij} = 1$ if $i = j$, and 0 otherwise. Then, $\Gamma(\rho)$ counts the number of times that a row object is mapped to the same column object by the permutation ρ. This is a problem of matching in the combinatorial probability literature for which a Poisson limit is more appropriate (David and Barton, 1962, pp. 102-118). In this particular case, the mean and variance are both 1, and thus the Hoeffding condition fails. A Poisson distribution with parameter $\lambda = 1$ would be a more reasonable approximation. In general, if C is dichotomous (zero-one) and has finite row and column sums as $n \to \infty$, a Poisson approximation would be expected and the Hoeffding sufficient condition would not be satisfied.

1.2.3 A Numerical Example (A)

An example may be helpful at this point to illustrate how the mechanics of the LA inference model can be carried out. Table 1.2.3a presents data, originally given by Phillips (1978), on the relationship between the month of birth and the month of death for 348 notable Americans. The conjecture of interest to Phillips, called the death-dip hypothesis, concerns the ability of people to postpone their deaths until after the occurrence of an important event, which in this case is a birthday. If this conjecture is reasonable, we would expect the circled entries in Table 1.2.3a to be relatively small compared to similar sets of entries that could be constructed from the table.

If Table 1.2.3a were treated as a 12×12 matrix $\underset{\sim}{C}$, then the death-dip hypothesis refers to a particular assignment of the row categories to the column categories, which can be represented by a permutation ρ_0 : $\rho_0(i) = i + 1$ for $1 \leqslant i \leqslant 11$, and $\rho_0(12) = 1$. The index $\Gamma(\rho_0)$ is 16, and our task is to decide whether this value is sufficiently small with respect to other comparable conjectures. When evaluated against all possible assignments that could be constructed, the mean and variance of $\Gamma(\rho)$ from the formulas given earlier are 29 and 24.2727, respectively, producing a Z statistic of -2.64. Consequently, compared to all equally likely assignments that could be formed, the one that represents the death-dip hypothesis produces an index that is significantly small, assuming a normal approximation to the distribution of Z. At least in relation to this data set, there is some evidence that a death-dip has occurred.

One of the technical points that should be raised for the data in Table 1.2.3a involves the varying row and column marginals. If we wished, a normalizing process could have been carried out initially to produce a table in canonical form in which the row and column sums are forced to be equal. Standardizations of this type are a fairly routine way of mitigating the effects

Table 1.2.3a Month of Birth and Month of Death for 348 Notable Americans

Death month	Birth month											
	Jan.	Feb.	Mar.	Apr.	May	June	July	Aug.	Sept.	Oct.	Nov.	Dec.
January	1	(1)	2	1	2	2	4	3	1	4	2	4
February	2	3	(1)	3	1	0	2	1	2	2	6	4
March	5	6	5	(3)	1	0	5	1	2	5	3	1
April	7	6	3	2	(1)	3	3	1	3	2	4	4
May	4	4	2	2	1	(2)	4	1	3	2	1	5
June	4	0	4	5	1	1	(1)	2	1	2	4	0
July	4	0	3	4	3	3	4	(1)	6	4	2	5
August	4	4	4	4	2	2	3	3	(1)	1	2	0
September	2	2	1	0	2	0	2	4	2	(0)	5	2
October	4	2	2	3	2	2	2	3	3	1	(4)	5
November	0	2	0	2	1	1	0	3	4	3	1	(0)
December	(1)	2	2	1	2	1	4	1	4	0	2	2

that varying row and column totals may have on an analysis. Typically, a process of iterative proportional fitting is used (e.g., see Bishop, Fienberg, and Holland, 1975, pp. 91-102) in which the entries in each row are, say, divided by the row sum. A similar operation is then performed on columns with column sums, which requires a renormalization of the rows, and so on until no changes in any of the table entries occur, i.e., until all the rows and columns sum to one.

In a slightly different sense, the assignment model has already compensated for differing row and column totals if we recall the equivalence of using the matrix $\{d_{ij}\}$ in equation (1.2.2b) in place of $\{c_{ij}\}$. In an analysis-of-variance interpretation, $\underset{\sim}{D}$ consists of interactions for which the row and column main effects have been removed. Consequently, because a standardization that compensates for row and column effects is part of the LA method automatically, the proportional fitting process was not implemented, although the analysis could obviously be repeated for such a matrix.

Use of the Interaction Matrix. The $\Gamma(\rho_0)$ statistic in conjunction with the LA model defines a very global assessment of the "uniqueness" of a conjectured permutation, ρ_0. To provide more detailed information, at least in an informal way, the interaction matrix, $\underset{\sim}{D}$, can be used to help identify how specific rows and/or columns contribute to the overall descriptive measure and to what degree. For example, it is apparent from Table 1.2.3b, which gives the interaction matrix corresponding to Table 1.2.3a, that certain death/birth month combinations contribute the most to an overall significant assignment, and some combinations even contribute oppositely (e.g., the two death/birth pairs of May/June and October/November). In general, a decision as to whether these differences have any substantive importance would require the study of other data sets. If these differences are relatively stable, then an alternative to (or at least a more detailed form of) the "death-dip" conjecture might be necessary to account for the differential patterning of entries in a birth/death table.

The data set in Table 1.2.3a and this particular analysis based on LA is very specialized to the death-dip hypothesis. As will be seen in later sections, however, alternative applications of the LA model are possible for data of exactly the same form. For this reason, the data in Table 1.2.3a will be used in a number of examples throughout this chapter and the next to illustrate various parts of the discussion.

1.2.4 Normalizing the Raw Index $\Gamma(\rho_0)$ (A)

Before various applications of the LA model are presented in the following sections, it may be appropriate to make a few general comments about normalizing a raw index $\Gamma(\rho_0)$. Or stated another way, how can we provide a final descriptive measure that is suitably standardized? Significance testing can always be carried out directly on the raw index $\Gamma(\rho_0)$, but it may be

TABLE 1.2.3b Interaction Matrix Corresponding to Table 1.2.3a

Death month	Birth month											
	Jan.	Feb.	Mar.	Apr.	May	June	July	Aug.	Sept.	Oct.	Nov.	Dec.
January	-2.0	-1.5	- .25	-1.33	+ .58	+ .75	+1.33	+1.17	-1.42	+2.0	- .83	+1.5
February	-1.0	+ .5	-1.25	+ .67	- .42	-1.25	- .67	- .83	- .42	0.0	+3.17	+1.5
March	+1.17	+2.67	+1.92	- .17	-1.25	-2.08	1.5	-1.67	-1.25	2.17	- .67	-2.33
April	+3.00	+2.5	- .25	-1.33	-1.42	+ .75	- .67	-1.83	- .42	-1.00	+ .17	+ .50
May	+ .67	+1.17	- .58	- .67	- .75	+ .42	1.00	-1.17	+ .25	- .33	-2.17	+2.17
June	+1.17	-2.33	+1.92	+2.83	- .25	- .08	-1.5	+ .33	-1.25	+ .17	+1.33	-2.33
July	.00	-3.5	- .25	+ .67	+ .58	+ .75	+ .33	-1.83	+2.58	+1.00	-1.83	+1.5
August	+ .75	+1.25	+1.5	+1.42	+ .33	+ .50	+ .08	+ .92	-1.67	-1.25	-1.08	-2.75
September	- .58	- .08	- .83	-1.92	1.00	- .83	- .25	+2.58	0.0	-1.58	+2.58	- .08
October	+ .50	-1.00	- .75	+ .17	+ .08	+ .25	-1.17	+ .67	+ .08	-1.50	+ .67	+2.00
November	-2.80	+ .42	-1.33	+ .58	+ .50	+ .67	-1.75	+2.08	1.50	+1.92	- .92	-1.58
December	-1.58	- .08	+ .17	- .92	+1.00	+ .17	+1.75	- .42	2.00	-1.58	- .42	- .08

15

inappropriate as a descriptive statistic because its size depends on what
may be an arbitrary scaling of the entries in C.

There are a number of very thorny issues in defining suitably normalized measures that probably will never be resolved satisfactorily with only
one approach. In the ideal situation, no additional standardization would
be necessary. For instance, in the simple illustration of Section 1.2.3
dealing with the death-dip conjecture, it may be sufficient to report the
obvious fact that 16 people were in cells that defined the death-dip conjecture and 29 were expected if such an assignment had been chosen at random.
The difference of 13 between the observed and expected has meaning without any further transformation. Unfortunately, the exact interpretation of
an observed index may not be as apparent in other applications. In these
cases, one strategy would be to choose between two versions of a general
transformation of $\Gamma(\rho_0)$, depending on whether relatively large or small
values of $\Gamma(\rho_0)$ are of primary interest. In particular, if relatively large
values are of concern, then we consider

$$\Gamma^*(\rho_0) = \frac{\Gamma(\rho_0) - E(\Gamma(\rho))}{\max_{\rho} \Gamma(\rho) - E(\Gamma(\rho))}$$

and alternatively, for relatively small values of $\Gamma(\rho_0)$

$$\Gamma^{**}(\rho_0) = \frac{E(\Gamma(\rho)) - \Gamma(\rho_0)}{E(\Gamma(\rho)) - \min_{\rho} \Gamma(\rho)}$$

In both cases, the normalized indices have zero expectations, are bounded
above by one, and invariant under linear transformations of the index in
which the multiplying constant is positive. Obviously, the upper bound can
be achieved by a particular assignment. Because the normalizing process
does not vary for different permutations, statements of significance for the
raw and normalized measures are the same.[†]

Two special cases of $\Gamma^*(\rho_0)$ and $\Gamma^{**}(\rho_0)$ will be encountered in various
applications of the assignment model. When $\min_{\rho} \Gamma(\rho)$ is 0, $\Gamma^{**}(\rho_0)$ takes
the simple form:

$$\Gamma^{**}(\rho_0) = 1 - \frac{\Gamma(\rho_0)}{E(\Gamma(\rho))}$$

[†]Throughout this book we shall continually use the simple formulas for the
expectations and variance of linear combinations of random variables, i.e.,
if $Y = aX + b$, where a and b are constants, and X and Y are random variables, then $E(Y) = aE(X) + b$ and $V(Y) = a^2V(X)$. Typically, X will correspond to some raw index, and Y to some normalized version.

and, when $E(\Gamma(\rho))$ is zero:

$$\Gamma^*(\rho_0) = \frac{\Gamma(\rho_0)}{\max_\rho \Gamma(\rho)} \qquad \Gamma^{**}(\rho_0) = \frac{\Gamma(\rho_0)}{\min_\rho \Gamma(\rho)}$$

For some applications, the $\min_\rho \Gamma(\rho)$ and $\max_\rho \Gamma(\rho)$ terms can be found by a simple process of inspection, but whenever necessary, they could be obtained through the optimization methods mentioned in Section 1.2.1.

Many of the indices that have been proposed in the literature for a variety of data analysis problems have the same general form given by $\Gamma^*(\rho_0)$ and $\Gamma^{**}(\rho_0)$, although this may not be at all apparent when the measure is discussed in isolation. Almost invariably, the $\min_\rho \Gamma(\rho)$ and $\max_\rho \Gamma(\rho)$ terms are approximated by various bounds to obtain convenient descriptive measures that do not require the solution of an explicit optimization problem. These bounds are usually obtained by reference to the data generation process and to what types of patterns could conceivably arise. We will see this in detail in Section 1.3.3, where the comparison of two sequences is discussed using a variety of correlation coefficients. In general, our presentation will also follow this practice of using bounds because it greatly simplifies the task of constructing appropriate descriptive indices. Thus, we may, at times, equivocate slightly in referring to measures of the form $\Gamma^*(\rho_0)$ and $\Gamma^{**}(\rho_0)$, even though bounds are being used in place of the optimization task. Hopefully, the context in each case will make this clear if it is not mentioned explicitly.[†]

In a somewhat broader sense, the use of a bound in place of $\max_\rho \Gamma(\rho)$ or $\min_\rho \Gamma(\rho)$ may lead to a descriptive measure that can be given meaning in terms of other indices that are more well known. The latter are typically average correlation coefficients or similar descriptive statistics that are familiar from other contexts. The dependence on such bounds implies that some external reference outside the given data is being relied on to provide the intuitive meaning for a measure. In general, whenever an index can be defined that has inherent meaning for a given problem, it should be used. This applies to simple averages, proportions, and the like. We will make use of this option at various points throughout the book.

With respect to the two transformations introduced above, it is an obvious nuisance to require two different coefficients, depending on whether

---

[†]Many of the normalizations used throughout this book as convenient descriptive statistics may also have natural population analogs that have been discussed in the literature. Because we prefer to deal strictly with the descriptive aspects of the problem and with a comprehensive strategy that encompasses a wide variety of applications, we will not stop to pursue these population analogs in any specific instance.

relatively large or small values of $\Gamma(\rho_0)$ are of primary interest, but there does not seem to be any convenient alternative that is universally acceptable. There are some situations, however, in which $\Gamma*(\rho_0)$ and $\Gamma**(\rho_0)$ have a very nice dual relationship, and only one of them will be needed. For example, if the distribution of $\Gamma(\rho)$ is symmetric about $E(\Gamma(\rho))$, then $\Gamma*(\rho_0) = -\Gamma**(\rho_0)$. Thus, if only the $\Gamma*(\rho_0)$ index were used, it would be bounded below by -1, and each negative value for $\Gamma*(\rho_0)$ would correspond to a comparable positive value on the index $\Gamma**(\rho_0)$. A similar comment holds if only $\Gamma**(\rho_0)$ were used; it would be bounded below by -1, and negative values would correspond to positive values on $\Gamma*(\rho_0)$. Consequently, in instances such as these, only one index will be needed with the understanding that negative values have meaning and are comparable to the corresponding positive values for the other.

Along these same lines but relaxing the assumption of symmetry, suppose a value, say B, is found such that

$$E(\Gamma(\rho)) - \min_{\rho} \Gamma(\rho) \leqslant B$$

and

$$\max_{\rho} \Gamma(\rho) - E(\Gamma(\rho)) \leqslant B$$

Then, if we want large raw measures to correspond to positive normalized values, the single index

$$\frac{\Gamma(\rho_0) - E(\Gamma(\rho))}{B} \tag{1.2.4a}$$

could be used. This quantity is bounded in the following way:

$$-1 \leqslant -\Gamma**(\rho_0) \leqslant \frac{\Gamma(\rho_0) - E(\Gamma(\rho))}{B} \leqslant \Gamma*(\rho_0) \leqslant +1$$

Thus, even though the distribution of $\Gamma(\rho)$ may not be symmetric around $E(\Gamma(\rho))$, we still might find it convenient to use the single index in expression (1.2.4a) in place of the other two. Again, negative values on this index would correspond to positive values on $\Gamma**(\rho_0)$. If we wished to have relatively small values on the raw index correspond to positive values on the normalized measure, the numerator in expression (1.2.4a) would be replaced by $E(\Gamma(\rho)) - \Gamma(\rho_0)$:

$$\frac{E(\Gamma(\rho)) - \Gamma(\rho_0)}{B} \tag{1.2.4b}$$

Then,

$$-1 \leqslant -\Gamma^*(\rho_0) \leqslant \frac{E(\Gamma(\rho)) - \Gamma(\rho_0)}{B} \leqslant \Gamma^{**}(\rho_0) \leqslant +1$$

As one final comment about standardization, this time in the choice of the conjectured permutation, we note that ρ_0 could always be taken as the identity permutation, ρ_I, without loss of any generality. If $\rho_0(i) = j$, $1 \leqslant i$, $j \leqslant n$, then a "new" assignment matrix could be constructed from $\underset{\sim}{C} = \{c_{ij}\}$ by the interchange of columns. The index $\Gamma(\rho_I)$ based on an assignment matrix of the form $\{c_{i\rho_0(j)}\}$ is the same as the index $\Gamma(\rho_0)$ based on the original definition. Furthermore, the reference distributions would be identical starting from either matrix. In some applications, it is convenient to use a nonidentity permutation, ρ_0, to characterize a conjectured matching, and we will continue to do so. Nevertheless, we should keep in mind the equivalence to using ρ_I and an appropriately reorganized matrix of assignment scores. Such flexibility will be of great notational help in this chapter and later on.

1.3 APPLICATIONS OF THE LA MODEL[†]

1.3.1 Nominal Scale Agreement: Weighted Kappa (B)

Weighted kappa (κ), proposed in a very general form by Cohen (1972), is appropriate for measuring nominal-scale agreement between two raters, using an R × C contingency table of the type given in Table 1.3.1a. The term "nominal" has its usual meaning and implies that the row and column categories have no inherent order, or possibly, any such order that exists is to be ignored. Cohen's index is explicitly defined by the expression:

$$\kappa = \frac{P_0 - P_e}{1 - P_e}$$

where

$$P_0 = \sum_{u=1}^{R} \sum_{v=1}^{C} w_{uv} p_{uv}$$

[†]For historical reference, we mention that some of the connections between the LA model and the applications given in Section 1.3 have been noted in the literature by the author, e.g., see Hubert (1976, 1978, 1984). Parts of this first chapter rely heavily on these sources.

Table 1.3.1a Illustrative Contingency Table for
n Objects and Two Raters

| Rater 1 category | Rater 2 category | | | Row sum |
	b_1	b_2	b_C	
a_1	n_{11}	n_{12} $\cdots$	n_{1C}	$n_{1 \cdot}$
a_2	n_{21}	n_{22} $\cdots$	n_{2C}	$n_{2 \cdot}$
$\cdot$	$\cdot$	$\cdot$	$\cdot$	$\cdot$
$\cdot$	$\cdot$	$\cdot$	$\cdot$	$\cdot$
a_R	n_{R1}	n_{R2} $\cdots$	n_{RC}	$n_{R \cdot}$
Column sum	$n_{\cdot 1}$	$n_{\cdot 2}$ $\cdots$	$n_{\cdot C}$	$n_{\cdot \cdot} = n$

$$P_e = \sum_{u=1}^{R} \sum_{v=1}^{C} w_{uv} p_{u \cdot} p_{\cdot v}$$

$$p_{uv} = \frac{n_{uv}}{n} \qquad p_{u \cdot} = \frac{n_{u \cdot}}{n} \qquad p_{\cdot v} = \frac{n_{\cdot v}}{n}$$

and the weights $\{w_{uv}\}$ are chosen to represent the relative seriousness of rater 1 placing an object in category a_u, and of rater 2 placing the same object in category b_v.

In the typical applications discussed by Cohen (1968) or Fleiss, Cohen, and Everitt (1969), it is assumed that R = C and that the placement of an object in category a_u by rater 1 or in category b_u by rater 2 denotes the same decision. Because the more general case sketched above is no more difficult to consider and is needed for some recent work on confirmatory hypothesis testing in contingency tables that is mentioned in the next section, this limitation to an equal number of rater categories will not be imposed. Nevertheless, it is of some historical interest to keep in mind that most of the literature on nominal-scale agreement has required this equal category condition and has specified the weights to count only those entries along the main diagonal, i.e., $w_{uu} = 1$ for $1 \leq u \leq R = C$, and 0 otherwise. This restriction is understandable, given the ambiguity inherent in choosing the a priori weights for defining κ. For instance, in introducing this index, Cohen (1968) gave only a very brief example, using three psychiatric diagnoses—and stated that a clinician would consider a diagnostic disagreement between neurosis and psychosis to be more serious than a disagreement between neurosis and personality disorder. Although no explicit procedure was given for actually assigning numerical values to the required weights, it was assumed that an expert could do so realistically.

Under the null hypothesis of independence defined in the usual manner for a two-way contingency table with variable marginal totals, Cohen (1972) gave the large-sample variance of κ in the following form:

$$V(\kappa) = \frac{1}{n(1 - P_e)^2} \left\{ \sum_{u,v} p_{u\cdot} p_{\cdot v} [w_{uv} - (\overline{w}_{u\cdot} + \overline{w}_{\cdot v})]^2 - P_e^2 \right\}$$

where

$$\overline{w}_{u\cdot} = \sum_v w_{uv} p_{\cdot v}$$

$$\overline{w}_{\cdot v} = \sum_u w_{uv} p_{u\cdot}$$

Because the expectation of κ is zero under the independence assumption, a rejection of this latter independence hypothesis is justified when the quantity $\kappa/\sqrt{V(\kappa)}$ is sufficiently large. Assuming the adequacy of the usual large-sample approximation, this ratio could be compared to a normal distribution with mean zero and variance one.

Although Cohen (1972) derived the large-sample variance of κ under the assumption that only the total sample size n is fixed, that is, for a contingency table defined by multinomial sampling and having variable marginal totals, many of the historically important applications of an index of nominal-scale agreement are more appropriate for a model that assumes fixed marginals. For instance, a fixed-marginals model would clearly be of most concern if κ were to be applied to the experimental paradigm of matching, as discussed by Mosteller and Bush (1954), Vernon (1936), and McHugh and Apostolakos (1959), among others. Here, a subject is given two sets of n objects and explicitly told of the marginal (row and column) frequencies in both. The task is to match the n objects from one set to the n objects from the other according to some stated criterion.

Even though the most natural sampling model may be multinomial, a justification for the fixed-marginal notion can be found in conditional inference. This point was made recently by Lehmann (1975) in some detail, but the general utility of conditional inference has been recognized for a very long time. For instance, Hotelling (1940, p. 276) summarized this position well when he stated:

The practice now accepted (after a controversy) for testing independence of two modes of classification, such as classification of persons according as they have or have not been vaccinated, and again according as they

live through an epidemic or die, is to compare the observed contingency table , not with all possible contingency tables of the same numbers of rows and columns, but only with the possible contingency tables
having exactly the same marginal totals as the observed table.

Under the fixed-marginal assumption, the variance term for κ is
relatively easy to obtain once the correspondence between κ and the assign-
ment statistic $\Gamma(\rho_0)$ is pointed out. Explicitly, define a $n \times n$ matrix, $\underset{\sim}{C}$,
as follows:

$$
\underset{\sim}{C} =
\begin{array}{c}
\\
\\
n_{1\cdot}\ \{ \\
\\
n_{2\cdot}\ \{ \\
\\
\vdots \\
\\
n_{R\cdot}\ \{
\end{array}
\overbrace{}^{n_{\cdot 1}}\quad \overbrace{}^{n_{\cdot 2}}\quad \cdots \quad \overbrace{}^{n_{\cdot C}}
\begin{bmatrix}
-w_{11}- & -w_{12}- & \cdots & -w_{1C}- \\
-w_{21}- & -w_{22}- & \cdots & -w_{2C}- \\
\vdots & \vdots & \cdots & \vdots \\
-w_{R1}- & -w_{R2}- & \cdots & -w_{RC}-
\end{bmatrix}
$$

where the symbol $-w_{uv}-$ refers to a rectangular matrix of size $n_{u\cdot} \times n_{\cdot v}$
in which each entry is w_{uv}. The n objects contained in the contingency
table form the n rows and columns of $\underset{\sim}{C}$; the n objects are grouped according to the two categorizations used by rater 1 (the rows) and rater 2 (the
columns).

Each possible permutation or matching of the row and column objects
defines one possible realization of the R × C contingency table with the
given marginals; conversely, all possible realizations of the table can be
represented in terms of such permutations. The observed table itself can
be obtained using ρ_0, defined as that assignment matching identical row
and column objects. Thus, $\Gamma(\rho_0)$ is equal to $\Sigma_{u,v}\, w_{uv}n_{uv} = nP_0$, which
forms the variable part of κ under a fixed marginal assumption, and
$E(\Gamma(\rho)) = (1/n)\Sigma_{u,v}\, w_{uv}n_{u\cdot}.n_{\cdot v} = nP_e$. By considering all possible such
permutations as equally likely under the LA model, the usual hypergeo-
metric distribution is generated for the entries in a contingency table with
fixed marginals (see David and Barton, 1962; Agresti, Wackerly, and Boyett, 1979).

As a final normalization that assumes large values of $\Gamma(\rho_0)$ to be desirable, κ and $\Gamma^*(\rho_0)$ have the same form if we define $\max_\rho \Gamma(\rho)$ to be n.

The latter condition is made more realistic by assuming, as is typical, that all weights are nonnegative and less than or equal to 1, because then $\max_\rho \Gamma(\rho) \leqslant n$. It is always true that $\kappa \leqslant \Gamma^*(\rho_0)$, but depending on the weight structure, it is possible that this upper bound cannot be achieved by any permutation ρ. In this case, κ would be bounded away from 1 for all possible realizations of the contingency table with the given marginals.

The measure κ, as defined by Cohen (1972), is a linear transformation of $\Gamma(\rho_0)$, and thus, relying on well-known formulas from elementary statistics, one can use $V(\Gamma(\rho))$ to obtain $V(\kappa)$ in the fixed-marginal framework. Simple algebra shows that $V(\kappa)$ in this latter context is $n/(n-1)$ times that given earlier for variable margins. Obviously, any discrepancy between the two formulas is negligible for moderate values of n, implying that either approach should lead to the same decision in a significance test of κ. For a more complete discussion of alternative inference models for indices of nominal-scale agreement, the reader is referred to Hubert (1977, 1980) and to Sections 4.3.5 and 4.3.7.

We have implicitly assumed that weighted kappa is intended for use with nominal categories, but it may be applied as well in the ordered-category case through an appropriate choice of weights. For instance, following Goodman and Kruskal (1954), agreement indices appropriate for ordered categories when R = C could be defined by a set of weights with the property

$$w_{uv} = 0 \text{ for } |u - v| \geqslant e$$

where e is some integer chosen by the researcher. As an example, if $w_{uv} = 1/n$, when $|u - v| \leqslant 1$ and 0 otherwise, then e = 2 and

$$\Gamma(\rho_0) = \sum_{|u-v| \leqslant 1} \frac{n_{uv}}{n}$$

In words, $\Gamma(\rho_0)$ is the proportion of entries that fall along the main diagonal or in cells that define categories one unit apart on the assumed ordering. This last interpretation corresponds to one particular index mentioned explicitly by Goodman and Kruskal. A further use of the weighted kappa notion is developed in the next section.

1.3.2 Prediction Analysis (B)

We have phrased the discussion of weighted kappa for the task of evaluating nominal-scale agreement between two raters. More generally, however, this same strategy is appropriate for $R \times C$ contingency tables in which the raters are replaced by two attributes that cross-classify n objects. For example, in the prediction analysis system developed by Hildebrand, Laing, and Rosenthal (1977), it is assumed that the numerical weights, $\{w_{uv}\}$, when considered as a set, characterize some relevant hypothesis that we may wish to confirm. Usually, the weights are dichotomous (zero-one) values, where the one's formally identify "error" cells according to the given conjecture, i.e., those cells that should be empty or at least those that should contain relatively few observations. To measure the degree of confirmation based on the given set of weights, Hildebrand et al. define an index $\hat{\nabla}$:

$$\hat{\nabla} = 1 - \frac{\sum_{u,v} w_{uv} p_{uv}}{\sum_{u,v} w_{uv} p_{u.} p_{.v}}$$

where in terms of our earlier notation, $\Sigma_{u,v}\, w_{uv} p_{uv}$ is the observed weighted proportion of all entries, and $\Sigma_{u,v}\, w_{uv} p_{u.} p_{.v}$ defines the expected weighted proportion under the assumption of independence for the row and column categorizations. Because small values of $\Gamma(\rho_0)$ are now of interest and we assume $\min_\rho \Gamma(\rho)$ is 0, or at least zero provides a lower bound, the $\hat{\nabla}$ index of Hildebrand et al. is of the same form as $\Gamma^{**}(\rho_0)$. In words, $\hat{\nabla}$ is one minus the number of observations in the error cells divided by the number expected under independence. For a more complete discussion of prediction analysis in the context of evaluating developmental hypotheses, the reader is referred to Froman and Hubert (1980, 1981).

As a simple example of how this approach could be carried out with the data in Table 1.2.3a, the death-dip hypothesis can be reevaluated with the assignment at the level of the individual rather than at the level of categories as in our previous analysis. Table 1.2.3a can be viewed as a contingency table with 12 error cells corresponding to the death-dip conjecture. If the weights for these twelve are assumed to be ones and all the remaining cells are given weights of zero, the observed index is again 16, the expected value is now 28.30, and the variance under independence is 25.8266. Thus, $\hat{\nabla} = 1 - (16.00/28.30) = .43$, with an associated Z statistic of -2.42. Even though the assignment is now at the individual level, the death-dip hypothesis is again confirmed, and in fact, with about the same relative size as before for the associated Z statistic.

1.3.3 The Comparison of Two Numerical Sequences (B)

To move into a context that is correlational in orientation, suppose $\{x_1, x_2, \ldots, x_n\}$ and $\{y_1, y_2, \ldots, y_n\}$ denote two numerical sequences in which the corresponding elements x_i and y_i are matched in some manner. Clearly, this framework is the natural situation for developing a measure of association based on the n bivariate pairs (x_1, y_1), (x_2, y_2), $\ldots$, (x_n, y_n). Because no restriction is imposed on the sequences other than they be numerical, an initial transformation of the x's and y's to ranks may be used when rank correlations are desired. Or, if we wished, an additional transformation could be carried out from ranks to normal scores of some type. For instance, the ith rank could be replaced by the expected value of the ith smallest observation in a sample of size n chosen from a standard normal distribution (see Lehmann, 1975). In general, any of the analysis procedures proposed throughout this book for a given data set apply automatically to normal scores or ranks merely by treating the latter as the original data.

The assignment score, c_{ij}, is constructed for the pair x_i and y_j and is used to define a summary index of correspondence between the two sequences based on ρ_0 as the identity permutation, ρ_I, i.e.,

$$\Gamma(\rho_I) = \sum_i c_{ii}$$

Two special cases of score specification are of particular interest. If the sequences are ranks, then $c_{ij} = |x_i - y_j|$ provides the basis of a measure of rank correlation called Spearman's footrule; if $c_{ij} = (x_i - y_j)^2$, the basis of Spearman's more common rank correlation is obtained. In either case, if $\min_\rho \Gamma(\rho)$ is assumed to be zero, as it would be for untied ranks, and assuming that small values of $\Gamma(\rho_I)$ are desirable, then

$$\Gamma^{**}(\rho_I) = 1 - \frac{\Gamma(\rho_I)}{E(\Gamma(\rho))}$$

provides one normalized form for each statistic.

To be somewhat more explicit for the case of untied ranks, the specification $c_{ij} = (x_i - y_j)^2$ leads to $E(\Gamma(\rho)) = n(n^2 - 1)/6$, and thus,

$$\Gamma^{**}(\rho_I) = 1 - \frac{6 \sum_i (x_i - y_i)^2}{n(n^2 - 1)}$$

as given by Kendall (1970, p. 8). If $c_{ij} = |x_i - y_j|$, then $E(\Gamma(\rho)) = (n^2 - 1)/3$, and

$$\Gamma^{**}(\rho_I) = 1 - \frac{3 \sum_i |x_i - y_i|}{n^2 - 1}$$

again, as defined by Kendall (1970, p. 32). If relatively large values of $\Gamma(\rho_I)$ were of interest, as they would be when negative correlation is expected, then an index of the form $\Gamma^*(\rho_I)$ would be more appropriate. Because the distribution is symmetric around $E(\Gamma(\rho))$ when $c_{ij} = (x_i - y_j)^2$,

$$\Gamma^*(\rho_I) = -\Gamma^{**}(\rho_I)$$

Thus, exactly the same index can be used either for relatively small or large values of $\Gamma(\rho_I)$ in the usual Spearman rank correlation context, if one remembers that small values lead to positive normalized indices and large values lead to negative normalized indices. The bounds are ± 1, and these can be achieved for untied ranks. Unfortunately, this nice dual relationship does not exist for $c_{ij} = |x_i - y_j|$. Because $\max_\rho \Gamma(\rho)$ equals $(n^2 - 1)/2$ for n odd and $n^2/2$ for n even, these latter values could be used to define $\Gamma^*(\rho_I)$ when large values of $\Gamma(\rho_I)$ are of concern. Or possibly, we could look for a bound and define an index of the type given in expression (1.2.4b).

The problems identified in standardizing the raw footrule index to provide a normalized statistic extend beyond the obvious need to consider two separate statistics $\Gamma^*(\rho_I)$ and $\Gamma^{**}(\rho_I)$. Even if this separation is carried out consistently, depending on whether large or small values of the raw index are of interest, an anomaly still arises in the related significance statements. In particular , suppose we have two sequences of untied ranks $\{x_1, \ldots, x_n\}$ and $\{y_1, \ldots, y_n\}$, and our concern is with negative association. Instead of obtaining $\Gamma(\rho_I)$ and normalizing to obtain $\Gamma^*(\rho_I)$, suppose we arbitrarily reverse the ranking in one sequence and consider positive association instead. Unfortunately, and in contrast to the use of squared differences between ranks, the distribution of the raw footrule index is not symmetric; more specifically, each ranking does not have a natural conjugate or reversal that would lead to an index as much below the mean as the original index was above the mean (or conversely). Thus, the reversal of the ranking and a consideration of positive association does not bear any nice relationship to the original strategy.

A simple example may help clarify this problem. Suppose the x sequence is $\{1, 2, 3, 4\}$ and the y sequence is $\{4, 2, 3, 1\}$, and our concern is with a negative relationship. The raw footrule measure is $|1 - 4| + |2 - 2| + |3 - 3| + |4 - 1| = 6$ with an upper-tail significance of $13/24$. If we reversed the y sequence to generate $\{1, 3, 2, 4\}$, the footrule measure is $|1 - 1| + |2 - 3| + |3 - 2| + |4 - 4| = 2$ and gives a much different lower-tail significance of $4/24$. As we will see, this problem arises in other contexts as well. Thus, what could be an arbitrary keying of the direction for one set of ranks or scores could have a major effect, not only in terms of normalized indices, but on the significance statements themselves.

If there are tied ranks (resolved, say, by using midranks) or if the original sequences are used without a rank transformation, significance testing for the index $\Gamma(\rho_I)$ remains fairly straightforward, but the task of defining a final descriptive index becomes even more complicated and open to debate. For example, suppose $c_{ij} = (x_i - y_j)^2$ and consider the index

$$\Gamma^{**}(\rho_I) = \frac{E(\Gamma(\rho)) - \Gamma(\rho_I)}{E(\Gamma(\rho)) - \min_{\rho} \Gamma(\rho)}$$

Simple algebra shows this to be equivalent to:

$$\frac{\sum_i (x_i - \overline{x})(y_i - \overline{y})}{\max_{\rho} \sum_i (x_i - \overline{x})(y_{\rho(i)} - \overline{y})}$$

where $\overline{x}$ and $\overline{y}$ are the means of the x and y sequences, respectively. The term in the denominator can be found by merely ordering the deviations of the form $(x_i - \overline{x})$ and $(y_i - \overline{y})$ from smallest to largest and summing the n cross-products (the corresponding minimum could be found by inverting the order for the y sequence). However, this is not the common measure that is used; instead, the denominator is replaced by an upper bound. In particular,

$$\max_{\rho} \sum_i (x_i - \overline{x})(y_{\rho(i)} - \overline{y}) \leq \sqrt{\sum_i (x_i - \overline{x})^2 \sum_i (y_i - \overline{y})^2}$$

and when the term on the right is used instead of the left-hand term, the standard Pearson correlation coefficient is obtained:

$$r_{xy} = \frac{\sum_i (x_i - \overline{x})(y_i - \overline{y})}{\sqrt{\sum_i (x_i - \overline{x})^2 \sum_i (y_i - \overline{y})^2}}$$

This latter expression must always be less than or equal to $\Gamma^{**}(\rho_I)$. If we had started with $\Gamma^{*}(\rho_I)$, a lower bound would have been appropriate that would result in an index of this same type, i.e.,

$$\min_{\rho} \sum_i (x_i - \overline{x})(y_{\rho(i)} - \overline{y}) \geq -\sqrt{\sum_i (x_i - \overline{x})^2 \sum_i (y_i - \overline{y})^2}$$

In other words, r_{xy} is a measure of the form presented in expression 1.2.4b), where $B = \sqrt{\sum_i (x_i - \overline{x})^2 \sum_i (y_i - \overline{y})^2}$; thus $-\Gamma^{*}(\rho_I) \leq r_{xy} \leq \Gamma^{**}(\rho_I)$.

This simple demonstration can be used to make several important points. First of all, the LA paradigm as a significance testing strategy includes the usual randomization models for the Pearson correlation and Spearman's two indices in both the tied and untied cases. Secondly, the usual descriptive indices in the literature can be interpreted through the use of a bound either on $\max_\rho \Sigma_i (x_i - \bar{x})(y_{\rho(i)} - \bar{y})$ or on $\min_\rho \Sigma_i (x_i - \bar{x})$ $\times (y_{\rho(i)} - \bar{y})$. In fact, because the same expression is employed for these bounds, the use of the correlation coefficient r_{xy} effectively sidesteps the issue of a possible asymmetry in the randomization distribution merely by using one index instead of two. Third, if we use ranks that are tied, the correlation r_{xy} is equivalent to what is usually called Kendall's ρ_b. If we continue even further with this notion of bounding, the denominator in the untied case of $n(n^2 - 1)/12$ is larger than $\sqrt{\Sigma_i (x_i - \bar{x})^2 \Sigma_i (y_i - \bar{y})^2}$. Thus, if the former quantity were used, an alternative normalization is constructed, referred to as Kendall's ρ_a, which in absolute value is always less than or equal to the absolute value of ρ_b.

To mention two further technical points, we first note that the bivariate specification of $(x_i - y_j)^2$ could have been replaced by $x_i y_j$ without any real loss of generality, i.e., because

$$\Gamma(\rho_I) = \sum_i (x_i - y_i)^2 = \sum_i x_i^2 + \sum_i y_i^2 - 2 \sum_i x_i y_i$$

$\Gamma(\rho_I)$ is a simple linear transformation of an observed statistic based on the assignment score $c_{ij} = x_i y_j$. This same relationship was shown earlier in the reduction of the normalized measures to a multiplicative form. Because the Pearson correlation between the two sequences is also a simple linear transformation of the observed statistic $\Gamma(\rho_I) = \Sigma_{i=1}^n x_i y_i$, i.e.,

$$r_{xy} = \frac{\Gamma(\rho_I) - (1/n) \sum_i x_i \sum_i y_i}{\sqrt{\sum_i (x_i - x)^2 \sum_i (y_i - y)^2}}$$

and

$$E(\Gamma(\rho)) = \frac{1}{n} \sum_i x_i \sum_i y_i$$

$$V(\Gamma(\rho)) = \left(\frac{1}{n-1}\right) \sum_i (x_i - \bar{x})^2 \sum_i (y_i - \bar{y})^2$$

the mean and variance of r_{xy} under random assignment can be obtained very easily. Specifically,

$$E(r_{xy}) = 0 \quad \text{and} \quad V(r_{xy}) = \frac{1}{n-1}$$

Secondly, we have restricted our discussion in this section to numerical sequences because, in effect, we have already dealt with the problem of relating two nonnumerical sequences in the earlier applications of $\Gamma(\rho_I)$. For example, in the contingency table scheme of the last two sections, the sequence $\{x_1, \ldots, x_n\}$ could be used to represent the labels for the R row categories, and $\{y_1, \ldots, y_n\}$ could be used to represent the labels for the C column categories. If we define $c_{ij} = w_{uv}$ when x_i is placed in the row category labeled u and when y_j is placed in the column category labeled v, the analysis of contingency tables could be interpreted as a correlational problem of relating two nonnumerical sequences.

The assignment index in its product form, $\Sigma_i\, x_i y_i$, has been discussed extensively in the literature under the general title of a linear permutation statistic. We reserve discussion of one particular special case for the next section, where the y sequence consists of zeros and ones denoting membership in one of two groups. For now, we briefly introduce a related application, referred to as a quadrant measure of association between the two sequences $\{x_1, \ldots, x_n\}$ and $\{y_1, \ldots, y_n\}$. To develop this idea formally, suppose

$$\text{sign}(u) = \begin{cases} +1 & u > 0 \\ 0 & u = 0 \\ -1 & u < 0 \end{cases}$$

and define

$$c_{ij} = \text{sign}(x_i - M_X)\, \text{sign}(y_j - M_Y)$$

where M_X and M_Y are constants (see Kendall and Stuart, 1979, p. 538). The index $\Gamma(\rho_I)$ is a raw measure of association between the two sequences and may be represented as U − V, where U is the number of objects for which the associated values in the two sequences are both strictly above or both strictly below their respective values, M_X and M_Y; V refers to the number of objects for which the associated observations are strictly above the given constant in one sequence and strictly below for the second.

Because $-n \leqslant \Gamma(\rho_I) \leqslant n$, the values of $\pm n$ can be used as bounds to define normalized indices of the form $\Gamma^*(\rho_I)$ or $\Gamma^{**}(\rho_I)$. In the special instance when $\Sigma_i\, \text{sign}(x_i - M_X) = 0$ or $\Sigma_i\, \text{sign}(y_i - M_Y) = 0$, as would typically be the case when M_X and M_Y denote medians, the measure of the

$\Gamma^*(\rho_I)$ type based on the bound of n would be represented as $U/n - V/n$. Here, U/n denotes the probability that in a random selection from the n bivariate pairs, the one chosen has both values strictly above or strictly below their respective values M_X and M_Y; V/n is the probability that one value in the pair is strictly above and the other value strictly below. Thus, in this context only one normalized measure is necessary; moreover, it can be given a convenient operational interpretation with respect to the original data. A measure of the $\Gamma^{**}(\rho_I)$ form would merely be $V/n - U/n$ when based on the bound of $-n$.

1.3.4 The Linear Permutation Statistic: Two-Independent Sample Problems (B)

The linear permutation statistic mentioned in the context of the cross-product index $\Sigma_i x_i y_i$ of the last section includes many of the well-known nonparametric two-independent sample tests for location and scale. Here, we are given n "scores" $s_1, s_2, \ldots, s_n$, e.g., ranks, normal scores, the original observations, and so on, where the first n_1 are in group I and the remaining $n - n_1$ are in group II. The observed test statistic $\Gamma(\rho_I)$ is defined as the sum of the scores in group I, $\Sigma_{i=1}^{n_1} s_i$. This measure can be generated using the $\underset{\sim}{C}$ matrix defined as

$$
\{c_{ij}\} =
\begin{array}{c}
n_1 \left\{ \vphantom{\begin{array}{c} a \\ b \\ c \\ d \end{array}} \right. \\
\\
n - n_1 \left\{ \vphantom{\begin{array}{c} a \\ b \\ c \\ d \end{array}} \right.
\end{array}
\overbrace{}^{n_1} \quad \overbrace{}^{n - n_1}
\begin{bmatrix}
s_1 & \cdots & s_1 & 0 & \cdots & 0 \\
\vdots & & \vdots & \vdots & & \vdots \\
s_{n_1} & \cdots & s_{n_1} & 0 & \cdots & 0 \\
s_{n_1+1} & \cdots & s_{n_1+1} & 0 & \cdots & 0 \\
\vdots & & \vdots & \vdots & & \vdots \\
s_n & \cdots & s_n & 0 & \cdots & 0
\end{bmatrix}
$$

In the notation of the last section, the sequence $\{s_1,\ldots,s_n\}$ is identified with $\{x_1,\ldots,x_n\}$, whereas $\{y_1,\ldots,y_n\}$ is defined by

$$y_i = \begin{cases} 1, & \text{if } i \leqslant n_1 \\ 0, & \text{otherwise} \end{cases}$$

Because ρ_I is the identity permutation, and the entry c_{ij} in the above matrix could be represented as $x_i y_j$,

$$\Gamma(\rho_I) = \sum_{i=1}^{n} x_i y_i = \sum_{i=1}^{n_1} s_i$$

In any case, after some simple algebra, we obtain

$$E(\Gamma(\rho)) = \frac{n_1}{n} \sum_i s_i$$

$$V(\Gamma(\rho)) = \frac{n_1(n-n_1)}{n(n-1)} \cdot \left[\sum_i s_i^2 - \frac{1}{n}\left(\sum_i s_i\right)^2 \right]$$

These formulas assume that the $n!$ possible assignments based on the matrix $\underset{\sim}{C}$ are equally likely, or equivalently, that the $\binom{n}{n_1}$ possible allocations of n_1 scores to group I are equally likely. This particular form of $\underset{\sim}{C}$ induces the latter distribution over the $\binom{n}{n_1}$ subsets because $n_1!\,(n-n_1)!$ different assignments lead to each possible allocation of n_1 scores to group I.

The most well-known example of this two-group index is Wilcoxon's rank sum statistic (see Lehmann, 1975). The n scores are the n ranks, and we wish to detect location differences between the groups through an extreme value for $\Gamma(\rho_I)$. When the ranks are untied, the mean and variance formulas are very simple and reduce to

$$E(\Gamma(\rho)) = \frac{1}{2}n_1(n+1)$$

$$V(\Gamma(\rho)) = \frac{1}{2}(n-n_1)n_1(n+1)$$

Because the distribution of $\Gamma(\rho)$ is symmetric when there are no ties, for a normalized statistic, we have $\Gamma^*(\rho_I) = -\Gamma^{**}(\rho_I)$, and thus, only $\Gamma^*(\rho_I)$ needs to be considered. In fact, $\Gamma^*(\rho_I)$ can be represented as $2h/[n_1(n - n_1)] - 1$, where h is the number of pairs of observations constructed from group I and group II in which the group I observation is higher. Thus, $h/[n_1(n - n_1)]$ is the probability that in a randomly selected pair, the group I observation is larger. The transformation provided by $\Gamma^*(\rho_I)$ merely scales this probability from -1 to +1. Somewhat more generally, the index $\Gamma^*(\rho_I)$ is defined, in words, as the quotient of (the mean of the scores in group I minus the grand mean)/(the maximum possible mean in group I minus the grand mean). The index $\Gamma^{**}(\rho_I)$ could be given a similar interpretation, merely by replacing the maximum possible mean for group I by the minimum.

As a second slightly different application that could just as well have been discussed in the contingency table context, suppose the n scores are dichotomous (zero-one). Then, the index $\Gamma(\rho_I)$ simply counts the number of ones in group I, and the data can be represented as frequencies in the well-known fourfold table appropriate for Fisher's exact test for a 2×2 contingency table:

		Group I	Group II	
	1	a	b	a + b
Score				
	0	c	d	c + d
		$a + c = n_1$	$b + d = n - n_1$	n

With this notation,

$$\Gamma(\rho_I) = a$$

$$E(\Gamma(\rho)) = \frac{(a + c)(a + b)}{n}$$

and

$$V(\Gamma(\rho)) = \frac{(a + c)(b + d)(a + b)(c + d)}{n^2(n - 1)}$$

There are now, however, four possible normalized statistics:

If $a + b \leqslant a + c$, then

$$\Gamma^*(\rho_I) = \frac{ad - bc}{(a + b)(b + d)} \qquad \Gamma^{**}(\rho_I) = \frac{bc - ad}{(a + b)(a + c)}$$

and if $a + b \leqslant a + c$, then

$$\Gamma^*(\rho_I) = \frac{ad - bc}{(a + c)(c + d)} \qquad \Gamma^{**}(\rho_I) = \frac{bc - ad}{(b + d)(c + d)}$$

It is interesting to note that the geometric means of the two $\Gamma^*(\rho_I)$ measures and of the two $\Gamma^{**}(\rho_I)$ measures are the same:

$$\frac{|ad - bc|}{\sqrt{(a + b)(b + d)(a + c)(c + d)}}$$

This latter expression is equal in absolute value to the usual fourfold point correlation.

Although we have defined the two sample index as the sum of the scores in group I, other possibilities exist that would be equivalent for purposes of significance testing. These would be defined by monotone transformations of the sum with forms that are invariant over all permutations. For example, the difference between the means in the two groups is one possibility, and the standard two-independent sample t-statistic is another. In terms of simplicity, however, the sum of scores in group I is probably the best representative. For some applications of these ideas to measuring properties of a free-recall sequence in psychology, the reader is referred to Hubert and Levin (1978).

1.3.5 Association Between Spatially Defined Variables (B)

In a recent article, Tjøstheim (1978) proposed an index of association between two variables observed over the same n geographic locations, say $O_1, O_2, \ldots, O_n$, that explicitly uses spatial information to help characterize the observed degree of correspondence. As notation, suppose the two variables, F and G, are both defined by the untied ranks from 1 to n. A general form for the observed value of Tjøstheim's index (in an unnormalized form) is given by

$$\Gamma(\rho_0) = \sum_i c_{i\rho_0(i)}$$

where c_{ij} is some appropriate measure of spatial separation or distance between locations O_i and O_j, and $\rho_0(i) = j$ if the rank on F observed in location O_i is the same as the rank on G observed in location O_j. In other words, $\Gamma(\rho_0)$ is the sum of distances between those locations that contain identical ranks on F and G, and thus specifies the degree to which identical rank values on F and G occupy geographical positions that are close spatially.

Measures of association that employ spatial information have as their primary use the evaluation of the correspondence between two variables when each is spatially autocorrelated, e.g., when similar values on F (and G) tend to be in close spatial locations (see Cliff and Ord, 1981, for a more explicit discussion of the topic). The usual measures of association between two variables (e.g., Kendall's tau, Spearman's rank order correlation, and so on), ignore the actual spatial positioning of the observation pairs and are concerned with indexing the similarity between two variables over identical spatial positions. The index given above essentially does the opposite. It indexes similarity between spatial positions for identical observations. The reader is referred to Glick (1982) for a Monte Carlo study that makes these same points empirically through several spatially defined processes that generate the variables F and G.

Typically, c_{ij} represents some standard distance function, although no explicit interpretation is really necessary for developing the methodology. The LA interpretation, for example, could be applied even when the locations are regions and where c_{ij} is a zero-one dichotomous variable that identifies simple spatial contiguity, although in this case a Poisson approximation, as mentioned in Section 1.2.2, may be more appropriate than the normal. Irrespective of how the spatial separation measures are keyed, either relatively large or small values of $\Gamma(\rho_0)$ may be of interest. For instance, suppose c_{ij} has the interpretation of a distance value in which small numbers indicate close locations. A notion of <u>positive</u> association would be indicated by a relatively small value of $\Gamma(\rho_0)$ in which identical ranks on F and G are in close locations; conversely, <u>negative</u> association would be indicated by a relatively large value for $\Gamma(\rho_0)$ in which identical ranks on F and G are in relatively more distant locations. A reversal in the keying of c_{ij} would result in a reversal of the interpretation for the extremes of $\Gamma(\rho_0)$. We should keep in mind, however, that the keying or directionality of the rankings is important, just as it was in the case of Spearman's footrule. Again, there is no natural concept of a conjugate ranking that would produce a symmetric distribution for $\Gamma(\rho)$ around its mean.

Our usual inference model essentially assumes that one set of ranks and their spatial locations are fixed (e.g., variable F) and the ranks for the second (e.g., variable G) are assigned at random to the n spatial locations. The effect of this can be seen very clearly for one special case of a separation measure defined by squared euclidean distances. In particular, if (x_i, y_i) and (x_j, y_j) denote the two coordinates for locations O_i and O_j, respectively, then the squared euclidean distance between O_i and O_j is

$$(x_i - x_j)^2 + (y_i - y_j)^2$$

and $\Gamma(\rho_0)$ can be written as

$$\text{constant} - 2 \sum_i (x_i x_{\rho_0(i)} + y_i y_{\rho_0(i)})$$

Thus, without loss of generality, we can define

$$c_{ij} = x_i x_j + y_i y_j \qquad (1.3.5a)$$

on the basis of the "active" part of the spatial separation measure. If we assume, also as a convenience, that the x and y coordinates have been standardized to mean zero and variance 1, then

$$E(\Gamma(\rho)) = 0$$

and

$$V(\Gamma(\rho)) = \left(\frac{2n^2}{n-1}\right)(1 + r_{xy}^2)$$

where r_{xy} is the Pearson correlation between the x and y coordinates over the n spatial locations. The observed index $\Gamma(\rho_0)$ is $n(r_{x_F x_G} + r_{y_F y_G})$, where $r_{x_F x_G}$ and $r_{y_F y_G}$ are the correlations between the x and y coordinates for identically ranked observations on F and G.

The extension to more than two dimensions is immediate. For example, if the location O_i is characterized by K coordinates,

$$V(\Gamma(\rho)) = \left(\frac{n^2}{n-1}\right)\left(K + 2 \sum_{i<j} r_{x_{(i)} x_{(j)}}^2\right)$$

where $r_{x_{(i)} x_{(j)}}$ denotes the Pearson correlation between the ith and jth coordinates over the spatial locations. The observed index is n times the sum of the K correlations between the coordinates (one for each dimension) for identically ranked observations on the two variables. Given the keying of the separation measures, the normalized version $\Gamma^*(\rho_0)$ (in two dimensions) would have the very simple form:

$$\Gamma^*(\rho_0) = \frac{\Gamma(\rho_0)}{\max_{\rho} \Gamma(\rho)}$$

$$= \frac{n(\Gamma_{x_F x_G} + \Gamma_{y_F y_G})}{\sum_i (x_i^2 + y_i^2)}$$

$$= \frac{r_{x_F x_G} + r_{y_F y_G}}{2}$$

That is, $\Gamma^*(\rho_0)$ is the average of the correlations defined over identically ranked observations on F and G between the x and y coordinates. The extension to K dimensions should be obvious because it just requires the average of the K correlations between coordinates. Unfortunately, there does not appear to be a comparable simple reduction for $\Gamma^{**}(\rho_0)$, even though $\min_\rho \Gamma(\rho)$ is fairly easy to find: The x coordinates are first ranked from smallest to largest and then the reverse; when the two sums of cross-products for the x and y coordinates are added together, this quantity defines the minimum. We do note, however, that because $-2n$ provides a lower bound to $\min_\rho \Gamma(\rho)$ in this two-dimensional case, $\Gamma^{**}(\rho_0)$ would have the same form as $\Gamma^*(\rho_0)$ if this bound were used, but would be opposite in sign. Thus, $\Gamma^*(\rho_0)$ may be relied on as a single index much as the correlation coefficient of Section 1.3.3, and again with the same understanding that negative values are meaningful.

The notion of association in a spatial context has a number of other applications beyond those that relate to two ranked variables. For example, suppose we are given two different sets of n spatial locations, $\{O_1, O_2, \ldots, O_n\}$ and $\{p_1, p_2, \ldots, p_n\}$, and some conjectured matching, ρ_0, defined from the first set of locations to the second. If c_{ij} is a spatial separation measure between O_i and p_j, then $\Gamma(\rho_0)$ is an index of spatial bias for the conjectured matching. For example, Tobler (1964) considers a problem in which the first n locations are residences of brides and the second set of n locations are residences of the respective grooms. When the matching ρ_0 defines the actual married couples, $\Gamma(\rho_0)$ would be a raw index of spatial bias in the selection of spouse. Or possibly, the first n locations could specify the origins of movement for n individuals, with the second set containing the n destinations, e.g., we could be given residences of criminals along with the corresponding places of crime occurrence. The index $\Gamma(\rho_0)$ then provides a measure of spatial bias in the commission of the crimes when ρ_0 matches origin and destination.

Some further applications of this type are discussed in Chapter 2 along with several extensions of the association notion. In particular, we develop indices for variables that do not define complete sets of untied ranks and/ or that rely on more information than just the spatial separation between identical ranks on F and G. A discussion is also given in Section 2.3.1, as to how this association measure in a spatial framework relates to Spearman's rank order correlation.

1.4 TECHNICAL ISSUES IN THE LA MODEL

1.4.1 Moments of the Index $\Gamma(\rho)$ (C)

When the first two moments of $\Gamma(\rho)$ were presented, a discussion of their derivation was delayed. We now illustrate a very general strategy for finding moments that is used implicitly throughout the book. The approach is based on a zero-one indicator function for an arbitrary permutation ρ defined as

$$I_\rho(i,j) = \begin{cases} 1, & \text{if } \rho(i) = j \\ 0, & \text{otherwise} \end{cases}$$

Using this indicator, $\Gamma(\rho)$ can be rewritten as

$$\Gamma(\rho) = \sum_{i,j} c_{ij} I_\rho(i,j)$$

Indicators have the one important property that their expectation equals the probability of the defining event, e.g., $E(I_\rho(i,j)) = P(\rho(i)=j)$. Thus, to obtain the mean of $\Gamma(\rho)$:

$$E(\Gamma(\rho)) = E\left\{\sum_{i,j} c_{ij} I_\rho(i,j)\right\}$$

$$= \sum_{i,j} c_{ij} E\{I_\rho(i,j)\}$$

$$= \sum_{i,j} c_{ij} P(\rho(i)=j) = \frac{1}{n} \sum_{i,j} c_{ij}$$

Here, we use the fact that $P(\rho(i)=j) = 1/n$ when all permutations are assumed equally likely.

The second moment follows in a slightly more complicated way through index restriction:

$$E([\Gamma(\rho)]^2) = E\left[\sum_{i,j,r,s} c_{ij}c_{rs}I_\rho(\rho(i)=j)I_\rho(\rho(r)=s)\right]$$

$$= \sum_{i,j,r,s} c_{ij}c_{rs}P(\rho(i)=j \text{ and } \rho(r)=s)$$

$$= \sum_{i,j} c_{ij}^2 P(\rho(i)=j)$$

$$+ \sum_{\substack{i,j,r,s \\ i\neq r \\ j\neq s}} c_{ij}c_{rs}P(\rho(i)=j \text{ and } \rho(r)=s)$$

Because permutations are one-to-one functions, either $i = r$ and $j = s$, giving the first term; or $i \neq r$ and $j \neq s$, giving the second. This pair of terms reduce further when the statements that $P(\rho(i)=j$ and $\rho(r)=s) = 1/[n(n-1)]$ and $P(\rho(i)=j) = 1/n$ are used, i.e.,

$$\frac{1}{n}\sum_{i,j} c_{ij}^2 + \frac{1}{n(n-1)}\sum_{\substack{i,j,r,s \\ i\neq r \\ j\neq s}} c_{ij}c_{rs}$$

The second term itself can now be written:

$$\sum_{\substack{i,j,r,s \\ i\neq r \\ j\neq s}} c_{ij}c_{rs} = \left(\sum_{i,j} c_{ij}\right)^2 - \sum_i c_{i\cdot}^2 - \sum_j c_{\cdot j}^2 + \sum_{i,j} c_{ij}^2$$

Finally, because $V(\Gamma(\rho)) = E([\Gamma(\rho)]^2) - [E(\Gamma(\rho))]^2$, and after some simplification, our earlier formula for the variance is achieved.

In this same manner, the raw third moment can be found:

$$E(\{\Gamma(\rho)\}^3) = \frac{n}{(n-1)(n-2)}\sum_{i,j} d_{ij}^3$$

where $d_{ij} = c_{ij} - \overline{c}_{\cdot j} - \overline{c}_{i\cdot} + \overline{c}_{\cdot\cdot}$ in accordance with earlier notation in (1.2.2b). In turn, an explicit formula for the skewness parameter, γ, can be obtained as

$$\gamma = \frac{n(n-1)^{1/2}}{n-2} \frac{\sum\limits_{i,j} d_{ij}^3}{\left(\sqrt{\sum\limits_{i,j} d_{ij}^2}\right)^3}$$

where $\gamma = \{ E(|\Gamma(\rho)|^3) - 3E(\Gamma(\rho))V(\Gamma(\rho)) - [E(\Gamma(\rho))]^3 \}/[V(\Gamma(\rho))]^{3/2}$. This last expression is important for one approximate significance testing strategy mentioned in the next section.

Various mean and variance expressions are given throughout the book, usually without an explicit derivation. In principle, however, all of these formulas could be obtained using the indicator method and a lot of very tedious algebra. We will leave it to the reader to carry out the necessary mechanics. An alternative discussion of the moments of the LA statistic, including an explicit formula for the fourth moment, is given by Hajek and Sidak (1967, p. 82).

1.4.2 Significance Testing Procedures for the LA Model (A)

The most theoretically appropriate significance testing method is through complete enumeration of $\Gamma(\rho)$ over the set of n! equally likely permutations. When n is even moderate in size, however, the computational burden that complete enumeration imposes is enormous. For this reason alone, it is necessary to develop alternatives. The one obvious possibility of using an asymptotic distribution, such as the normal, to approximate the results obtained with an exhaustive listing has already been relied on in the earlier sections. We mention three other alternatives here and illustrate their use with the data of Table 1.2.3a.

The first alternative is based on what is called <u>Cantelli's inequality</u> and is very crude. At times, however, it may be sufficient to generate a significance level that is small enough to make the more laborious methods redundant. In particular, for any distribution with a finite second moment, $P(Z > k) \le 1/(k^2 + 1)$ and $P(Z < -k) \le 1/(k^2 + 1)$. Thus, an observed Z value of, say, 4.36 would lead to an upper-tail significance level of at most .05. Obviously, if approximate normality holds, the use of Cantelli's inequality is very conservative. Nevertheless, for very large Z values, it still may be all that is necessary to convince someone that $\Gamma(\rho_0)$ is significant at some reasonable level.

A second strategy for constructing a significance test, and one that we will rely on heavily throughout the book, can be developed using the Monte Carlo paradigm discussed by Hope (1968), Edgington (1969), and others. In this approach, M permutations are selected at random and with replacement from the complete enumeration. The one-tailed lower (upper) significance level of the observed statistic is identified with the proportion of M + 1 index values (with M selected randomly and with the single

observed measure which is assumed to be another random draw under the
null conjecture) that are as small (large) or smaller (larger) than the obtained value. Thus, if M = 999 and 50 indices in the total set of 1000 are
as small or smaller than the observed statistic, an .05 significance level
is declared. In fact, on the basis of an M of 999, we are assured, with
probability .99, that an obtained test statistic judged significant at the .01
(.05) level if the complete distribution were available will be given a significance level no larger than .018 (.066) by use of the Monte Carlo significance test (see Edgington, 1969, pp. 152-155, for a proof and extended
discussion). In general, Monte Carlo significance testing will be relied on
heavily throughout this book and could be considered our method of choice.
Although Z statistics will be obtained routinely, any significance level generated from an assumed normal approximation will typically be "verified"
by random sampling from the exact distribution. This process is particularly important for several nonpathological applications, mentioned in
Chapter 4, where asymptotic normality is very troublesome.

   Our final significance testing strategy is based on fitting a function of
a very specific form with the use of the first three moments of the exact
permutation distribution. Following the lead of Mielke, Berry, and Brier
(1981) in a closely related context, a Pearson type III distribution (the
gamma) is used as an approximation to the distribution of Z. The type III
distribution is standardized to have mean zero, variance 1, and skewness
parameter γ. Thus, because tables are available, e.g., Salvosa (1930)
and Harter (1969), all that is required operationally is to compare the
observed value of Z to the tabled percentage points of the appropriate distribution, where the latter is characterized by a particular value of γ.

   To be more explicit, the functional form that is used to approximate
the exact permutation distribution, for positive γ, can be given as follows:

$$F(y) = \frac{(2/\gamma)^{4/\gamma^2}}{\Gamma(4/\gamma^2)} \cdot \left(\frac{2 + y\gamma}{\gamma}\right)^{(4/\gamma^2)-1} \cdot e^{-2(2+y\gamma)/\gamma^2}$$

where $-2/\gamma < y < \infty$, and $\Gamma(\cdot)$ is the usual gamma function (and not our LA
index). This distribution has mean zero, variance 1, and skewness γ. As
$\gamma \longrightarrow 0$, f(y) approaches the normal density; for $\gamma < 2$, f(y) is unimodal,
and for $\gamma \geq 2$, it is J-shaped. For example, to calculate an upper-tailed
significance level for an observed value, Z, we would find the area from
Z to ∞ under the density f(y). For negative skewness and lower-tailed
probabilities, we could merely calculate the area from $-Z$ to ∞ under the
same density f(y), using the absolute value of γ.

   Table 1.4.2a gives a few of the values that Z would have to have for
upper-tail significance levels of .10, .05, .01, and .001, and for positive
skewness values of .0, .1, .2, ..., 2.0. Obviously, for positive skewness, the use of a normal table would lead to the assertion of smaller
upper-tail significance probabilities.

Table 1.4.2a Critical Values for Z, Given Different Positive
Skewness Values—Upper-Tail Probabilities

	Significance level			
Skewness	.10	.05	.01	.001
.0 (normal)	1.28	1.64	2.33	3.00
.1	1.29	1.67	2.40	3.23
.2	1.30	1.70	2.47	3.38
.3	1.31	1.73	2.54	3.52
.4	1.32	1.75	2.62	3.67
.5	1.32	1.77	2.69	3.81
.6	1.33	1.80	2.76	3.96
.7	1.33	1.82	2.82	4.10
.8	1.34	1.84	2.89	4.24
.9	1.34	1.86	2.96	4.39
1.0	1.34	1.88	3.02	4.53
1.1	1.34	1.89	3.09	4.67
1.2	1.34	1.91	3.15	4.81
1.3	1.34	1.92	3.21	4.96
1.4	1.34	1.94	3.27	5.10
1.5	1.33	1.95	3.33	5.23
1.6	1.33	1.96	3.39	5.37
1.7	1.32	1.97	3.44	5.51
1.8	1.32	1.98	3.50	5.64
1.9	1.31	1.99	3.55	5.78
2.0	1.30	2.00	3.61	5.91

More complete tables are given by Salvosa (1930) and Harter (1969),
or they can be constructed rather easily by using a common subroutine for
evaluating cumulative probabilities for the gamma distribution in the Inter-
national Mathematical-Statistical Library series available at most computer
facilities. In particular, for upper-tail probabilities corresponding to an
observed value Z and positive skewness, we merely evaluate

Table 1.4.2b   Sample Cumulative Probability for $\Gamma(\rho)$ and Z
Based on a Sample Size of 1000—Evaluation of
the Death–Dip Conjecture

Sample cumulative probability	Γ	Z
.001	15	-2.84
.002	17	-2.43
.005	18	-2.23
.010	19	-2.02
.040	21	-1.62
.070	22	-1.42
.100	23	-1.22
.200	25	- .81
.300	27	- .41
.400	28	- .20
.500	29	.00
.600	30	.20
.700	32	.61
.800	33	.81
.900	36	1.42
.930	37	1.62
.960	38	1.83
.990	41	2.44
.995	41	2.44
.998	45	3.25
.999	46	3.45
1.000	46	3.45

$$1 - \int_0^b \frac{x^{\theta-1}}{\Gamma(\theta)} e^{-x} dx$$

where $\theta = 4/\gamma^2$ and $b = [Z + (2/\gamma)](2/\gamma)$. When the user specifies θ and b, the integral from 0 to b can be directly evaluated by a single call to the IMSL subroutine MDGAM.

As an example of how these procedures operate in practice, we reconsider the data in Table 1.2.3a and the test of the death-dip hypothesis based on the 12! possible assignments of row to column categories. As discussed in Section 1.2.3, $\Gamma(\rho_0) = 16$, $E(\Gamma(\rho)) = 29$, $V(\Gamma(\rho)) = 24.2727$, and the Z statistic is -2.64. If we assume a normal approximation, a significance level (lower-tail) of .004 would be obtained. The skewness parameter is $+.046$ and the Pearson type III approximation would also result in a significance value of .004, up to three decimal places. Table 1.4.2b gives a sample cumulative distribution based on 1000 randomly selected permutations. (A sample size of 1000 is used rather than, say 999, to give an easy way of obtaining the cumulative probabilities without rounding. In the remainder of the book, however, we will use 999 and provide cumulative frequencies rather than cumulative probabilities.) As can be seen from this sample distribution, a Monte Carlo significance level of $(1 + 1)/(1000 + 1) =$ .002 would be declared because only 1 permutation out of the 1000 that were randomly generated gave an index less than or equal to the observed value of 16. In general, all these significance values are fairly close. The very conservative Cantelli inequality, however, would not be very appropriate here because $1/(k^2 + 1) = .125$ when $k = -2.64$.

1.4.3 Correlations Between LA Indices (C)

In addition to the study of a single assignment statistic, it is relatively straightforward to calculate the correlation between two such indices over the n! possible permutations. Exactly the same indicator argument presented in Section 1.4.1 would be used. In general, if c_{ij} and c'_{ij} denote two possible C matrices, and d_{ij} and d'_{ij} are obtained from c_{ij} and c'_{ij} as in equation (1.2.2b), then the covariance between the corresponding measures, $\Gamma(\rho)$ and $\Gamma'(\rho)$, can be given as

$$\text{cov}(\Gamma(\rho), \Gamma'(\rho)) = \left(\frac{1}{n-1}\right) \sum_{i,j} d_{ij} d'_{ij}$$

Normalizing by the square roots of the variances, we obtain

$$\text{Correlation} \left(\Gamma(\rho), \ \Gamma'(\rho) \right) = \frac{\displaystyle\sum_{i,j} d_{ij} d'_{ij}}{\sqrt{\displaystyle\sum_{i,j} d^2_{ij} \sum_{i,j} d'^2_{ij}}}$$

Consequently, because $\Sigma_{i,j} d_{ij} = \Sigma_{i,j} d'_{ij} = 0$, the correlation between the indices is identical to the correlation between the entries in the two interaction matrices.

As one very obvious application of this correlation idea, suppose c_{ij} and c'_{ij} are defined in a product form from the numerical sequences $\{x_1, \ldots, x_n\}$; $\{y_1, \ldots, y_n\}$ and $\{x'_1, \ldots, x'_n\}$; $\{y'_1, \ldots, y'_n\}$, respectively:

$$c_{ij} = x_i y_j$$

$$c'_{ij} = x'_i y'_j$$

Then, $d_{ij} = (x_i - \overline{x})(y_j - \overline{y})$; $d'_{ij} = (x'_i - \overline{x}')(y'_j - \overline{y}')$; and correlation $\left(\Gamma(\rho), \Gamma'(\rho) \right)$ equals

$$\frac{\displaystyle\sum_{i,j} (x_i - \overline{x})(y_j - \overline{y})(x'_i - \overline{x}')(y'_j - \overline{y}')}{\sqrt{\displaystyle\sum_{i,j} (x_i - \overline{x})^2 (y_j - \overline{y})^2 \sum_{i,j} (x'_i - \overline{x}')^2 (y'_j - \overline{y}')^2}} = r_{xx'} r_{yy'}$$

where $r_{xx'}$ and $r_{yy'}$ are the Pearson correlations between the indicated sequences, i.e., between x and x' or y and y'. Thus, the degree to which the LA indices are correlated is a direct function of the degree to which the sequences used to define the assignment scores are correlated.

In addition to this general result, which is appropriate for assignment scores having a product form, we give three more specific examples as representative illustrations. First, if, as in Section 1.3.3, we compare two numerical sequences $\{x_1, \ldots, x_n\}$ and $\{y_1, \ldots, y_n\}$ consisting of untied ranks and let $c_{ij} = (i - j)^2$ and $c'_{ij} = |i - j|$, then

$$\text{Correlation} \left(\Gamma(\rho), \ \Gamma'(\rho) \right) = \frac{3}{\sqrt{10}} \left\{ \frac{n^2 + 1}{\sqrt{(n^2 - 1)[n^2 + (7/2)]}} \right\}$$

Surprisingly, this correlation is bounded away from 1 for large n and approaches $3/\sqrt{10}$ as $n \longrightarrow \infty$. For the quadrant and squared difference specifications, the correlation is $n^2(3/4)/(n^2 - 1)$ for n even and $(n + 1)(3/4)/n$

for n odd, and both values converge to $3/4$ as $n \rightarrow \infty$. Given the size of these two correlations, one would expect a greater degree of consistency between the squared and absolute difference measures than between those constructed by the squared difference and quadrant definitions. This situation contrasts with an asymptotic correlation of 1.00 between the usual definition of Spearman's rank order correlation based on $c_{ij} = (i - j)^2$ and Kendall's tau statistic (Kendall, 1970, p. 80).

As a second example, suppose we are in the framework of linear permutation statistics and consider two sets of scores $s_1, s_2, \ldots, s_n$, and $s_1', \ldots, s_n'$. One set is given by the untied ranks to give Wilcoxon's rank sum statistic, i.e., $s_i = i$ for $1 \leqslant i \leqslant n$. The second set is dichotomous (zero-one) and gives the "a" term in the fourfold table if $s_i' = 1$ for $1 \leqslant i \leqslant T$, and 0 otherwise. In the notation of Section 1.3.4, the point of dichotomy T would be equal to $a + c$. The correlation between these two statistics is

$$- \left[\frac{3T(n - T)}{(n - 1)(n + 1)} \right]^{1/2}$$

which, surprisingly, depends only on T and not on the sizes of the two groups (see David and Barton, 1962, p. 190). When T is close to the median object, this correlation is close to its maximum absolute value. Thus, these two statistics are most consistent when the row sums in the fourfold table are nearly equal.

Finally, in the association framework of Section 1.3.5, suppose c_{ij} defines a symmetric spatial separation matrix between the n geographical locations and $c_{ij}' = (i - j)^2$. In other words, our concern is with the relationship between Spearman's rank order correlation and the measure based on c_{ij}. In general,

$$\text{Correlation}(\Gamma(\rho), \Gamma'(\rho)) = \frac{- \frac{(n+1)^2}{2} \sum_{i,j} c_{ij} + 2(n + 1) \sum_i i \sum_j c_{ij} - 2 \sum_{i,j} ijc_{ij}}{\left(\frac{n^3 - n}{6} \right) \sqrt{\frac{1}{n} \left(\sum_{i,j} c_{ij} \right)^2 - \frac{2}{n} \sum_i \left(\sum_j c_{ij} \right)^2 + \sum_{i,j} c_{ij}^2}}$$

If c_{ij} is again the reduced squared euclidean measure, this expression simplifies to:

$$- \frac{r^2_{Fx_F} + r^2_{Fy_F}}{\sqrt{2(1 + r^2_{xy})}}$$

where r_{Fx_F} and r_{Fy_F} are the correlations between the ranks on F and the
x and y coordinates that correspond to these ranks, respectively; r_{xy} is
again the correlation of the x and y coordinates over the n spatial locations.

There are two points to be made regarding this last correlation. First,
it depends on which variable we arbitrarily assume is fixed (F and G) because a direct analog is possible when the ranks of G are fixed in constructing the reference distribution for $\Gamma(\rho)$ merely by replacing $r_{Fx_F}^2$ by

$r_{Gx_F}^2$ and $r_{Fy_F}^2$ by $r_{Gy_G}^2$. Secondly, there is in general a nonzero correlation between the Spearman index and the Tjøstheim measure, suggesting
that the assignment inference model for the latter is contaminated to some
extent with an association statistic that is aspatial in design. Some of the
later chapters will address this problem again and offer alternative inference models to resolve this apparent difficulty.

Besides discussing the correlation between two indices descriptively,
we may be able to go one step further and assert asymptotic joint normality.
In particular, Fraser (1956) shows that if the two indices satisfy the usual
regularity condition for asymptotic normality separately and if their correlation has a limit, the pair of indices $(\Gamma(\rho), \Gamma'(\rho))$ are asymptotically bivariate normal. The same type of statement can be extended to more than two
indices and to the assertion of multivariate normality as long as all the
intercorrelations converge to limits.

REFERENCES

Agresti, A., D. Wackerly, and J. Boyett. Exact conditional tests for
    cross-classifications: Approximation of obtained significance levels,
    Psychometrika 44 (1979), 75-83.
Bishop, Y. M. M., S. E. Fienberg, and P. W. Holland. Discrete
    Multivariate Analysis: Theory and Practice, The MIT Press, Cam-
    bridge, Mass., 1975.
Cliff, A. D., and J. K. Ord. Spatial Processes: Models and Applications,
    Pion, London, 1981.
Cohen, J. Weighted kappa: Nominal scale agreement with provision for
    scaled disagreement or partial credit, Psychol. Bull. 70 (1968), 213-
    220.
Cohen, J. Weighted chi-square: An extension of the kappa method,
    Educ. Psychol. Meas. 32 (1972), 61-74.
David, F. N., and D. E. Barton. Combinatorial Chance, Hafner, New
    York, 1962.
Edgington, E. S. Statistical Inference: The Distribution-Free Approach,
    McGraw-Hill, New York, 1969.

Fraser, D. A. S. A vector form of the Wald-Wolfowitz-Hoeffding theorem, Ann. Math. Stat. 27 (1956), 540-543.

Fliess, J. L., J. Cohen, and B. S. Everitt. Large sample standard errors of kappa and weighted kappa, Psychol. Bull. 72 (1969), 323-327.

Froman, T., and L. J. Hubert. Application of prediction analysis to developmental priority, Psychol. Bull. 87 (1980), 136-146.

Froman, T., and L. J. Hubert. A reply to Moshman's critique of prediction analysis and developmental priority, Psychol. Bull. 88 (1981), 188.

Garfinkel, R. S., and G. L. Nemhauser. Integer Programming, Wiley, New York, 1972.

Glick, B. J. A spatial rank-order correlation measure, Geogr. Analysis 14 (1982), 178-181.

Goodman, L. A., and W. H. Kruskal. Measures of association for cross classifications, J. Am. Stat. Assoc. 49 (1954), 732-764.

Hájek, J., and Z. Šidák. Theory of Rank Tests, Academic Press, New York, 1967.

Harter, H. L. A new table of percentage points of the Pearson type III distribution, Technometrics 11 (1969), 177-187.

Hildebrand, D., M. Laing, and A. Rosenthal. Prediction Analysis of Cross Classifications, Wiley, New York, 1977.

Hoeffding, W. A combinatorial central limit theorem, Ann. Math. Stat. 22 (1951), 558-566.

Hope, A. C. A. A simplified Monte Carlo significance test procedure, J. R. Stat. Soc. [B] 30 (1968), 582-598.

Hostelling, H. The selection of variates for use in prediction with some comments on the general problem of nuisance parameters, Ann. Math. Stat. 11 (1940), 271-283.

Hubert, L. J. A relationship between the assignment problem and some simple statistical techniques, Qual. Quant. 10 (1976), 341-348.

Hubert, L. J. Kappa revisited, Psychol. Bull. 84 (1977), 289-297.

Hubert, L. J. A general formula for the variance of Cohen's weighted kappa, Psychol. Bull. 85 (1978), 183-184.

Hubert, L. J. Alternative inference models based on matching for a weighted index of nominal scale response agreement, Qual. Quant. 14 (1980), 711-725.

Hubert, L. J. Statistical applications of linear assignment, Psychometrika 49 (1984), 449-473.

Hubert, L. J., and J. R. Levin. Evaluating priority effects in free recall, Br. Math. Stat. Psychol. 31 (1978), 11-18.

Kendall, M. G. Rank Correlation Methods, 4th ed., Griffin, London, 1970.

Kendall, M. G., and A. Stuart. The Advanced Theory of Statistics, vol. 2, 4th ed., Macmillan, New York, 1979.

Lehmann, E. Nonparametrics: Statistical Methods Based on Ranks, Holden-Day, San Francisco, 1975.

McHugh, R. B., and P. C. Apostolakos. Methodology for the comparison of clinical with actuarial predictions, Psychol. Bull. 56 (1959), 301-308.

Mielke, P. W., K. J. Berry, and G. W. Brier. Application of multi-response permutation procedures for examining seasonal changes in monthly mean sea-level pressure patterns, Monthly Weather Rev. 109 (1981), 120-126.

Mosteller, F., and R. R. Bush. Selected quantitative techniques, in G. Lindsey (Ed.), Handbook of Social Psychology, vol. 1, Addison-Wesley, Cambridge, Mass., 1954.

Phillips, D. P. Deathday and birthday: An unexpected connection, in Tanur, J. M., et al. (Eds.), Statistics: A Guide to the Unknown, Holden-Day, San Francisco, 1978, pp. 71-85.

Puri, M. L., and P. K. Sen. Nonparametric Methods in Multivariate Analysis, Wiley, New York, 1971.

Salvosa, L. R. Tables of Pearson's type III function, Ann. Math. Stat. 1 (1930), 191-198.

Thorndike, R. L. The problem of classification of personnel, Psychometrika 15 (1950), 215-235.

Tjøstheim, D. A measure of association for spatial variables, Biometrika 65 (1978), 109-114.

Tobler, W. Computation of the correspondence of geographical patterns, Papers Regional Sci. Assoc. 12 (1964), 131-139.

Vernon, P. E. The matching method applied to investigations of personality, Psychol. Bull. 33 (1936), 149-177.

Ward, J. H. The counseling assignment problem, Psychometrika 23 (1958), 55-65.

2

Evaluating the Correspondence Between Two Rectangular Matrices of the Same Size

2.1 INTRODUCTION (A)[*]

This chapter presents a generalization of the LA model to the task of comparing two rectangular data matrices of the same size. As background and as a way of motivating the extensions to be pursued, it may be helpful if we start with some of the same basic assignment ideas developed in Chapter 1 but rephrased using a slightly different notation. A strategy for the comparison of rectangular matrices is then suggested merely by relaxing several of the restrictions that need to be imposed to obtain the usual LA interpretation.

The LA index $\Gamma(\rho)$ is written in its typical form through the single matrix $\underset{\sim}{C} = \{c_{ij}\}$:

$$\Gamma(\rho) = \sum_i c_{i\rho(i)}$$

However, suppose a second $n \times n$ matrix $\underset{\sim}{B} = \{b_{ij}\}$ is defined as follows:

$$b_{ij} = \begin{cases} 1, & \text{when } i = j \\ 0, & \text{otherwise} \end{cases}$$

Then trivially,

$$\Gamma(\rho) = \sum_i c_{i\rho(i)} = \sum_i \sum_j b_{ij} c_{i\rho(j)}$$

---

[*] The reader is referred to the Preface for an explanation of the use of (A), (B), and (C) in text headings.

In words, $\Gamma(\rho)$ is a cross-product measure between two matrices, $\{b_{ij}\}$ and $\{c_{i\rho(j)}\}$, where $\{b_{ij}\}$ tells us which elements in the matrix $\{c_{i\rho(j)}\}$ to sum, i.e., those down the main left-to-right diagonal. The matrix $\{c_{i\rho(j)}\}$ contains the same entries as did $\underset{\sim}{C}$, but the columns have been interchanged by the permutation ρ. The $\rho(j)$ column in $\underset{\sim}{C}$ is now the jth column of $\{c_{i\rho(j)}\}$.

Given this framework, the present chapter is concerned with two extensions. First, the restriction is removed that $\{b_{ij}\}$ is necessarily dichotomous with a single 1 in each row and column. Second, the condition that $\underset{\sim}{B}$ and $\underset{\sim}{C}$ are square is relaxed as long as both have the same size. As notation, $\underset{\sim}{B}$ and $\underset{\sim}{C}$ are two $p \times q$ matrices, and the cross-product index will be denoted by $\Lambda(\rho)$:

$$\Lambda(\rho) = \sum_{i,j} b_{ij} c_{i\rho(j)}$$

Various applications of this measure will again require the identification of some particular matching or permutation ρ_0. The observed index $\Lambda(\rho_0)$ is then compared to all q! equally likely values of $\Lambda(\rho)$.

The topic of significance testing for an observed index $\Lambda(\rho_0)$ will turn out to be fairly simple because a convenient reduction actually makes it part of the assignment strategy of Chapter 1. Because these results are helpful in demonstrating the utility of the $\Lambda(\rho_0)$ measure later on, we begin with this discussion in the next section before a number of applications are presented.

2.2 REDUCTION OF THE RECTANGULAR MATRIX COMPARISON TASK TO A LINEAR ASSIGNMENT PROBLEM (A)

Although it should be obvious that the assignment index $\Gamma(\rho)$ is a special case of $\Lambda(\rho)$, we can also show the converse. Given the definition of the index $\Lambda(\rho)$, we merely have to follow through a number of identities:

$$\Lambda(\rho) = \sum_{i,j} b_{ij} c_{i\rho(j)}$$

$$= \sum_{i,j} \sum_{k} b_{ij} c_{ik} I_\rho(j,k)$$

$$= \sum_{j} \sum_{k} I_\rho(j,k) \sum_{i} b_{ij} c_{ik}$$

$$= \sum_j \sum_k s_{jk} I_\rho(j,k)$$

$$= \sum_j s_{j\rho(j)}$$

where

$$I_\rho(j,k) = \begin{cases} 1, & \text{if } \rho(j) = k \\ 0, & \text{otherwise} \end{cases}$$

and

$$s_{jk} = \sum_i b_{ij} c_{ik}$$

Stated another way, $\Lambda(\rho)$ is a linear assignment (LA) index of the form $\Gamma(\rho)$, where the (reduced) matrix $\underset{\sim}{R}_c = \{s_{jk}\}$ is of size $q \times q$. Each entry in the latter matrix represents a raw measure of correspondence between a particular pair of columns, one chosen from $\underset{\sim}{B}$ and one chosen from $\underset{\sim}{C}$. The subscript c on $\underset{\sim}{R}_c$ refers to the fact that a matching was done between the columns of $\underset{\sim}{B}$ and $\underset{\sim}{C}$. Obviously, significance testing and the corresponding descriptive normalizations can proceed exactly as in Chapter 1, but through the reduced matrix $\underset{\sim}{R}_c$.

As a very simple numerical illustration, suppose $\underset{\sim}{B}$ and $\underset{\sim}{C}$ are defined as the two 4×3 matrices:

$$\underset{\sim}{B} = \begin{bmatrix} 1 & 7 & 3 \\ 4 & 2 & 1 \\ 3 & 5 & 2 \\ 4 & 5 & 6 \end{bmatrix} \qquad \underset{\sim}{C} = \begin{bmatrix} 1 & 2 & 5 \\ 2 & 3 & 7 \\ 4 & 1 & 6 \\ 3 & 5 & 4 \end{bmatrix}$$

and suppose ρ_0 is the identity permutation: $\rho_0(1)=1$, $\rho_0(2)=2$, $\rho_0(3)=3$. Then, $\Lambda(\rho_0) = (1\cdot1 + 7\cdot2 + 3\cdot5) + (4\cdot2 + 2\cdot3 + 1\cdot7) + (3\cdot4 + 5\cdot1 + 2\cdot6) + (4\cdot3 + 5\cdot5 + 6\cdot4) = 141$. Alternatively, the 3×3 matrix $\underset{\sim}{R}_c = \{s_{jk}\}$, which is

$$\begin{bmatrix} 33 & 37 & 67 \\ 46 & 50 & 99 \\ 31 & 41 & 58 \end{bmatrix}$$

and $\Gamma(\rho_0) = 33 + 50 + 58 = 141$.

The mean and variance formulas for $\Lambda(\rho)$ follow directly from those given for $\Gamma(\rho)$:

$$E(\Lambda(\rho)) = \frac{1}{q} \sum_k b_{k \cdot} c_{k \cdot}$$

where $b_{k \cdot}$ and $c_{k \cdot}$ are the sums in row k for $\underset{\sim}{B}$ and $\underset{\sim}{C}$, respectively;

$$V(\Lambda(\rho)) = \frac{1}{q(q-1)} \left[\frac{1}{q} E_1 - (E_2 + E_3) + q E_4 \right]$$

where

$$E_1 = \left(\sum_k b_{k \cdot} c_{k \cdot} \right)^2$$

$$E_2 = \sum_j \left(\sum_k b_{k \cdot} c_{kj} \right)^2$$

$$E_3 = \sum_i \left(\sum_k c_{k \cdot} b_{ki} \right)^2$$

$$E_4 = \sum_i \sum_j \left(\sum_k b_{ki} c_{kj} \right)^2$$

In comparing two rectangular matrices, however, it may be just as convenient to obtain $\underset{\sim}{R}_c$ and use the assignment moments in (1.2.2a) directly.

Because the matrices being compared are rectangular, the choice is not arbitrary as to whether the reference distribution is constructed over all permutations of the columns, as we assumed, or alternatively, over all permutations of the rows. This contrasts with the situation in Chapter 1, where instead of the simple assignment index $\Gamma(\rho) = \Sigma_i c_{i\rho(i)}$, we could have used an index of the form $\Sigma_j c_{\phi(j)j}$ throughout. Here, ϕ merely denotes a permutation that assigns columns to rows. Both statistics would have exactly the same distribution over all n! permutations.

In general, then, a comparison of two rectangular matrices requires a decision as to whether the rows or the columns are to be matched at random in constructing the reference distribution, i.e., whether we use

$$\sum_{i,j} b_{ij} c_{i\rho(j)} \quad \text{or} \quad \sum_{i,j} b_{ij} c_{\phi(i)j}$$

Obviously, the moment formulas can be adapted easily for the matching of rows by simply taking the transposition of $\underset{\sim}{B}$ and $\underset{\sim}{C}$ and defining the problem on columns instead.

When rows are matched, we use the symbol $\Lambda(\phi)$ to make the distinction:

$$\Lambda(\phi) = \sum_{i,j} b_{ij} c_{\phi(i)j}$$

The matrix $\underset{\sim}{R}_r$ needed for a reduction to an LA problem would be of size $p \times p$, where

$$\underset{\sim}{R}_r = \{t_{jk}\} = \left\{ \sum_i b_{ji} c_{ki} \right\}$$

The subscript r on $\underset{\sim}{R}_r$ indicates that the matching is being carried out between rows. Usually, the context of the problem will also make it clear whether the assignment is to be defined on rows or on columns.

There is one equivalence between row and column matching that should be noted. If ρ_0 and ϕ_0 are both the identity permutation, then $\phi(\rho_0) = \Lambda(\phi_0)$, and this common value can be obtained by summing the main left-to-right diagonal entries in either of the reduced matrices $\underset{\sim}{R}_c$ or $\underset{\sim}{R}_r$. The major operational differences in matching by rows or by columns are in the reference distributions, and consequently, in the expectation and variance terms. Irrespective of whether the matching is defined on rows or columns, however, both reduced matrices should be obtained, or better, the interaction matrices constructed from each. The latter show explicitly which particular rows and columns are contributing to the raw index, $\Lambda(\rho_0)$ or $\Lambda(\phi_0)$, and to what extent.

In summary, when a particular permutation ρ_0 is identified and is defined on columns, the raw index $\Lambda(\rho_0)$ indicates how similar the q columns are between $\underset{\sim}{B}$ and $\underset{\sim}{C}$ as matched by ρ_0. When ϕ_0 is defined on rows, the raw index $\Lambda(\phi_0)$ measures how similar the p rows are between $\underset{\sim}{B}$ and $\underset{\sim}{C}$ as matched by ϕ_0. Depending on where the matching is defined, the results may vary because of different reference distributions against which the observed indices are compared. We do note, however, that there is one special case discussed in the next section where the row and column matchings must lead to equivalent results. This occurs when $\underset{\sim}{B}$ and $\underset{\sim}{C}$ are both square and symmetric.

An Example. For an illustration of how these reductions could be carried out on a real data analysis problem, we return to the matrix in Table 1.2.3a used to evaluate the death-dip hypothesis. In addition to this latter conjecture, Phillips (1978) was also interested in the possibility of a rise in deaths in the birth month and the three successive months thereafter. Thus, if the Table 1.2.3a matrix defines $\underset{\sim}{B}$, then $\underset{\sim}{C}$ can be given as Table

2.2a. The two matrices in Tables 2.2b and c define the reduced matrices R_c and R_r, depending on whether the matching is to be done on columns or rows, respectively. In both cases, the entries in the reduced matrices, R_c and R_r, define all possible sums of four consecutive monthly entries either within columns (Table 2.2b) or within rows (Table 2.2c). Both of these tables also include the interaction terms for each of the main diagonal entries (in parentheses).

Given the way in which B is defined, ρ_0 and ϕ_0 are both the identity permutation and generate the same observed statistic $\Lambda(\rho_I) = \Lambda(\phi_I) = 140$.

For matching by columns, $E(\Lambda(\rho)) = 116$ and $\sqrt{V(\Lambda(\rho))} = 8.9628$, which produces $Z = 2.68$ (the skewness parameter is .013). Based on a random sample of size 999, six indices were as large or larger than 140, giving a Monte Carlo significance level of .007. Matching by rows produced very comparable results: $E(\Lambda(\phi)) = 116$, $\sqrt{V(\Lambda(\phi))} = 9.1791$, $Z = 2.61$ (the skewness parameter is .053). The Monte Carlo significance level also turned out to be .007, based on a sample of size 999. In this case, the two expectations are the same, even though B and C are not symmetric because the row and column sums of C are all equal to the same value. Descriptively, Table 2.2b suggests that the birth months of May, June, and July contribute the least in terms of interaction (and even negatively for May and July) to the overall death-rise index (a slight death-dip due to summer?). Table 2.2c indicates that for whatever reason, the death months of May, October, and December all contribute negatively.

Although we have approached the death-rise conjecture at the category level, the assessment of this hypothesis could have been carried out at the individual level through the use of the Hildebrand et al. prediction analysis strategy. Here, the zeros in Table 2.2a define the error cells that should contain relatively few entries if the death-rise conjecture is supported by the data. The observed number of observations in the error cells is 208, with 231.84 expected under independence. Thus, the prediction index $\hat{\nabla}$ is .10 with an associated Z value of 2.73. Clearly, for this problem the assignments at either the individual or the category level lead to very similar results.

This first example has been used mainly as a numerical illustration of how two rectangular matrices could be compared, but it also suggests a general strategy for comparing a given contingency table against a second, where the latter may be constructed by the researcher. Obviously, when the assignment is defined at the level of the individual, these assessment problems can be rephrased in the prediction analysis context of Hildebrand et al., discussed in Section 1.3.2. However, the same evaluation tasks could also be approached through a row or column matching strategy as in our example dealing with the death-rise conjecture. The given contingency table is treated as a rectangular matrix that is to be compared to a second matrix of the same size. We might wish to impose some further normalization on the empirically generated contingency table to make the terms of

Table 2.2a Second Matrix C~ Used to Define the Death–Rise Conjecture

Death month	Birth month											
	Jan.	Feb.	Mar.	Apr.	May	June	July	Aug.	Sept.	Oct.	Nov.	Dec.
Jan.	1	0	0	0	0	0	0	0	0	1	1	1
Feb.	1	1	0	0	0	0	0	0	0	0	1	1
Mar.	1	1	1	0	0	0	0	0	0	0	0	1
Apr.	1	1	1	1	0	0	0	0	0	0	0	0
May	0	1	1	1	1	0	0	0	0	0	0	0
June	0	0	1	1	1	1	0	0	0	0	0	0
July	0	0	0	1	1	1	1	0	0	0	0	0
Aug.	0	0	0	0	1	1	1	1	0	0	0	0
Sept.	0	0	0	0	0	1	1	1	1	0	0	0
Oct.	0	0	0	0	0	0	1	1	1	1	0	0
Nov.	0	0	0	0	0	0	0	1	1	1	1	1
Dec.	0	0	0	0	0	0	0	0	1	1	1	1

Table 2.2b Reduced R_c Matrix to Define Column Matching for the Death–Rise Conjecture[a]

Birth month	Birth month											
	Jan.	Feb.	Mar.	Apr.	May	June	July	Aug.	Sept.	Oct.	Nov.	Dec.
Jan.	15(1.17)	18	20	19	16	14	14	10	7	6	4	9
Feb.	16	19(6.88)	16	10	8	6	8	10	8	7	8	12
Mar.	11	11	14(3.00)	12	13	12	10	7	5	6	5	10
Apr.	9	10	12	13(1.50)	15	13	11	9	6	7	7	8
May	5	4	4	6	7(-.08)	8	9	7	7	7	6	6
June	5	5	6	9	8	6(.33)	7	5	4	6	4	3
July	14	14	13	12	12	10	11(-1.00)	7	8	10	10	15
Aug.	6	4	5	5	7	10	11	13(6.25)	11	10	8	6
Sept.	8	10	9	13	11	10	12	9	12(3.58)	11	10	9
Oct.	13	11	11	10	9	7	6	5	4	8(.83)	9	11
Nov.	15	14	12	11	9	13	13	12	12	9	11(1.00)	13
Dec.	13	14	10	14	10	7	12	7	9	11	10	11(.58)

[a]Interactions are given in parentheses for the main diagonal entries.

Table 2.2c Reduced $\underset{\sim}{R}_r$ Matrix to Define Row Matching for the Death-Rise Conjecture[a]

Death month	Death month											
	Jan.	Feb.	Mar.	Apr.	May	June	July	Aug.	Sept.	Oct.	Nov.	Dec.
Jan.	11(.67)	8	8	5	6	7	9	11	10	12	10	11
Feb.	14	15(4.17)	10	9	8	5	6	4	5	7	11	14
Mar.	14	15	17(3.42)	19	15	9	9	7	8	13	11	11
Apr.	17	21	20	18(3.92)	12	9	9	8	10	9	10	13
May	12	14	15	12	9(-.83)	7	9	8	10	10	7	11
June	10	8	8	13	10	11(4.42)	8	5	5	6	9	7
July	15	11	12	11	10	13	14(2.33)	11	14	15	13	17
Aug.	7	10	12	16	14	12	11	10(1.83)	9	8	7	4
Sept.	9	11	7	5	5	3	4	8	8(1.50)	8	11	9
Oct.	14	15	13	11	9	9	9	9	10	9(-1.92)	11	13
Nov.	4	3	2	4	5	4	4	5	7	9	10(4.58)	7
Dec.	5	7	7	6	7	6	8	8	10	9	7	8(-.08)

[a]Interactions are given in parentheses for the main diagonal entries.

the cross-product insensitive to row or column marginals (e.g., using conditional probabilities within columns when the matching is carried out between columns), but otherwise, the comparison of rectangular matrices provides one possible alternative to the system of prediction analysis discussed in Section 1.3.2. It should be noted, however, that matching at the level of the individual may be the more reasonable strategy to follow whenever the number of rows or columns on which the matching is defined is small. In these cases, the usual significance levels may be unobtainable because of the limited number of rows or columns actually present.

2.3 APPLICATIONS OF THE INDICES $\Lambda(\rho)$ AND $\Lambda(\phi)$

2.3.1 Association Between Spatially Defined Variables (B)

Recalling the discussion and notation of Section 1.3.5 on Tjøstheim's (1978) index of association between two spatially defined variables, suppose we wish to include more information in such a statistic than is provided by the distances between locations with identical ranks on F and G. Assume that the n locations $O_1, O_2, \ldots, O_n$ are labeled by the ranks on F, define c_{ij} as the symmetric spatial separation measure between locations O_i and O_j, and let $b_{ij} = (i - j)^2$. If $\rho_0(i) = j$ when the ith rank on G is in location O_j, then $\Lambda(\rho_0)$ is obtained by first weighting the distance c_{ij} by the squared difference between the F and G ranks at locations O_i and O_j, respectively. If small (large) values of c_{ij} denote spatial closeness, then positive association would be represented by relatively large (small) values for $\Lambda(\rho_0)$. (Because $\underset{\sim}{B}$ and $\underset{\sim}{C}$ are both symmetric, row or column matchings lead to equivalent methods.)

The approach to spatial association just described can be used to show explicitly the information on which Spearman's rank order correlation is based. If the distance measure c_{ij} is a (zero-one) dichotomous variable in which 1 is assigned to identical spatial locations and 0 is assigned to different spatial locations and $b_{ij} = (i - j)^2$, then $\Lambda(\rho_0)$ is the "active" part of Spearman's measure and $\Lambda^{**}(\rho_0)$ is the usual normalized index. In short, Tjøstheim's original statistic is based on a general distance matrix $\{c_{ij}\}$ and a zero-one dichotomous matrix $\{b_{ij}\}$ that matches identical ranks; Spearman's index is essentially the converse because it is based on a zero-one dichotomous distance matrix $\{c_{ij}\}$ and a more general weight matrix $\{b_{ij}\}$. Thus, the index mentioned above that uses both matrices in nondichotomous forms is a hybrid of Tjøstheim's and Spearman's indices which relies on information that both use in turn but that the other then ignores.

In an exactly parallel sense, suppose c_{ij} is again a zero-one dichotomous variable that identifies identical spatial locations and the values on F and G are standardized to mean zero and variance 1, and denoted by $f_1, f_2, \ldots, f_n$ and $g_1, g_2, \ldots, g_n$, respectively, over the n spatial locations. Then, by using squared differences for b_{ij}, i.e., $b_{ij} = (f_i - g_j)^2$, $\Lambda(\rho_I)$ is $2n(1 - r_{FG})$, where r_{FG} is the Pearson correlation between F and G. Because the ith location O_i contains the values f_i and g_i, we can now use the identity permutation ρ_I to define the observed index. Furthermore, the index itself has a minimum value of zero, at least in terms of a bound; consequently, the normalized index of the form $\Lambda^{**}(\rho_I)$ is appropriate and turns out to be r_{FG} using this value of 0 for the raw index minimum. In effect, this is exactly what we did earlier in discussing the comparison between two numerical sequences in Section 1.3.3.

The measures we have just discussed are obviously of interest in the explicitly spatial framework of concern to Tjøstheim. Nevertheless, the strategy of measuring association between two variables, as represented in our generalized index $\Lambda(\rho_0)$, has significance beyond the strictly geographic context. The matrix $\{c_{ij}\}$ merely represents information against which we wish to compare the relationship between F and G. The distance interpretation between geographic locations in obvious, but any matrix could be used. For example, $\{c_{ij}\}$ could represent the interpoint distances between spatial locations constructed from a multidimensional scaling, or be ultrametric values reconstructed from a hierarchical clustering. In all cases, our interests are in using the information provided by $\{c_{ij}\}$ to help assess the correspondence between the two variables F and G (see Section 4.3.9).

Finally, we might mention in passing, several possible variations on the association problem that will not be pursued either here or for Tjøstheim's original context. First, the spatial separation measures could be asymmetric, and second, functions other than squared differences considered for $\underset{\sim}{B}$, e.g., absolute differences. Given the nice parallels to the usual correlation coefficient, however, only squared differences are discussed, and because spatial separation measures are typically symmetric, this assumption is made explicitly for convenience. Carrying out generalizations of this type would be straightforward merely by constructing the $\underset{\sim}{B}$ and $\underset{\sim}{C}$ matrices in the appropriate manner. For some further comments on the problem of assessing spatial association, the reader is referred to Hubert and Golledge (1982b). Much of the material in this section as well as in Section 1.3.5 is based on this latter source.

2.3.2 Comparing Object by Attribute Data Tables (B)

Object by attribute data tables are frequently encountered in the literature. The term "object" usually refers to some sampling unit, such as subjects, groups, cities, or regions; the term "attribute" typically denotes a measured variable such as crime rate, achievement score, time to complete a task, or rainfall. To be specific in terms of notation, suppose $\underset{\sim}{B}$ and $\underset{\sim}{C}$ are two $p \times q$ object by attribute data matrices, where the rows and columns denote the same p objects and q attributes, respectively, and where the matrices themselves could represent, say, two different points in time. Our interest is in comparing $\underset{\sim}{B}$ and $\underset{\sim}{C}$ in a correlational sense, but we would like to carry out the process with the matrices more or less intact. Because there is a natural one-to-one correspondence between the rows of $\underset{\sim}{B}$ and $\underset{\sim}{C}$ that defines the identity permutation on the first p integers, ϕ_I, and we have tacitly assumed that the rows represent sampling units of some type, the comparison task is approached through the observed statistic:

$$\Lambda(\phi_I) = \sum_{i,j} b_{ij} c_{ij}$$

Given the way object by attribute tables are usually represented, it is natural to define the matching on rows rather than columns to obtain an analog of the usual correlation coefficient; nevertheless, the converse could also be carried out.

One rather complete illustration of how such a strategy of row matching can be used is available in the literature. In an example presented by Tsutakawa and Yang (1974), but rephrased in our notation, two 14×16 matrices, $\underset{\sim}{B}$ and $\underset{\sim}{C}$, were defined, with the 14 rows corresponding to farms and the 16 columns to patterns of antibiotic drug resistance. The two matrices referred to humans and farm animals, respectively, with the entries being a function of the number of times that a particular resistance pattern was observed. In this framework, the raw cross-product statistic, $\Lambda(\phi_I)$, indicated the degree to which the same resistance patterns are found within the same farms. Its relative size was evaluated with respect to the distribution of $\Lambda(\phi)$ over all 14! possible permutations of the rows of one matrix against the fixed second. The reader interested in a more thorough discussion of this particular application can consult the original source.

Given the typical form of object by attribute data matrices, the entries in $\underset{\sim}{B}$ and $\underset{\sim}{C}$ may not be strictly comparable. Thus, it is usually necessary to carry out some appropriate transformation on the matrix entries to assure a reasonable commensurability between all the terms making up the cross-product index. As one convenient method that we will follow, it will be assumed that the data in each column of $\underset{\sim}{B}$ and $\underset{\sim}{C}$ represent z scores; that is, within each column, the sample mean is zero and the sample variance is one. When q is 1, $\underset{\sim}{B}$ and $\underset{\sim}{C}$ are single numerical (column) sequences. In this case, one classical comparison strategy is obvious—the Pearson correlation could be obtained between the two variables and tested for significance by the standard randomization technique. As in Section 1.3.3, the obtained correlation is compared to what would be expected under the null model that all n! permutations of the entries in the single $\underset{\sim}{C}$ column were equally likely compared to the fixed $\underset{\sim}{B}$ column (or conversely). In fact, the matrix $\underset{\sim}{R}_C$ that allows a reduction to the assignment model would have the form $s_{jk} = x_j y_k$, where x_j and y_k are the jth and kth elements in the single $\underset{\sim}{B}$ and $\underset{\sim}{C}$ columns, respectively. This leads directly back to the framework of Section 1.3.3.

Because the entries within each column of B and C are assumed to be z scores, the raw measure $\Lambda(\phi_I)$ between $\underset{\sim}{B}$ and $\underset{\sim}{C}$ is p times the sum of the q Pearson correlations between all pairs of columns. Thus, if r_j denotes the correlation between the jth columns, then $\Lambda(\phi_I)$ is actually $p \sum_{j=1}^{q} r_j$.

The inference model that compares $\Lambda(\phi_I)$ to the distribution of $\Lambda(\phi)$ maintains the integrity of the relationships among the variables, because we assume as the null conjecture that the rows of one of the matrices have been matched at random to those of the second. In other words, as a natural generalization of the permutation model for a single correlation, complete vectors of observations attached to the sampling units are now permuted as a whole. The z score constraint, by itself, reduces the moment formulas considerably:

$$E(\Lambda(\phi)) = 0$$

$$V(\Lambda\underset{\sim}{(\phi)}) = \left(\frac{p^2}{p-1}\right) \sum_{i,j} r\underset{ij}{\overset{B}{\sim}} r\underset{ij}{\overset{C}{\sim}}$$

where $r\underset{ij}{\overset{B}{\sim}}$ and $r\underset{ij}{\overset{C}{\sim}}$ are the correlations between the ith and jth columns in $\underset{\sim}{B}$ and $\underset{\sim}{C}$, respectively.

If large values of $\Lambda(\phi_I)$ are of interest and we replace $\max_\rho \Lambda(\rho)$ by its upper bound of pq, then $\Lambda^*(\phi_I)$ has the form

$$\frac{1}{q} \sum_{j=1}^{q} r_j$$

i.e., the average correlation between the corresponding columns of $\underset{\sim}{B}$ and $\underset{\sim}{C}$. In terms of this normalized measure, $E(\Lambda^*(\phi)) = 0$, and

$$V(\Lambda^*(\phi)) = \frac{1}{q^2(p-1)} \sum_{i,j} r\underset{ij}{\overset{B}{\sim}} r\underset{ij}{\overset{C}{\sim}} \qquad (2.3.2a)$$

There are several comments that should be made about the comparison strategy just described and the moments of the normalized measure $\Lambda^*(\phi)$ given above. First of all, if $q = 1$, then the index $\Lambda^*(\phi_I)$ reduces to a single correlation. The mean of 0 and variance from equation (2.3.2a) of $1/(p-1)$ are what they should be. Second, if all q column pairs between the two matrices were compared independently, then the appropriate expectation of an index like $\Lambda^*(\phi)$ would again be zero, but the variance would now be $1/[q(p-1)]$. The formula we use differs from this latter value by a factor of $\alpha = (1/q) \sum_{i,j} r\underset{ij}{\overset{B}{\sim}} r\underset{ij}{\overset{C}{\sim}}$. Typically, α will be greater than 1 but must

be less than q (as can be shown from the Cauchy-Schwarz inequality).
Stated in other words, the variance in equation (2.3.2a) will usually be
larger than what would be obtained if independence were assumed between
all the column comparisons; moreover, the value in equation (2.3.2a) must
always be less than or equal to $1/(p - 1)$. This latter bound would be at-
tained when all of the columns of $\underset{\sim}{B}$ are the same (in z scores) and all the
columns of $\underset{\sim}{C}$ are the same (again, in z scores). When this occurs, we in a
sense have only one pair of columns, and the index $\Lambda(\phi_I)$ is really a single
correlation, and thus it is reasonable to have a variance of $1/(p - 1)$. In
any case, the index α (or $\sqrt{\alpha}$ if we wish to consider the change in the
standard deviation rather than the variance) is a measure of how the relationship among the columns in the two matrices is compensated for by the
comparison process over and above the baseline generated if the q pairs of
columns were to be compared independently.

An Example. As a simple illustration of how the mechanics of the comparison process can be carried out, Tables 2.3.2a and b present, respectively, the 1976 and 1977 rates per 100,000 population for the seven standard
index crimes (murder and nonnegligent manslaughter, forcible rape, robbery, aggravated assault, burglary, larceny-theft, motor vehicle theft) for
20 standard metropolitan statistical areas (SMSAs): Akron (a), Bakers-
field (b), Bridgeport (c), Columbia (d), Davenport-Rock Island-Moline (e),
El Paso (f), Fresno (g), Gary-Hammond-East Chicago (h), Grand Rapids (i),
Harrisburg (j), Honolulu (k), Jersey City (l), Little Rock-North Little
Rock (m), Mobile (n), New Haven-West Haven (o), Omaha (p), Peoria (q),
Springfield-Chicopee-Holyoke (r), Syracuse (s), and Tacoma (t). Once the
entries in each column have been standardized to z scores, our concern is
with a comparison of these two 20 × 7 matrices.

If we match on rows, the observed index $\Lambda(\phi_I) = 131.15$, which is 20
times the sum of the seven between-column correlations. Thus, the average correlation between the seven columns, $\Lambda^*(\phi_I)$, is a very substantial .94.
The variance of $\Lambda^*(\rho)$ is .0168, and the corresponding Z statistic is 7.23
(with a skewness parameter of .106). In a sample of 999 random assignments, no index larger than the observed was found; thus, a Monte Carlo
significance level of .001 can be reported. In short, there appears to be
substantial commonality between the two tables, as measured by the average
correlation between columns.

Going one step further in this example, Tables 2.3.2c and d
present the interaction matrices for matching by columns and rows,
respectively. All the main diagonal terms are positive in each matrix,
suggesting a general consistency between variables as well as between
SMSAs (the diagonal terms are circled in Tables 2.3.2c and d). The
column by column interaction matrix in Table 2.3.2c displays rather
markedly the distinctness of the motor vehicle theft variable. The

Table 2.3.2a 1976 Crime Rates per 100,000 Population for the 20
          SMSAs Listed in the Text

| | Index crime | | | | | | |
SMSA	Murder	Rape	Robbery	Assault	Burglary	Larceny	Auto theft
a	4.7	32.0	139.7	250.0	1412.1	3788.7	372.6
b	14.1	38.3	191.7	287.8	2383.1	4662.5	512.5
c	3.9	10.4	110.9	106.0	1293.6	2906.1	649.5
d	10.7	43.0	165.4	370.6	2247.5	4172.2	346.8
e	4.6	17.5	126.7	171.6	1556.5	3745.7	237.0
f	6.6	37.2	205.5	175.2	1751.1	4022.9	522.5
g	12.2	37.7	242.6	333.6	2906.2	4460.7	695.7
h	12.4	36.2	259.7	266.1	1322.7	2971.3	867.1
i	2.7	22.5	74.7	179.6	1187.7	3016.3	165.4
j	3.1	24.1	160.6	189.2	1265.3	2650.1	273.2
k	5.6	22.8	154.8	52.9	1910.8	3630.4	593.0
l	7.7	18.9	357.5	185.3	1617.7	2315.7	952.6
m	12.5	56.6	253.5	327.5	1951.3	4513.0	395.8
n	17.6	40.0	198.7	345.8	2482.3	3150.8	303.3
o	3.1	14.3	119.3	81.7	1625.0	3070.4	655.6
p	4.6	40.1	155.4	227.8	1168.5	3591.5	487.1
q	4.0	24.9	123.0	336.4	1197.8	2806.7	235.2
r	3.0	18.6	95.3	245.1	1969.2	2449.8	711.0
s	3.3	12.9	92.9	110.0	1433.0	2939.8	262.3
t	5.9	50.6	141.7	232.5	1923.0	2894.0	317.1

large diagonal entries in the 20 × 20 interaction matrix for SMSAs in
Table 2.3.2d identify extreme profiles that are stable from year to
year. For example, Fresno (g) has consistently high rates of murder
and burglary, and Jersey City (l) has high robbery and auto theft but
relatively low larceny.

Table 2.3.2b 1977 Crime Rates per 100,000 Population for the 20
            SMSAs Listed in the Text

| SMSA | Index crime | | | | | | |
	Murder	Rape	Robbery	Assault	Burglary	Larceny	Auto theft
a	5.8	37.9	129.8	284.2	1310.9	3362.8	358.1
b	13.9	41.9	196.1	370.9	2240.3	4394.5	606.9
c	4.6	14.0	102.5	96.4	1287.5	2738.6	670.7
d	13.1	54.2	207.4	552.1	2488.9	3351.7	387.9
e	2.4	18.4	134.7	163.1	1309.3	3005.6	260.2
f	8.4	30.3	203.9	156.7	1696.2	3459.9	620.2
g	21.9	35.5	306.9	366.8	2783.3	3932.3	671.9
h	16.7	49.8	256.6	275.3	1296.7	2645.8	876.9
i	3.5	32.0	87.1	183.1	1270.5	2944.4	187.7
j	3.3	24.8	164.4	167.0	1054.0	2349.6	248.2
k	6.4	24.4	149.7	49.4	1840.9	3917.8	519.0
l	11.2	22.9	321.0	187.7	1596.4	1969.8	985.4
m	11.7	81.9	301.2	375.5	2053.1	3800.4	460.7
n	22.2	46.6	206.2	371.6	1905.7	2493.4	300.9
o	3.3	17.5	137.6	113.0	1611.4	2941.6	661.7
p	6.5	36.8	165.7	221.8	1342.2	3380.0	536.3
q	4.2	22.9	89.2	359.9	1225.3	2659.2	213.2
r	4.7	23.3	72.0	298.8	1514.4	2429.2	587.3
s	2.5	13.0	90.7	102.1	1531.4	2923.4	224.7
t	6.7	54.4	135.4	247.3	2034.4	2952.7	333.0

The topic of object by attribute data matrices is taken up again
in Section 3.2.6, but in the context of two-group concordance. The
reader may also be interested in consulting Hubert and Golledge
(1982a); this reference forms the basis of the current presentation.

Table 2.3.2c Column Matching Interaction Matrix for the Crime Index Data of Tables 2.3.2a and b

Index crime (1976)	Index crime (1977)						
	Murder	Rape	Robbery	Assault	Burglary	Larceny	Auto theft
Murder	3.85	.77	.36	.62	-.25	-2.50	-2.85
Rape	-1.93	7.61	-1.94	2.38	-.59	1.69	-7.22
Robbery	.97	-1.82	7.54	-4.47	-3.66	-6.18	7.62
Assault	.95	3.86	-2.86	9.90	-1.21	-3.34	-7.29
Burglary	1.28	-3.38	-2.85	-.02	6.57	1.75	-3.34
Larceny	-4.17	.12	-3.32	-1.12	2.27	11.91	-5.69
Auto theft	-.94	-7.15	3.08	-7.29	-3.14	-3.34	18.78

Table 2.3.2d Row Matching Interaction Matrix for the Crime Index Data
of Tables 2.3.2a and b

SMSA	a	b	c	d	e	f	g	h	i
a	1.72	-.09	.05	-.44	1.40	-.37	-2.94	-1.73	1.91
b	-.92	8.41	-5.16	7.15	-4.45	.75	11.05	1.11	-4.08
c	.06	-5.20	6.60	-8.35	3.57	.97	-7.16	-1.03	2.40
d	.37	6.35	-5.82	8.60	-3.31	-1.06	7.19	-.41	-2.04
e	1.41	-1.67	1.95	-3.26	3.59	.04	-4.13	-4.56	3.33
f	-.02	1.99	-.87	-.06	-.70	1.20	1.85	.53	-.86
g	-2.51	9.96	-6.21	9.87	-6.39	1.12	14.96	2.25	-6.88
h	1.74	.75	-.69	.66	-3.80	.72	4.14	8.05	-4.42
i	2.50	-5.50	3.38	-4.99	5.10	-1.33	-10.07	-4.79	5.88
j	1.11	-5.54	2.41	-4.08	3.59	-.83	-7.38	-1.96	3.56
k	-.86	-.98	3.33	-5.07	1.39	2.01	-1.12	-1.29	.45
l	-3.82	-2.35	1.74	-2.79	-2.31	2.01	3.92	7.63	-5.40
m	.11	7.56	-7.62	8.61	-4.90	-0.02	9.43	3.07	-3.66
n	-1.78	4.78	-6.47	9.36	-4.80	-1.70	10.24	2.30	-3.62
o	-.40	-4.02	5.86	-7.35	2.97	1.36	-5.48	-1.61	1.80
p	1.64	-.97	.42	-1.48	.78	-.13	-3.98	.61	1.52
q	2.16	-3.65	.58	-.17	2.76	-2.35	-6.53	-2.24	3.49
r	-.71	-3.10	3.71	-1.87	.90	-.56	-3.67	-.80	.48
s	1.15	-5.76	4.88	-6.88	5.21	-.25	-8.59	-4.75	4.91
t	.53	-.97	-2.08	2.56	-.59	-1.48	-1.74	-.39	1.24

j	k	l	m	n	o	p	q	r	s	t
	.26	-3.40	-.06	-1.55	-.22	.87	1.89	.33	1.11	.26
-6.61	.90	-2.72	7.41	3.74	-3.50	-.69	-4.24	-4.07	-4.79	.71
3.73	2.32	2.88	-9.33	-4.96	4.93	.87	1.79	3.55	4.72	-2.34
-4.59	-1.58	-4.87	7.34	4.20	-4.39	-.82	-.91	-2.37	-3.93	2.06
2.56	2.32	-3.95	-4.34	-2.67	1.46	.50	2.62	.67	4.48	-.34
-1.34	1.73	-.34	3.10	-1.19	-.20	.66	-2.36	-2.19	-1.09	.15
-8.66	.18	.66	9.37	3.73	-3.38	-1.47	-6.19	-4.24	-6.97	.79
-1.83	-2.81	7.97	3.20	2.83	-1.04	.15	-3.26	-.98	-5.61	-2.27
5.47	.53	-4.64	-6.56	-2.71	1.81	.74	5.55	3.06	6.08	.50
(5.00)	-.75	-.11	-4.40	-1.53	1.38	.31	3.66	1.98	3.80	-.11
.35	(4.26)	1.22	-3.42	-3.79	3.31	.46	-2.14	-.19	2.67	-.61
.21	-1.67	(14.31)	-.33	.06	2.01	-.47	-4.42	-.73	-3.73	-3.88
-5.40	-1.52	-2.90	(12.42)	5.03	-5.79	-.08	-3.86	-5.53	-6.81	1.81
-4.39	-3.66	-.70	6.70	(9.42)	-5.38	-2.56	-1.92	-2.68	-5.00	1.87
2.60	3.20	2.48	-7.73	-5.26	(4.89)	.64	.44	2.70	4.38	-1.48
1.27	-.29	-1.45	1.06	-1.69	-.13	(1.36)	1.05	.25	-.03	.21
3.89	-3.05	-2.75	-3.21	.22	-.54	.14	(5.71)	2.96	2.41	.14
1.14	-.71	2.80	-6.00	-2.34	3.10	-.54	1.89	(4.58)	1.97	-.27
5.19	2.18	-2.03	-8.74	-3.61	3.46	.37	3.97	2.92	(6.85)	-.47
.43	-1.84	-2.46	3.52	2.05	-1.77	-.44	.66	.01	-.52	(3.27)

2.3.3 The Comparison of Two Vector Sequences: Directional Data (B)

The task of comparing two p × q matrices, $\underset{\sim}{B}$ and $\underset{\sim}{C}$, can be rephrased in
a way that may help emphasize its connection to the correlation problem of
comparing two numerical sequences discussed in Section 1.3.3. Assuming
that the matching is to be done on columns, each of the matrices B and C
can be represented as sequences of q vectors, $\{\underset{\sim}{b}_1, \ldots, \underset{\sim}{b}_q\}$ and $\{\underset{\sim}{c}_1, \ldots, \underset{\sim}{c}_q\}$.
The vectors $\underset{\sim}{b}_i$ and $\underset{\sim}{c}_i$ have p components and represent the ith columns of
B and C, respectively. (Obviously, matching by rows would involve two
sequences containing p vectors—each is defined by q components.) In
direct analogy to the correlation problem of Section 1.3.3, where the q
vectors would correspond to q numerical values, our task is one of assessing
the degree to which the sequences are similar. Here, we use the LA model
based on the q × q assignment matrix $R_c = \{s_{jk}\} = \{\Sigma_i\, b_{ij} c_{ik}\}$, which
merely relies on a particular measure of proximity between the vectors
from the two sets $\{\underset{\sim}{b}_1, \ldots, \underset{\sim}{b}_q\}$ and $\{\underset{\sim}{c}_1, \ldots, \underset{\sim}{c}_q\}$.

As one application of this vector correlation idea, suppose that
$\{\underset{\sim}{b}_1, \ldots, \underset{\sim}{b}_q\}$ and $\{\underset{\sim}{c}_1, \ldots, \underset{\sim}{c}_q\}$ represent two sets of n directions in p-dimen-
sional space, i.e., $\underset{\sim}{b}_i$ and $\underset{\sim}{c}_i$ are unit vectors containing p direction cosines
that are matched in some a priori way. The task is one of assessing
correlation between matched directions, using the measure of proximity
between b_j and c_k of $\Sigma_i\, b_{ij} c_{ik} = \underset{\sim}{b}_j^t \underset{\sim}{c}_k$, where t denotes the transposition.
For two dimensions (p = 2), $\underset{\sim}{b}_j^t \underset{\sim}{c}_k$ reduces to the cosine of the difference
between the two angles. In any case, normalized indices in the directional
context can be defined very easily by use of the upper and lower bounds
of ± n, i.e., $-n \leq \min_\rho \Lambda(\rho)$ and $\max_\rho \Lambda(\rho) \leq n$. For some further discussion, the reader is referred to Epp, Tukey, and Watson (1971), where
this same LA model for the problem of evaluating directional correspondence
is discussed in some detail.

2.3.4 Index Generalizations: Generalized Rater Agreement (B)

Although the index for comparing two rectangular matrices, $\Lambda(\rho)$, has been
defined in a cross-product form, an extension to more general measures is
possible and relatively easy to carry out. For example, the product
$b_{ij} c_{i\rho(j)}$ can be replaced by an arbitrary bivariate function $f(b_{ij}, c_{i\rho(j)})$ and
an index defined by the form

$$\Lambda(\rho) = \sum_{i,j} f(b_{ij}, c_{i\rho(j)})$$

The moment formulas remain the same as before with $s_{jk} = \Sigma_i\, f(b_{ij}, c_{ik})$. Our discussion has been limited to the special case of $f(b_{ij}, c_{i\rho(j)}) = b_{ij}c_{i\rho(j)}$, but several other possibilities may be considered. For example, $f(b_{ij}, c_{i\rho(j)})$ could be defined as $|b_{ij} - c_{i\rho(j)}|$ in analogy with Spearman's footrule, or if the entries in the matrix represent nominal category information, $f(b_{ij}, c_{i\rho(j)})$ could be defined as 1 if b_{ij} and $c_{i\rho(j)}$ denote the same category and 0 otherwise (or possibly, these dichotomous values could be replaced by preassigned weights).

What is required in all of these extensions is some alternative way of defining $\underset{\sim}{R}_c$, which contains measures of relationship between the columns of $\underset{\sim}{B}$ and the columns of $\underset{\sim}{C}$. Thus, even the restriction to sums of the form $\Sigma_i\, f(b_{ij}, c_{ik})$ can be removed, although this structure by itself offers a substantial degree of generality. Irrespective of how $\underset{\sim}{R}_c$ is specified, the assignment model can still be used to evaluate the relative degree of correspondence observed between $\underset{\sim}{B}$ and $\underset{\sim}{C}$. Obviously, these same comments are true if we match the rows between $\underset{\sim}{B}$ and $\underset{\sim}{C}$ instead of the columns. For example, in the illustration of the last section based on the matching of rows, the measure of correspondence between two SMSA profiles could have been redefined as the correlation. This latter measure is concerned with the "shapes" of the profiles over the seven index crimes and is unaffected by relative "level," i.e., the average z score within each row. The measure actually used, based on the sum of raw products across two profiles, was heavily dependent on the relative magnitude of the scores within each SMSA. Consequently, whenever this property is undesirable or shape is also of interest, the analysis could be redone with an alternative measure of profile similarity. Obviously, the interpretation of the descriptive measure, $\Lambda(\phi_I)$, would have to change to reflect the similarity measure actually used.

When reduced to the task of defining a square matrix $\underset{\sim}{R}_c$, the problem of comparing two rectangular matrices suggests a way of assessing rater agreement in some very complicated situations (see Hubert and Golledge, 1983). The raters define the two matrices $\underset{\sim}{B}$ and $\underset{\sim}{C}$, with the columns of each matrix corresponding to the q objects to be rated. A given column contains information provided by the rater regarding that particular object. For example, for each of the q objects, the raters may be asked to check for the presence or absence of symptoms or the appropriateness of certain adjectives, give a rank order of aptness for a set of statements, provide estimates of the degree to which certain facets that could characterize an object are present, and so on. All that is then required is some summary measure defined for all possible pairs of assignment profiles. The summary measure could be a correlation, a count of consistencies in the placement of ones and zeros, and the like. The raw agreement statistic, $\Lambda(\rho_I)$, is based on the final matrix $\underset{\sim}{R}_c$ and can be evaluated in the usual way with the LA model. Finally, a normalized descriptive statistic of the form $\Lambda^*(\rho_I)$ or $\Lambda^{**}(\rho_I)$ could be reported as the overall index of agreement between the two raters.

It should be clear how the usual weighted kappa measure could be rephrased in this structure. All of the q objects are assigned a given category by each rater, and the measure of correspondence between the two assessments of an object is defined by the weight w_{uv}, if the first rater assigns the single category labeled u in his/her system of categories and the second rater assigns the single category labeled v in his/her system of categories. In this case, as well as in the possible extensions beyond simple categories to more complicated assessment schemes, we in effect have a strategy for studying a very general correlational problem. Irrespective of the form these assessments take, all that is required is some specification of proximity that can be used to replace the set of weights $\{w_{uv}\}$. The reader may be interested in consulting a recent substantive use of the LA model in the work of Buffington, Martin, and Becker (1981) and Martin, Buffington, and Becker (1981). Here, the assessments take the form of visual evoked responses (EEGs) matched between or within subjects (e.g., same subject, twins, relatives, and so on); proximity is obtained through the Pearson product-moment correlation across time.

2.3.5 Unweighted Kappa: Incomplete Selection and Conditional Kappa (C)

Although rater agreement was discussed in Chapter 1 as well as in the previous section, it is instructive to show how the matrix comparison strategy clarifies some recent issues in the context of an unweighted kappa measure of nominal-scale response agreement. Suppose two raters place q objects into one of p categories. Rater 1 defines a p × q matrix $\underset{\sim}{B} = \{b_{ij}\}$, where $b_{ij} = 1$ if object j is placed into category i; a second matrix $\underset{\sim}{C}$ is defined similarly for rater 2. A matching by columns with the usual cross-product measure produces a dichotomous assignment matrix $\underset{\sim}{R_c}$ of size q × q, where the entry s_{jk} is 1 if objects j and k are placed in the same category, and 0 otherwise. The index, $\Lambda(\rho_I)$, is the sum of the main diagonal frequencies in the usual p × p contingency table, where the p rows define the categories used by rater 1 and the p columns define the categories used by rater 2. The expected value of $\Lambda(\rho)$ is the well-known expression:

$$E(\Lambda(\rho)) = \frac{1}{q} \sum_{i=1}^{p} b_i \cdot c_i.$$

If the p rows are matched instead, the reduced matrix $\underset{\sim}{R_r}$ would actually be the p × p contingency table mentioned above. As always, the observed index, $\Lambda(\phi_I)$, is the same as $\Lambda(\rho_I)$. However, the expected value differs:

$$E(\Lambda(\phi)) = \frac{q}{p}$$

This last expression has been discussed by several authors (e.g., see Brennan and Prediger, 1982) as an appropriate correction term in a normalized kappa index of the form $\Lambda^*(\phi_I)$. The argument for its use centers on the information in the margins of the $p \times p$ contingency table itself and how the lack or presence of marginal homogeneity should be included in a final descriptive measure. A reliance on the more common expression given earlier essentially eliminates the latter. As we see here, an alternative justification could be based on a model of category matching as a replacement for the more typical matching over the rated objects.

Finally, one special case of the matrix comparison task deserves explicit mention, in which the matrices $\underset{\sim}{C}$ and $\underset{\sim}{B}$ are both of order $n \times n$, and $\underset{\sim}{C}$ contains the usual assignment scores as in Chapter 1, but B restricts the number of assignments of row to column objects to K:

$$b_{ij} = \begin{cases} 1, & \text{if } 1 \leqslant i = j \leqslant K \\ 0, & \text{otherwise} \end{cases}$$

In other words, $\Lambda(\rho)$ is a sum of K entries in $\underset{\sim}{C}$ chosen from its first K rows. Within these rows, however, the entries may come from any of the n columns. Using the reduction to a simple LA model in Section 2.2, we in effect define a new assignment matrix, say, $\underset{\sim}{C}^o = \{c_{ij}^o\}$, which is identical to $\underset{\sim}{C}$ in its first K rows but is entirely zeros in the remaining $n - K$ rows.

The obvious application of this incomplete selection idea is to a conditional kappa index (see Light, 1971; Hubert, 1977). Specifically, suppose our concern is with nominal-scale agreement between two raters when rater 1 uses category a_r. Defining $\underset{\sim}{C}^o$ as the raw index $\Gamma(\rho_0)$ would have the form

$$\sum_v w_{rv} n_{rv}$$

and a conditional version of κ could be defined as

$$\frac{\sum_v w_{rv} n_{rv} - \sum_v w_{rv}(n_{r \cdot} n_{\cdot v}/n)}{n_{r \cdot} - \sum_v w_{rv}(n_{r \cdot} n_{\cdot v}/n)}$$

The LA inference model based on $\underset{\sim}{C}^o$ and n! equally likely permutations reduces to considering only $n!/(n - K - 1)!$ permutations. The usual moment formulas would remain appropriate, however, using $\underset{\sim}{C}^o$ as the assignment matrix. In the unweighted case, conditional kappa reduces to:

$$\frac{n_{rr} - (n_{r \cdot} n_{\cdot r}/n)}{n_{r \cdot} - (n_{r \cdot} n_{\cdot r}/n)}$$

Obviously, we could conditionalize on any number of categories for rater 1 merely by using the relevant categories for the initial rows of $\underset{\sim}{C}^o$.

2.4 AN INFERENCE MODEL BASED ON ROW AND COLUMN MATCHING (C)

Up to now, the comparison of two p × q rectangular matrices $\underset{\sim}{B}$ and $\underset{\sim}{C}$ has relied on an inference model that matches either on rows or on columns but not on both. Following Klauber (1971), a strategy for matching on both rows and columns simultaneously can be developed fairly easily. For the raw descriptive measure, two permutations, ρ_0 and ϕ_0, must be identified. As notation, the cross-product statistic is now denoted by

$$\Lambda(\phi_0, \rho_0) = \sum_{i,j} b_{ij} c_{\phi_0(i) \rho_0(j)}$$

and is compared to all p!q! equally likely realizations of $\Lambda(\phi, \rho)$.

The moment formulas need modification and are given by the following expressions:

Let

$$B_1 = \sum_{i,j} b_{ij} \quad B_2 = \sum_j \left(\sum_i b_{ij} \right)^2 \quad B_3 = \sum_i \left(\sum_j b_{ij} \right)^2 \quad B_4 = \sum_{i,j} b_{ij}^2$$

and

$$C_1 = \sum_{i,j} c_{ij} \quad C_2 = \sum_j \left(\sum_i c_{ij} \right)^2 \quad C_3 = \sum_i \left(\sum_j c_{ij} \right)^2 \quad C_4 = \sum_{i,j} c_{ij}^2$$

Then,

$$E(\Lambda(\phi,\rho)) = \frac{B_1 C_1}{pq}$$

$$\begin{aligned}
V(\Lambda(\phi,\rho)) = & \left\{ \frac{1}{pq(p-1)(q-1)} \right\} \times \left\{ (p-1)(q-1)B_4 C_4 \right. \\
& + (p-1)(B_3 - B_4)(C_3 - C_4) + (q-1)(B_2 - B_4)(C_2 - C_4) \\
& \left. + [B_1^2 - (B_2 + B_3) + B_4][C_1^2 - (C_2 + C_3) + C_4] \right\} - \left(\frac{B_1 C_1}{pq} \right)^2
\end{aligned}$$

Unfortunately, there does not appear to be a nice way of simplifying the row and column matching strategy to obtain an LA model of the form discussed in Chapter 1. As we will see in Chapter 7, however, the row and column matching strategy can be considered a special case of another model. The latter is based on a notion of quadratic assignment with a restriction on the admissible assignments. Consequently, except for the short discussion on spatial association in the next section, we delay a further development of this topic until later.

2.4.1 Row and Column Matching in a Spatial Context (C)

There are several obvious applications of an inference strategy based on row and column matching in areas that have already been mentioned. For example, the task of assessing association between spatially defined variables as discussed in Section 2.3.1 required an arbitrary fixing of the values for one variable, F, in given spatial locations. The values on the second variable were then permuted to obtain the reference distribution for $\Lambda(\rho_0)$. In Tjøstheim's original context in which the spatial separation measures were summed for identical ranks only, the LA model of Chapter 1 is appropriate, and either variable could have been fixed without loss of generality. The reference distributions would be identical. In fact, when the second matrix $\underset{\sim}{B}$ is the usual dichotomous matrix with a single one in each row and each column, then the reference distributions based on $\Lambda(\phi,\rho)$, and $\Lambda(\phi)$, and $\Lambda(\rho)$ are all the same.

In a more general context where the entries in $\underset{\sim}{B}$ are, say, defined as squared differences between ranks, this complete equivalence does not necessarily hold. Even when $\underset{\sim}{B}$ and $\underset{\sim}{C}$ are square and symmetric, as they are in the generalization we offered to Tjøstheim's problem in Section 2.3.1, the distributions for $\Lambda(\phi)$ and $\Lambda(\rho)$ are the same but they differ in general from $\Lambda(\phi,\rho)$. This fact can be seen very easily in the expectations. If $\underset{\sim}{B}$ and $\underset{\sim}{C}$ are symmetric, then

$$E(\Lambda(\phi)) = E(\Lambda(\rho)) = \frac{1}{n} \sum_{i,j,k} b_{ij} c_{ik}$$

$$E(\Lambda(\phi,\rho)) = \left(\frac{1}{n^2}\right)\left(\sum_{i,j} b_{ij}\right)\left(\sum_{i,j} c_{ij}\right)$$

where $n = p = q$. Even though the distributions of $\Lambda(\phi)$ and $\Lambda(\rho)$ are identical, the fixing of one variable in given spatial locations may seem arbitrary, in which case an inference model based on $\Lambda(\phi,\rho)$ could be used.

Even in Tjøstheim's original problem in which spatial separations were summed between identical ranks, there is some theoretical advantage to considering the inference model to be based on row and column matching. Although all three distributions are the same and operationally there is no difference between their use, they are not equivalent theoretically when we try to relate the values of Tjøstheim's measure to that of Spearman's rank order correlation based on the same permutations. For example, in Section 1.4.3 an expression was given for the correlation between the Spearman and Tjøstheim measures. This correlation assumed a reduced squared euclidean distance measure and a fixed set of ranks on F that also labeled the n spatial locations. Obviously, this value depends on the correlation between the coordinates for the n locations and the fixed set of ranks on F. If an inference model based on row and column permutations were used, this correlation would be reduced to $-[2/(n-1)]\sqrt{2(1 + r_{xy}^2)}$. There is still some negative association between the Tjøstheim and the Spearman measures, but it does not seem as serious as before, given this latter expression based on row and column matching. In fact, this quantity converges to 0 as $n \longrightarrow \infty$.

Although we have emphasized the problem of association between two variables over one set of geographical locations, the spatial context was also of interest to Klauber (1971), but in the framework of two distinct sets of geographical locations. The first contained p locations, $\{O_1^{(1)}, \ldots, O_p^{(1)}\}$, and the second contained q locations $\{O_1^{(2)}, \ldots, O_q^{(2)}\}$. The two matrices $\underset{\sim}{B}$ and $\underset{\sim}{C}$ defined measures of similarity or proximity between the objects from the two sets. For instance, b_{ij} could denote spatial separation between $O_i^{(1)}$ and $O_j^{(2)}$, and c_{ij} could represent a corresponding absolute difference between the values on some given variable, e.g., the absolute difference between the times of occurrence for particular events. In this latter interpretation, the concern is with space-time clustering and the degree to which spatially close objects chosen from the two sets are also "close" in terms of the time variable.

Two examples mentioned explicitly by Klauber (1971) may help clarify this notion of space-time clustering. In one illustration, the two sets corresponded to the locations of dogs and cats, respectively, which were diagnosed as lymphoma cases in a given county over a three-year period. The entries in the matrix $\underset{\sim}{B}$ were actual spatial separation measures; $\underset{\sim}{C}$ contained functions of the absolute differences in the times of diagnosis. An evaluation of the measure $\Lambda(\phi_I, \rho_I)$ represented an assessment of the conjecture of disease contagion across the two sets, i.e., whether cases separated by short spatial distances were also separated by short absolute differences in times of diagnosis. In this context, both sets were considered random. A second example used two sets of geographic locations that corresponded to locations of underground nuclear explosions (considered fixed) and hypocenters of earthquakes, respectively. Both matrices contained zero-one entries; b_{ij} was 1 if and only if the location of the nuclear explosion $O_i^{(1)}$ was within 100 km of hypocenter $O_j^{(2)}$; c_{ij} was 1 if and only if the earthquake at $O_j^{(2)}$ occurred within 24 hr subsequent to the underground test at $O_j^{(1)}$. Obviously our concern here was with the degree to which underground nuclear explosions were related to subsequent earthquakes in the nearby area.

REFERENCES

Brennan, R. L., and D. J. Prediger. Coefficient kappa: Some uses, misuses, and alternatives, Educ. Psychol. Meas. 41 (1981), 687-699.

Buffington, V., D. C. Martin, and J. Becker. VER similarity between alcoholic probands and their first-degree relatives, Psychophysiology 18 (1981), 529-533.

Epp, R. J., J. W. Tukey, and G. S. Watson. Testing unit vectors for correlation, J. Geophys. Res. 76 (1971), 8480-8483.

Hubert, L. J. Kappa revisited, Psychol. Bull. 84 (1977), 289-297.

Hubert, L. J., and R. G. Golledge. Comparing rectangular data matrices, Environ. Planning [A] 14 (1981), 1087-1095.

Hubert, L. J., and R. G. Golledge. Measuring association between spatially defined variables: Tjostheim's index and some extensions, Geog. Anal. 14 (1982b), 273-278.

Hubert, L. J., and R. G. Golledge. Rater agreement for complex assessments, Br. J. Math. Stat. Psychol. 36 (1983), 207-216.

Klauber, M. R. Two sample randomization tests for space-time clustering. Biometrics 27 (1971), 129-142.

Light, R. J. Measures of agreement for qualitative data: Some generalizations and alternatives, Psychol. Bull. 76 (1971), 365-377.

Martin, D. C., V. Buffington, and J. Becker. Randomization test of paired data: Application to evoked responses, Psychophysiology 18 (1981), 524-528.

Phillips, D. P. Deathday and birthday: An unexpected connection, in Tanur, J. M., et al. (Eds.), Statistics: A Guide to the Unknown, Holden-Day, San Francisco, 1978, pp. 71-85.

Tjøstheim, D. A measure of association for spatial variables, Biometrika 65 (1978), 109-114.

Tsutakawa, R. K., and S. L. Yang. Permutation tests applied to antibiotic drug resistance, J. Am. Stat. Assoc. 69 (1974), 87-92.

3
Extensions of the LA Model

3.1 INTRODUCTION (A)[†]

The linear assignment (LA) model introduced in Chapter 1 has been the key to the applications discussed up to this point. Consequently, we might expect that any major generalization of this basic paradigm would, in turn, have a number of implications for the analysis of data. As always, the extensions we pursue are important in the recognition of problem commonality. The field of nonparametric statistics is very broad and diversified, and therefore, comprehensive organizing principles may be of immense pedagogical help to a student attempting to organize the field into some type of coherent whole.

This chapter is concerned with the notion of assignment based on scores that are coded by K subscripts, $c_{i_1 i_2 \cdots i_K}$. For example, if K is 3, then in its popular interpretation, $c_{i_1 i_2 i_3}$ may denote the value of assigning person i_1 to job i_2 at time i_3. For convenience, the K-dimensional matrix $\{c_{i_1 i_2 \cdots i_K}\}$ is denoted by C_K; thus, when K = 2, the simple assignment model of Chapter 1 is obtained for $\underset{\sim}{C}_2 = \underset{\sim}{C} = \{c_{ij}\}$.

As an optimization problem, the use of $\underset{\sim}{C}_K$ offers no real conceptual change. The objective function to be optimized has the form

$$\sum_{i_1, i_2, \ldots, i_K} c_{i_1 i_2 \cdots i_K} x_{i_1 i_2 \cdots i_K}$$

and the constraints are given by analogs to those of Section 1.2.1:

[†]The reader is referred to the Preface for an explanation of the use of (A), (B), and (C) in text headings.

1. The unknown indicator, $x_{i_1 \ldots i_K}$, is 0 or 1, $1 \leqslant i_1, \ldots, i_K \leqslant n$

2. $\Sigma_{i_1, \ldots, i_{k-1}, i_{k+1}, \ldots, i_K} \, x_{i_1} \cdots {}_{i_K} = 1$, for all k

where $1 \leqslant k \leqslant K$ (with i_0 ignored as a subscript when k = 1). Thus, when K = 3, the constraint in (2) implies that each person, i_1, is assigned to one job and one time ($\Sigma_{i_2, i_3} \, x_{i_1 i_2 i_3} = 1$); each job, i_2, is assigned to one person and one time ($\Sigma_{i_1, i_3} \, x_{i_1 i_2 i_3} = 1$); and finally, each time, i_3, is assigned to one person and one job ($\Sigma_{i_1, i_2} \, x_{i_1 i_2 i_3} = 1$). In general, there are only n subscript combinations (or K-tuples) that lead to a value of one for the corresponding indicator, and these combinations must satisfy the constraints given in (2).

When rephrased as a search over permutations, the objective function would take the form

$$\Gamma_K(\rho_1, \ldots, \rho_{K-1}) = \sum_i c_{i \rho_1(i) \cdots \rho_{K-1}(i)} \qquad (3.1a)$$

where, without loss of generality, the summation has been arbitrarily keyed off the first subscript as in Chapter 1. If we had wished, an additional permutation could have been applied to the first subscript as well. As a notational convenience, this objective function in equation (3.1a) is denoted by $\Gamma_K(\rho)$. Thus, the (maximization) optimization task can be rephrased as

$$\max_{\rho_1, \ldots, \rho_{K-1}} \Gamma_K(\rho_1, \ldots, \rho_{K-1})$$

or alternatively, by the shorthand expression

$$\max_{\rho_1, \ldots, \rho_{K-1}} \Gamma_K(\rho)$$

As before, our interest is not in this optimization problem per se but in the actual distribution of the objective function over all permutations $\rho_1, \ldots, \rho_{K-1}$. There are $(n!)^{K-1}$ such possibilities that will be considered equally likely and used to form a reference distribution against which a given index value can be compared. This latter value will typically be defined by K - 1 identity permutations for $\rho_1, \ldots, \rho_{K-1}$ and denoted by $\Gamma_K(\rho_I)$. Finally, normalized indices of the form given in Section 1.2.4 could be developed for the raw index $\Gamma_K(\rho_I)$. Again, bounds are typically used to construct these latter measures as a way of avoiding the need to solve the corresponding optimization problem explicitly.

Assuming that all $(n!)^{K-1}$ realizations of $\Gamma_K(\rho)$ are equally likely, the moment formulas can become very complicated. For example, following the work of Dwyer (1964), the mean and variance are given as

$$E(\Gamma_K(\rho)) = \left(\frac{1}{n^{K-1}}\right) \sum_{i_1,\ldots,i_K} c_{i_1 i_2 \cdots i_K}$$

$$V(\Gamma_K(\rho)) = \left[\frac{(n-1)^{K-1} + (-1)^K}{n^{K-1}(n-1)^{K-1}}\right]\left(\sum_{i_1,\ldots,i_K} c_{i_1 \cdots i_K}^2\right)$$

$$+ \left[\frac{n^{K-1} - (n-1)^{K-1}}{(n^{K-1})^2 (n-1)^{K-1}}\right]\left(\sum_{i_1,\ldots,i_K} c_{i_1 \cdots i_K}\right)^2$$

$$+ \left[\frac{1}{n^{K-1}(n-1)^{K-1}}\right] \sum_{r=1}^{K-1} (-1)^r D_r$$

where D_r is the sum of all squared terms that have $K - r$ subscripts removed by summation. Thus, when $K = 3$,

$$D_1 = \sum_{i_1} c_{i_1 \cdot \cdot}^2 + \sum_{i_2} c_{\cdot i_2 \cdot}^2 + \sum_{i_3} c_{\cdot \cdot i_3}^2$$

and

$$D_2 = \sum_{i_1,i_2} c_{i_1 i_2 \cdot}^2 + \sum_{i_1,i_3} c_{i_1 \cdot i_3}^2 + \sum_{i_2,i_3} c_{\cdot i_2 i_3}^2$$

In applications of this K-place assignment model, an inference strategy based on Monte Carlo sampling from all $(n!)^{K-1}$ possible realizations of $\Gamma_K(\rho)$ would be one possible significance testing method that could be followed. In addition, for a few important special cases in which the K-place assignment scores are defined as sums of two-place scores, the skewness parameter has been worked out explicitly (e.g., see Mielke and Iyer, 1982). Thus, in these instances a type III approximation could also be used.

Some proofs of asymptotic normality exist for various special cases of the index $\Gamma_K(\rho)$ (cf. David and Barton, 1962), as well as proofs for the convergence to a chi-square random variable for related statistics such as Friedman's (see Section 3.2.2 and Lehmann, 1975). However, all of these approximations are of varying adequacy, depending on the size of K and of n, and on the particular assignment scores being used and their patterning. In some cases, such as the Page test discussed in Section 3.2.5, the normal approximation is easy to demonstrate and is probably very good if K and n are reasonably large. In this latter case, the distribution of the index is generated by a sum of $K - 1$ independent random variables, and each individual term is itself asymptotically normal under mild regularity conditions. In general, however, we will rely on type III approximations and/or approximate permutation tests based on sampling from the complete distribution. The latter will be a very appropriate strategy to implement routinely in the years to come, both because of the continuing reduction in computer costs and the procedure's wide applicability.

3.2 APPLICATIONS[†]

3.2.1 Contingency Tables (B)

As one particular application of the index $\Gamma_K(\rho_I)$, suppose a threefold contingency table is given having R rows, S columns, and T layers, and let n_{rst} denote the number of observations in row r, column s, and layer t. If w_{rst} defines a fixed weight attached to the corresponding cell, then the raw index $\Gamma_3(\rho_I)$ can be written as

$$\Gamma_3(\rho_I) = \sum_{r,s,t} w_{rst} n_{rst}$$

where the assignment score $c_{i_1 i_2 i_3}$ is defined as w_{rst} if object i_1 belongs to row r, i_2 belongs to column s, and i_3 belongs to layer t. The moment formulas reduce in a similar manner by representing the necessary sums in terms of the weights and marginal frequencies. In particular, if we let $p_{r..} = n_{r..}/n$, $p_{.s.} = n_{.s.}/n$, and $p_{..t} = n_{..t}/n$, then

$$E(\Gamma_3(\rho)) = n \sum_{r,s,t} w_{rst} p_{r..} p_{.s.} p_{..t}$$

and

$$V(\Gamma_3(\rho)) = \left(\frac{n}{n-1}\right)^2 \left\{ (n-2) \sum_{r,s,t} w_{rst}^2 p_{r..} p_{.s.} p_{..t} \right.$$
$$+ (2n-1) \left(\sum_{r,s,t} w_{rst} p_{r..} p_{.s.} p_{..t} \right)^2$$
$$\left. - n \left[\sum_r p_{r..} \left(\sum_{s,t} w_{rst} p_{.s.} p_{..t} \right)^2 \right. \right.$$

[†]Parts of the subsections that follow depend heavily on Hubert (1979), Hubert, Golledge, Costanzo, and Richardson (1981), and Hubert, Golledge, Kenney, and Costanzo (1981).

$$+ \sum_s p_{\cdot s \cdot} \left(\sum_{r,t} w_{rst} p_{r \cdot \cdot} p_{\cdot \cdot t} \right)^2$$

$$+ \sum_t p_{\cdot \cdot t} \left(\sum_{r,s} w_{rst} p_{r \cdot \cdot} p_{\cdot s \cdot} \right)^2 \Big]$$

$$+ \left[\sum_{r,s} p_{r \cdot \cdot} p_{\cdot s \cdot} \left(\sum_t w_{rst} p_{\cdot \cdot t} \right)^2 \right.$$

$$+ \sum_{r,t} p_{r \cdot \cdot} p_{\cdot \cdot t} \left(\sum_s w_{rst} p_{\cdot s \cdot} \right)^2$$

$$\left. + \sum_{s,t} p_{\cdot s \cdot} p_{\cdot \cdot t} \left(\sum_r w_{rst} p_{r \cdot \cdot} \right)^2 \right] \Big\}$$

When $R = S = T$, and three raters are assumed to define the dimensions of the three-way contingency table, a number of raw indices of rater agreement can be obtained by varying the definition of w_{rst}. As indicated by David and Barton (1962) and developed in the rater context by Hubert (1977), many different concepts of agreement are possible. Several of these definitions are given below, but the reader is referred to David and Barton (1962) for a more complete discussion and an examination of asymptotic normality for these alternatives when the weights are restricted to be dichotomous (zero-one) or when they arise as sums of dichotomous weights.

1. DeMoivre: An agreement occurs if and only if all raters place an object in the same category. Thus, we let $w_{rst} = 1$ if $r = s = t$, and 0 otherwise. Alternatively, if a_{rs} denotes a weighting between raters 1 and 2, b_{rt} a weighting between raters 2 and 3; and c_{st} a weighting between raters 1 and 3; $w_{rst} = $ minimum (a_{rs}, b_{rt}, c_{st}). In the specific cases being considered here, $a_{rs} = 1$ if $r = s$ and 0 otherwise, $b_{rt} = 1$ if $r = t$ and 0 otherwise, and $c_{st} = 1$ if $s = t$ and 0 otherwise. This notation will be used below as well.

2. Target: If the first rater is considered a target, then an agreement occurs if and only if another rater places an object in the same category as the first. Here, we let $w_{rst} = a_{rs} + b_{rt}$.

3. Pairwise: An agreement occurs if and only if two raters categorize an object consistently, i.e., $w_{rst} = a_{rs} + b_{rt} + c_{st}$.

In each of these cases, the expectations can be obtained readily from the general formula for $E(\Gamma_K(\rho))$. For example, when $K = 3$,

DeMoivre: $E(\Gamma_3(\rho)) = n \sum_r p_{r \cdot \cdot} p_{\cdot r \cdot} p_{\cdot \cdot r}$

Target: $E(\Gamma_3(\rho)) = n \sum_r (p_{r..}p_{.r.} + p_{r..}p_{..r})$

Pairwise: $E(\Gamma_3(\rho)) = n \sum_r (p_{r..}p_{.r.} + p_{r..}p_{..r} + p_{.r.}p_{..r})$

Also, in each instance rather simple upper bounds can be obtained on $\Gamma_K(\rho_I)$ to construct a normalized index $\Gamma_K^*(\rho_I)$, i.e., DeMoivre: n; target: $(K-1)n$; and pairwise: $\binom{K}{2}n$. The normalized indices in the target and pairwise cases, however, are not the simple averages of the relevant pairwise kappa indices because the expectation comes as a complete sum in the denominator of $\Gamma_K^*(\rho_I)$ and the constituent terms are not the same. For a discussion of the simple average method, the reader is referred to Conger (1980).

Although these three interpretations given for w_{rst} are probably the most obvious, several other possibilities exist:

4. If $w_{rst} = 1$ when at least one nontarget rater matches the target and 0 otherwise, then $w_{rst} = \text{maximum}(a_{rs}, b_{rt})$.
5. If $w_{rst} = 1$ when at least one pair of raters match and 0 otherwise, then $w_{rst} = \text{maximum}(a_{rs}, b_{rt}, c_{st})$.
6. If w_{rst} denotes the number of matches for pairs of raters with consecutive labels, then $w_{rst} = a_{rs} + c_{st}$.
7. If $w_{rst} = 1$ when a majority of raters match and 0 otherwise, then $w_{rst} = 1$ if $a_{rs} + b_{rt} + c_{st} \geq 2$ and 0 otherwise.

These definitions or extensions will not be pursued any further for more than three raters (or in general, to a K-fold contingency table for K greater than three), but the approach would follow exactly that given above. Also, a variant of the prediction analysis methodology developed by Hildebrand et al. (1977) for a two-way contingency table could be extended to a K-way table, using the weights to specify error cells.

3.2.2 Measuring Concordance Among K Sequences (B)

One of the standard nonparametric data analyses problems is typically discussed under the title of "Friedman's test" or "Kendall's coefficient of concordance." Here, K judges assign numerical values to n objects (e.g., ranks), and our interest is in (i) testing whether the n objects can be considered equally "preferable," and (ii) measuring the actual degree of concordance among the K judges. Both of these problems will be approached with a particular assignment score, $c_{i_1 i_2 \ldots i_K}$, having a simple additive structure:

$$c_{i_1 i_2 \cdots i_K} = \sum_{k<k'} c_{i_k i_{k'}} \qquad (3.2.2a)$$

Thus, we reduce each of the K-place assignment scores to a simple sum of two-place assignment scores defined over all possible pairs of sequences.

For now, the two-place assignment scores in equation (3.2.2a) will be defined as squared differences, i.e.,

$$c_{i_k i_{k'}} = \left(x_{i_k}^{(k)} - x_{i_{k'}}^{(k')} \right)^2$$

where the sequence of values over the n objects assigned by judge k is denoted by

$$[x_1^{(k)}, x_2^{(k)}, \ldots, x_n^{(k)}]$$

In other words, $c_{i_1 i_2 \cdots i_K}$ is constructed by all pairwise comparisons between the K sequences, using as a definition of similarity the squared differences between two entries from a pair of sequences. Thus, the raw index takes the form

$$\Gamma_K(\rho_I) = \sum_i \left\{ \sum_{k<k'} [x_i^{(k)} - x_i^{(k')}]^2 \right\}$$

Although the definition of an assignment score as a squared difference is followed throughout, we do point out a simple equivalence, using an alternative cross-product definition. In particular, $\Gamma_K(\rho_I)$ may be rewritten as

$$(K - 1) \sum_k \left\{ \sum_i [x_i^{(k)}]^2 \right\} - 2 \sum_i \sum_{k<k'} x_i^{(k)} x_i^{(k')}$$

which is a constant linear transformation of $\sum_i \sum_{k<k'} x_i^{(k)} x_i^{(k')}$. Thus, our presentation could just as well be phrased for assignment scores of the form

$$c_{i_k i_{k'}} = x_{i_k}^{(k)} x_{i_{k'}}^{(k')}$$

As an essentially arbitrary choice, however, we will continue to assume the squared-difference definition in the discussion below.

The interpretation of sequence data in a K × n analysis-of-variance format may help to clarify the meaning of the raw measure $\Gamma_K(\rho_I)$. The sum of squares for such a table can be decomposed into three parts: the sums of squares for judges, objects, and a residual. Under the randomness model being assumed, which considers all permutations of the entries in each row except the first to be equally likely, the total sum of squares and the judges sum of squares are constant. Thus, over all $(n!)^{K-1}$ realizations of the K × n table, the residual and object sum of squares must vary monotonically with respect to each other and with respect to the usual F ratio for assessing object differences (the F ratio would be defined by the mean square for objects divided by the mean square for the residual). Consequently, any of the latter three measures, or constant functions of them, could be used as a test statistic. This includes $\Gamma_K(\rho_I)$, which is K times the sum of squares for judges plus the residual. The smaller the raw index $\Gamma_K(\rho_I)$, the smaller the residual and the larger the object sum of squares and the F ratio.

If untied ranks are used rather than the original observations, all row sums are equal and the sum of squares for judges is 0. Thus, $\Gamma_K(\rho_I)$ is merely K times the residual sum of squares. In this context, the statistic, Q, which is usually attributed to Friedman, is a simple linear transformation of $\Gamma_K(\rho_I)$, i.e.,

$$Q = (n - 1)K - \frac{12}{Kn(n + 1)} \; \Gamma_K(\rho_I) = \frac{12}{Kn(n + 1)} \sum_{j=1}^{n} R_j^2 - 3K(n + 1)$$

where R_j is the sum of ranks in column j. We note in particular that when $n = 2$, Q reduces in turn to

$$4K\left(\frac{b}{K} - \frac{1}{2}\right)^2$$

where b is the number of times that an entry in column 2 is larger than the entry in column 1 (Lehmann, 1975). Because $(b/K) - (1/2)$ is one statistic that could be considered for a sign test, Friedman's Q [and thus, $\Gamma_K(\rho_I)$] also provides the basis for the two-tailed sign test as a special case. The two-dependent sample problem in a directional context is discussed again in Section 3.3.

Even when there are ties, the various modifications and specializations of Friedman's statistic that have been proposed (see Lehmann, 1975) turn out to be constant linear transformations of our basic measure $\Gamma_K(\rho_I)$. This is true, for example, even when the responses are (zero-one) dichotomous. Here, we obtain Cochran's test for n correlated proportions, or equivalently, McNemar's test when $n = 2$. This latter reduction is particularly instructive, given its familiarity. If a, b, c, and d refer to the number of (0,0), (0,1), (1,0), and (1,1) sequences, respectively, then McNemar's statistic

$$\frac{(b - c)^2}{b + c}$$

can be written as a constant linear transformation of $\Gamma_K(\rho_I)$:

$$\left[\frac{-2}{b + c}\right]\Gamma_K(\rho_I) + \left[b + c + \frac{2a(K - a)}{b + c} + \frac{2d(K - d)}{b + c}\right]$$

Given the constancy of the $(0,0)$ and $(1,1)$ rows, the $(n!)^{K-1}$ realizations of the table could be reduced substantially. In effect, we need to consider only 2^{b+c-1} equally likely realizations, i.e., one of the nonconstant rows can be fixed and each of the remaining $b + c - 1$ can be reorganized in only one of two ways. In short, a test of $\Gamma_K(\rho_I)$ is equivalent to a two-tailed binomial test of $p = 1/2$, where the number of trials is $b + c$, and the number of observed heads is b. Similar reductions hold for Cochran's extension when n is greater than 2.

Moments of $\Gamma_K(\rho)$. Because $\Gamma_K(\rho)$ is constructed through two-place assignment scores, it is convenient to define the moments of $\Gamma_K(\rho)$ in terms of interaction matrices. Suppose $\{c_{i_k i_{k'}}\}$ is replaced by $\{d_{i_k i_{k'}}\}$ through the removal of row and column effects as in equation (1.2.2b). When the measure $\Gamma_K(\rho)$ is defined with these reduced matrices, it is numerically the same as the original index $\Gamma_K(\rho)$ defined with $\{c_{i_k i_{k'}}\}$ minus the sum of expectations for each of the $\binom{K}{2}$ assignment problems:

$$\sum_{k<k'}\left(\frac{1}{n}\sum_{i_k, i_{k'}} c_{i_k i_{k'}}\right)$$

Because this latter term is actually $E(\Gamma_K(\rho))$ and is constant, the variance and skewness parameters can be conveniently represented through their associated interaction matrices. In particular, the pairwise indices are uncorrelated (cf. David and Barton, 1962, p. 218), and thus, the variance for $\Gamma_K(\rho)$ is the corresponding sum over $k<k'$ of the variances for the $\binom{K}{2}$ pairwise LA problems:

$$\gamma(\Gamma_K(\rho)) = \sum_{k<k'}\left(\frac{1}{n - 1}\right)\sum_{i_k, i_{k'}} d^2_{i_k i_{k'}}$$

The skewness parameter is a little more complicated and is given by

$$\gamma(\Gamma_K(\rho)) = \left\{\left[\frac{n}{(n-1)(n-2)}\right] \sum_{k<k'} \sum_{i_k,i_{k'}} d_{i_k i_{k'}}^3 + \left[\frac{6}{(n-1)^2}\right]\right.$$

$$\left. \times \sum_{k<k'<k''} \sum_{i_k,i_{k'},i_{k''}} d_{i_k i_{k'}} d_{i_k i_{k''}} d_{i_{k'} i_{k''}} \right\} / V(\Gamma_K(\rho))^{3/2}$$

(3.2.2b)

This latter quantity will allow a convenient type III approximation to the distribution of $\Gamma_K(\rho)$ when the original K-place assignment scores have the appropriate additive structure (see Mielke and Iyer, 1982).

Because small values of $\Gamma_K(\rho_I)$ are desirable given our definition of $\{c_{i_k i_{k'}}\}$, a normalized index of the form $\Gamma_K^{**}(\rho_I)$ would be appropriate. For example, if each sequence consists of the untied ranks, then $\min_{\rho_1,\ldots,\rho_{K-1}} \Gamma_K(\rho)$ is 0 and $E(\Gamma_K(\rho)) = \binom{K}{2} n(n^2-1)/6$. Consequently, if $r_{kk'}$ denotes the Spearman correlation between the k and k' sequences, then

$$\Gamma_K^{**}(\rho_I) = \frac{1}{\binom{K}{2}} \sum_{k<k'} r_{kk'}$$

the average of all such pairwise correlations. Although this last average could be reported as the final descriptive measure of concordance, the alternative measure suggested by Kendall (1970) could also be obtained by a simple linear transformation. Because this average pairwise correlation has several nice extensions later on, however, it will be our descriptive measure of choice in this context.

If ranks are replaced by z scores within sequences (i.e., we align the values within each sequence for location and scale), and if $\min_\rho \Gamma_K(\rho)$ were assumed 0 (possibly as a bound), then $\Gamma_K^{**}(\rho_I)$ would be the average of all Pearson correlations. This latter value could then be reported as the final descriptive measure of concordance.

Given the general structure of the assignment strategy, it should be apparent that concordance could be approached in a variety of ways. For example, the footrule norm could be used in the same additive framework of equation (3.2.2a), or possibly, alternative nonadditive functions could be proposed. The latter would measure variability within a column of the K × n table (e.g., the range) and lead to different notions of concordance across sequences. It should also be clear at this point that the problem of nominal-scale agreement among K raters, as discussed in Section 3.2.1, can be rephrased in this generalized Friedman paradigm involving K sequences of n observations. Instead of using, ranks, however, category labels are attached to each of the n objects by each rater. The assignment scores are defined as 1 or 0, depending on whether the category labels are identical or not.

3.2.3 Concordance in a More General Context (B)

Although we explicitly introduced the notion of concordance for K sequences of n numerical values in Section 3.2.2, the basic notion is really more general than this. The K-place assignment score in the additive form of equation (3.2.2a) leads to an index of concordance $\Gamma_K(\rho)$ that can be represented as

$$\Gamma_K(\rho) = \sum_{0 \leqslant k < k' \leqslant K-1} \left(\sum_i c_{\rho_k(i)\rho_{k'}(i)} \right)$$

where for notational convenience, the permutation on the first sequence, ρ_0, is assumed to be the fixed identity permutation. Thus, each term $\Sigma_i\, c_{\rho_k(i)\rho_{k'}(i)}$ represents the objective function for a particular assignment based on a matrix $\{c_{i_k i_{k'}}\}$; e.g., person $\rho_k(i)$ is assigned to job $\rho_{k'}(i)$. There are $\binom{K}{2}$ such matrices that need to be considered in defining $\Gamma_K(\rho)$. (The moments we gave in Section 3.2.2 were phrased in terms of general two-place assignment matrices. Thus, these formulas are appropriate here as well and will not be restated.)

We will assume, without loss of generality, that the observed index of concordance is defined by $\Gamma_K(\rho_I)$, in which each of the permutations is the identity. For example, suppose in our usual spatial context that K ranked variables F_1, F_2, ..., F_K are observed over n different spatial locations and the ranks on F_1 are arbitrarily used to label the n geographical locations. Then, instead of defining the assignment scores as spatial separation measures between the locations as we did before, the entry in the ith row and jth column of the assignment matrix $\{c_{i_k i_{k'}}\}$ is the spatial separation measure between the locations that contain rank i on F_k and rank j on $F_{k'}$. The index $\Gamma_k(\rho_I)$ then measures the degree of concordance among the K variables in relation to the spatial separations between the n locations.

The concept of generalized concordance as defined by the index $\Gamma_K(\rho_I)$ has a variety of possible applications. All that is required is a set of $\binom{K}{2}$ assignment matrices, and the measure can be constructed. For instance, in the context of object by attribute data matrices of the previous chapter, suppose B_1, B_2, ..., B_K represent K such matrices, each of size p × q. If matching is to be done, say, over columns, then each pair of matrices generates a q × q assignment matrix based on some appropriate measure of correspondence between columns. Because these assignment matrices can in turn be used to define an index of concordance for the K matrices, we are essentially extending a concordance notion from sequences to matrices. Obviously, similar statements could be made for the K p × p

assignment matrices that would be obtained if the matching were done
by rows.

As an explicit numerical example of how a notion of generalized concordance can be used in a real data analysis context, consider Table 3.2.3a.
This 30 × 30 symmetric matrix contains measures of similarity or proximity
between the crime profiles for six cities (Chicago, Los Angeles, Philadel-
phia, San Francisco, Boston, and Pittsburgh) at five time points (1969,
1971, 1973, 1975, and 1977). Each entry in the matrix corresponds to a
euclidean distance measure between the profiles for two city-time pairs
over the seven index crimes (murder, rape, robbery, assault, burglary,
larceny, auto theft). The entries across crimes were first made comparable by a transformation to z scores, that is, normalizing the rates within
each particular crime type to mean 0 and variance 1. The final proximity
between profiles was then obtained by taking the square root of the sum of
the squared differences over the seven index crimes. Obviously, other
measures could be used as well, for example, Mahalanobis distances,
correlations, and so on. In all cases, however, the analysis would be carried out in the same manner.

Considering the complete 30 × 30 matrix of proximities in Table
3.2.3a, the assessment of a spatial effect should intuitively depend on the
(reduced) level of elevation for the entries in the six "same-city" blocks
(each of size 5 × 5) on the main diagonal. Similarly, if we would reorganize
the matrix to nest cities under time, a temporal effect should be reflected
in the (reduced) elevation for each of the five "same-time" blocks (each of
size 6 × 6) on the main diagonal. In either case, a natural evaluation
strategy would first define an index for the degree of elevation in the on-
diagonal blocks, and then specify a procedure for assessing the relative
size of the observed index compared to some chance model. In these
respects, the general concordance notion represented by $\Gamma_K(\rho_I)$ is a natural
tool for carrying out this assessment.

Because the roles of the city and time factors are essentially interchangeable, we can limit our discussion to the one factor of time. Or, in
the language of the usual concordance notion for K sequences over n objects,
the K judges would correspond to the six cities, and the n objects to the
five time periods; thus, our interest is in "blocking" on the spatial factor
when testing for a temporal effect. As mentioned above, such an effect
would be reflected in the five "same-time" blocks of size 6 × 6 if Table
3.2.3a were reorganized to nest cities under time. However, as the table
is written now, these entries are already present in the "diagonal strips" in
the 15 city-to-city submatrices off the main diagonal. Each of these 5 × 5
matrices defined by two different cities, k and k', gives a matrix $\{c_{i_k i_{k'}}\}$

that can be used in constructing the overall index $\Gamma_K(\rho_I)$ and its reference
distribution. In fact, $\Gamma_K(\rho_I)$ is merely the sum of all the entries in the 15
diagonal strips that are obtained from all possible city pairs.

Because the expectation and variance of $\Gamma_K(\rho)$ are additive with respect to the separate assignment tasks, they may be obtained rather easily; the skewness parameter requires a little more effort but can be found directly from the formula in Section 3.2.2. Carrying out these computations for testing the effect of the spatial and the temporal factors, we would obtain the following (the Monte Carlo significance levels are based on samples of size 999):

	Spatial	Temporal
$\Gamma_K(\rho_I)$	101.75	270.76
$E(\Gamma_K(\rho))$	209.36	285.01
$V(\Gamma_K(\rho))$	108.183	4.858
γ	-.706	-.520
Z	-10.35	-6.47
Monte Carlo significance	.001	.001
Type III significance	.001	.001

Both Z statistics given above are large in absolute value and indicate that both the spatial and temporal classifications could help explain some of the differences in the proximities.

The cross-classified matrix considered in this example included only two dimensions, but extensions to more than two are immediate and offer no major difficulties. For example, suppose we have three factors A, B, and C, with a, b, and c levels, respectively. If an evaluation of A is desired, then factors B and C are simply combined to produce an aggregate factor with bc levels. Otherwise, the evelution process remains unchanged. It should be apparent that any number of dimensions and levels could be handled by the simple expedient of constructing an aggregate factor from all dimensions except the particular one under test.

The major contribution of a general notion of concordance is in the use of arbitrary proximity measures and the development of a strategy for blocking on the levels of one (or more) a priori dimension(s) when evaluating the effects produced by a second. The strategy being proposed is really very general even though the illustration we have used in explaining the method contained the three explicit classification dimensions of space, time, and crime type. For instance, because any two of the dimensions could in fact have been considered the major classification facets of interest, proximity measures could have been obtained between profiles over the

Table 3.2.3a 30 × 30 Symmetric Matrix Containing Measures of Proximity Between the Profiles for Six Cities at Five Time Points

	Chicago					Los Angeles					Philadelphia				
	1969	1971	1973	1975	1977	1969	1971	1973	1975	1977	1969	1971	1973	1975	1977
Chicago															
1969	0														
1971	.59	0													
1973	2.08	1.83	0												
1975	2.81	2.60	.93	0											
1977	2.53	2.48	1.39	1.12											
Los Angeles															
1969	3.07	3.03	2.83	2.99	3.17	0									
1971	3.54	3.39	3.02	3.15	3.53	1.02	0								
1973	4.02	3.79	2.78	2.62	3.28	1.77	1.36	0							
1975	4.47	4.18	2.98	2.77	3.59	2.62	2.04	.94	0						
1977	5.07	4.76	3.66	3.59	4.38	3.08	2.53	1.59	1.19	0					
Philadelphia															
1969	2.86	3.32	4.34	4.66	3.87	4.56	5.36	5.90	6.54	7.22	0				
1971	1.46	1.82	2.98	3.42	2.74	3.62	4.23	4.76	5.30	6.00	1.67	0			
1973	1.50	1.77	2.39	2.71	1.96	3.16	3.81	4.15	4.70	5.40	2.06	.89	0		
1975	1.63	1.68	1.74	1.93	1.36	2.86	3.41	3.55	4.03	4.80	2.82	1.59	.87	0	
1977	2.15	2.45	2.64	2.72	2.05	3.07	3.84	4.05	4.66	5.42	2.12	1.66	1.04	1.16	0
San Francisco															
1969	3.39	3.22	3.16	3.21	3.44	1.96	1.69	2.44	3.04	3.71	5.00	3.81	3.52	3.13	3.63
1971	3.06	2.88	2.69	2.70	3.03	1.96	1.77	2.29	2.80	3.65	4.75	3.53	3.20	2.68	3.23
1973	3.88	3.71	2.52	1.91	2.47	2.41	2.42	1.71	2.10	3.05	5.31	4.28	3.58	2.84	3.27
1975	4.54	4.30	2.89	2.21	2.99	3.09	2.93	1.86	1.88	2.76	6.12	5.05	4.34	3.54	4.05
1977	4.52	4.31	2.92	2.30	3.04	2.82	2.68	1.55	1.69	2.51	6.10	5.06	4.33	3.58	4.03
Boston															
1969	3.52	3.97	4.77	4.97	4.15	4.57	5.25	5.94	6.63	7.39	1.94	2.46	2.74	3.37	2.74
1971	3.21	3.60	4.28	4.47	3.74	4.07	4.60	5.33	5.98	6.78	2.55	2.39	2.60	3.06	2.70
1973	3.20	3.48	3.77	3.87	3.26	3.61	3.99	4.63	5.25	6.04	3.40	2.81	2.73	2.93	2.81
1975	4.69	4.80	4.63	4.60	4.33	4.22	4.16	4.72	5.19	5.90	5.54	4.72	4.53	4.46	4.63
1977	3.84	4.10	3.99	3.98	3.57	3.43	3.71	4.28	4.89	5.67	4.28	3.74	3.50	3.52	3.35
Pittsburgh															
1969	3.34	3.84	4.74	4.99	4.26	4.70	5.48	6.08	6.77	7.51	1.21	2.34	2.67	3.29	2.45
1971	3.12	3.61	4.53	4.79	4.09	4.48	5.29	5.87	6.55	7.27	.96	2.12	2.43	3.05	2.19
1973	3.14	3.63	4.44	4.67	3.93	4.59	5.39	5.90	6.56	7.30	1.03	2.15	2.38	2.97	2.09
1975	2.66	3.11	3.66	3.82	3.09	3.96	4.73	5.12	5.75	6.54	1.40	1.81	1.76	2.22	1.29
1977	2.96	3.42	4.03	4.19	3.47	4.14	4.96	5.38	6.05	6.79	1.27	2.08	2.07	2.57	1.56

	San Francisco					Boston					Pittsburgh				
	1969	1971	1973	1975	1977	1969	1971	1973	1975	1977	1969	1971	1973	1975	1977
	0														
	.90	0													
	2.50	2.05	0												
	3.17	2.74	.93	0											
	3.07	2.71	.91	.45	0										
	4.58	4.54	5.30	6.19	6.13	0									
	3.82	3.80	4.72	5.61	5.56	1.02	0								
	3.15	3.24	4.04	4.89	4.82	2.09	1.23	0							
	3.28	3.64	4.40	5.06	4.95	4.15	3.24	2.16	0						
	3.17	3.27	3.83	4.63	4.48	3.01	2.23	1.32	1.74	0					
	5.00	4.77	5.41	6.27	6.23	1.38	2.08	3.03	5.14	3.79	0				
	4.87	4.61	5.22	6.06	6.02	1.65	2.27	3.16	5.28	3.91	.44	0			
	4.96	4.67	5.17	5.99	5.97	1.68	2.29	3.12	5.25	3.87	.53	.42	0		
	4.37	4.02	4.34	5.15	5.13	1.90	2.17	2.72	4.78	3.37	1.28	1.11	.91	0	
	4.62	4.31	4.63	5.44	5.40	1.91	2.36	2.99	5.07	3.62	1.06	.84	.66	.50	0

cities and our interest directed toward the two dimensions of crime type and time. The basic inference principles would remain the same, and the analyses would be carried out as before. Hopefully, this discussion will suggest a way that researchers may assess dimensional salience in data sets that are not easily studied by more standard analysis-of-variance schemes because of an unusual proximity measure.

The general concordance statistic is defined in an additive manner, and thus, one form of a normalized measure can be developed fairly easily, using the $\binom{K}{2}$ individual assignment matrices. As notation, suppose $\Gamma_{\rho_k \rho_{k'}}$ denotes the assignment statistic based on the matrix $\{c_{i_k i_{k'}}\}$, i.e.,

$$\Gamma_{\rho_k \rho_{k'}} = \sum_i c_{\rho_k(i) \rho_{k'}(i)}$$

The raw concordance index can then be defined as

$$\Gamma_K(\rho) = \sum_{0 \leq k < k' \leq K-1} \rho_k \rho_{k'}$$

and an index of the form $\Gamma_K^*(\rho_I)$ by

$$\frac{\Gamma_K(\rho_I) - \sum\limits_{0 \leq k < k' \leq K-1} E(\Gamma_{\rho_k \rho_{k'}})}{\max\limits_{\rho_1, \ldots, \rho_{K-1}} \Gamma_K(\rho) - \sum\limits_{0 \leq k < k' \leq K-1} E(\Gamma_{\rho_k \rho_{k'}})} \qquad (3.2.3a)$$

If the bound

$$\max\limits_{\rho_1, \ldots, \rho_{K-1}} \Gamma_K(\rho) \leq \sum\limits_{0 \leq k < k' \leq K-1} \max\limits_{\rho_k, \rho_{k'}} \Gamma_{\rho_k \rho_{k'}}$$

were used in the denominator of expression (3.2.3a), we would obtain

$$\frac{\sum\limits_{0 \leq k < k' \leq K-1} \left[\Gamma_{\rho_I \rho_{I'}} - E(\Gamma_{\rho_k \rho_{k'}}) \right]}{\sum\limits_{0 \leq k < k' \leq K-1} \left[\max\limits_{\rho_k, \rho_{k'}} \Gamma_{\rho_k \rho_{k'}} - E(\Gamma_{\rho_k \rho_{k'}}) \right]} \qquad (3.2.3b)$$

where $\Gamma_{\rho_I \rho_{I'}}$ refers to the observed index for the assignment task defined by k and k'. In other words, the numerators and denominators for each of the individual assignment tasks are added together separately, and the ratio of these two sums define the index in expression (3.2.3b). If we wish, further bounding could be considered for each of the individual maximum terms in the denominator. It should be obvious that a similar discussion could be developed for $\Gamma_K(\rho_I)$; in fact, the normalized measure of concordance of the previous section, given by the average of the $\binom{K}{2}$ Pearson correlations between the K sequences, could be obtained in just this way.

Measures of the type given in expressions (3.2.3a) and (3.2.3b) would be particularly appropriate in the two contexts mentioned earlier in this section: the association among K variables in a spatial sense and the extension of concordance from K sequences to K object by attribute data matrices. For example, given K variables F_1, F_2, ..., F_K and Tjøstheim's original notion, discussed in Section 1.3.5, of specifying association as sums of distances between identical ranks, the use of the squared euclidean distance measure in two dimensions would lead to an index $\Gamma_K^*(\rho_I)$ from expression (3.2.3a) or (3.2.3b) of

$$\frac{1}{\binom{K}{2}} \left[\sum_{k<k'} (r_{x_{F_k} x_{F_{k'}}} + r_{y_{F_k} y_{F_{k'}}})/2 \right] \qquad (3.2.3c)$$

In a similar way for K p × q object by attribute data matrices, $\underset{\sim}{B}_1$, ..., $\underset{\sim}{B}_K$, with columns standardized to z scores, each separate assignment task leads to a raw measure that is p times the sum of the q Pearson correlations between columns. Using the usual bound of +1 for each such correlation, the index of expression (3.2.3b) would have the form

$$\frac{1}{\binom{K}{2}} \sum_{k<k'} R_{kk'}/q \qquad (3.2.3d)$$

where $R_{kk'}$ is the sum of the q Pearson correlations between the columns of $\underset{\sim}{B}_k$ and $\underset{\sim}{B}_{k'}$.

Although the indices in (3.2.3a) and (3.2.3b) may be appropriate in the generalized concordance framework, in applications to cross-classified proximity matrices of the type discussed in this section and the next it may be sufficient to use a much more direct descriptive approach for reporting the relative size of an observed index. For each cross-classified proximity matrix, two concordance problems are present, e.g., one based on temporal effects and one based on spatial effects. This structure defines three natural average proximities: Two are based on the proximities within the same levels of each factor, and a third is based on proximities for different levels on both factors. For example, the data of Table 3.2.3a provide the information to construct the average same-SMSA[†] proximity (1.70), the average same-time proximity (3.61), and the average for different SMSAs and for different times (3.85). As is apparent, the spatial effect is more pronounced than the temporal, even though both are significant. In short, if the proximities themselves have inherent meaning, these latter quantities may provide a direct descriptive indication of the size of the effects that have been declared significant or not. In our example, for instance, the spatial effect is much more pronounced than the temporal, even though both are significant. We will use these same indices again in the next section when the proximities are actually correlations; thus, the averages are average correlations with some inherent meaning of their own.

3.2.4 Multitrait-Multimethod Matrices (C)

A classic paper by Campbell and Fiske (1959) on convergent and discriminant validity proposed a heuristic strategy for investigating what is called the multitrait-multimethod matrix. These matrices have the same form as Table 3.2.3a, but the proximities are intercorrelations among several traits, each measured by several methods (the correlations are possibly corrected for attenuation as a result of unreliability—see Lord and Novick, 1968, pp. 69-74). Thus, traits and methods form the two classification factors, and the intercorrelations contain some of the information considered necessary in carrying out a validation study of trait assessment.

In general, convergent validity can be confirmed by observing relatively large correlations between measures of the same trait based on different methods. Conversely, discriminant validity can be confirmed when the correlations are relatively low between tests, based on the same or different methods, that are intended to be indicators of distinct traits. In the original Campbell and Fiske discussion, however, very few formal assessment techniques were suggested for performing the necessary comparisons (but see Schmitt, Coyle, and Saari, 1977, for a review of later formal work). For that matter, the paper leaves the distinct impression that any such rigor may be undesirable. To the extent that a method of analysis can distort the search for the relatively straightforward properties or conditions defined by Campbell and Fiske, the simple inspection process they advocate may be the most appropriate. Nevertheless, it is possible to codify some of their heuristics in a very direct manner through the use of the concordance concepts developed in Section 3.2.3, and in the process, obtain a formal assessment framework that remains close to their original intent.

[†]SMSA, standard metroplitan statistical area.

Table 3.2.4a   A Multitrait-Multimethod Matrix from Kelley and Krey
(1934) (Sample Size of 311)

	Peer ratings				Association test			
	A_1	B_1	C_1	D_1	A_2	B_2	C_2	D_2
Peer ratings								
Courtesy A_1	X							
Honesty B_1	.74							
Poise C_1	.63	.65	X					
School drive D_1	.76	.78	.65	X				
Association test								
Courtesy A_2	.13	.14	.10	.14	X			
Honesty B_2	.06	.12	.16	.08	.27	X		
Poise C_2	.01	.08	.10	.02	.19	.37	X	
School drive D_2	.12	.15	.14	.16	.27	.32	.18	X

One of the easiest ways of discussing the multitrait-multimethod matrix
is to present a small example and then introduce some of the necessary
terminology in relation to this concrete illustration. With such a purpose
in mind, Table 3.2.4a gives a (trivial) 8 × 8 correlation matrix originally
obtained by Kelley and Krey (1934), and reanalyzed by Campbell and Fiske,
involving two methods—peer judgment and word association—that were used,
in turn, as techniques for measuring the four personality traits of courtesy,
honesty, poise, and school drive.

In developing their validation strategy, Campbell and Fiske specify
several conditions or criteria that should be satisfied by a multitrait-multimethod matrix. First of all, correlations among those tests that supposedly
measure the same trait should be relatively large (an assessment of convergent validity), and second, these same-trait values should be higher
than the correlations between tests of different traits based on different
methods (an assessment of discriminant validity). Together these two conditions can be evaluated directly, using the concordance notion of Section
3.2.3, by "blocking" on (or controlling for) method and evaluating the difference across traits. In Table 3.2.4a, this would be reflected in the relative size of the diagonal strip in the 4 × 4 submatrix that contains correlations between the association test and peer ratings. Because each possible
assignment picks one element from each row and column, the concordance

strategy compares the same-trait correlations to those matrix entries that
include one of the two original measures.

Finally, as a third condition, the same-method correlations should be
relatively small. This can be assessed again by our concordance idea if
the table is reordered to nest methods under traits. At least intuitively,
none of these three conditions appears confirmed to any degree, and consequently, we will not bother to carry out the details of the formal assessment. The average same-method correlation, which should be small, is a
substantial .48. However, the average for the same traits is only .13, and
this value is not markedly different from the average of .10 for different
traits and different methods. In fact, the same-trait assignment is not even
optimal, and thus we could not declare it significant at the usual .05 level.

The criteria suggested by Campbell and Fiske for evaluating convergent
and discriminant validity have an obvious intuitive appeal for the researcher
involved in assessing the merits of a test battery. As developed, however,
their analysis strategy is somewhat vague, especially as to how the varying
degrees of confirmation or nonconfirmation are to be measured. Although
inspection techniques are obviously important, it is also of value to have
some type of statistical framework within which such an evaluation could be
performed. We note that although the matrices typically employed in an
assessment of convergent and discriminant validity contain correlation
coefficients, the analysis techniques based on concordance are not restricted
to any particular measure of proximity between variables as long as a
cross-classification of traits and methods (or a similar nesting of two other
variables is present). This same point was made in Section 3.2.3.

3.2.5 Testing for an A Priori Pattern in K Sequences (B)

Instead of a general concordance test of the Friedman type, we can also
define a test procedure sensitive to a particular a priori pattern hypothesized
for the K sequences. Such an extension can be viewed as an analog of the
target-rater agreement problem mentioned in Section 3.2.1. In the notation
of Section 3.2.2, if an a priori set of weights is defined by the first
sequence $[x_1^{(1)}, \ldots, x_n^{(1)}]$, then an appropriate test statistic can be given in
a form equivalent to that used by Page (1963) and by Pirie and Hollander
(1972) based on the K-place assignment score

$$
\begin{aligned}
c_{i_1 i_2 \cdots i_K} &= \sum_{1 < k \leqslant K} c_{i_1 i_k} \\
&= \sum_{1 < k \leqslant K} [x_{i_1}^{(1)} - x_{i_k}^{(k)}]^2
\end{aligned}
$$

$$(3.2.5a)$$

where, again, we have assumed that the two-place assignment scores are defined by squared differences. The observed index

$$\Gamma_K(\rho_I) = \sum_i \left\{ \sum_{1 < k \leqslant K} [x_i^{(1)} - x_i^{(k)}]^2 \right\}$$

is based on the sum of all pairwise comparisons between the first sequence and the remaining $K - 1$. (The comments in Section 3.2.2 about the equivalence of using a cross-product index of the form $x_{i_1}^{(1)} x_{i_k}^{(k)}$ in place of the squared difference apply here in exactly the same way.)

Typically, the weights in the first sequence are integers from 1 to n, and the values $x_i^{(k)}$ for $k \geqslant 2$ are ranks (Page, 1963) or other transformed scores (Pirie and Hollander, 1972). The mean and variance formulas follow directly from the expressions in Section 3.1, or alternatively, because $\Gamma_K(\rho)$ can be defined as a sum of $K - 1$ independent terms:

$$\sum_{i=1}^{n} [x_i^{(1)} - x_{\rho_1(i)}^{(2)}]^2, \ldots, \sum_{i=1}^{n} [x_i^{(1)} - x_{\rho_{K-1}(i)}^{(K)}]^2$$

the mean and variance could also be obtained by the usual moment formulas for linear combinations of random variables. Obviously, the formulas for the mean and variance of each individual term can be found by the relatively simple expressions used for the LA model of Chapter 1. The skewness parameter can also be obtained through a fairly simple operation because the raw third moment is additive. Thus, $E([\Gamma_K(\rho)]^3$ is

$$\left[\frac{n}{(n-1)(n-2)} \right] \sum_{1 < k \leqslant K} \sum_{i_1, i_k} d_{i_1 i_k}^3$$

where $\{d_{i_1 i_k}\}$ is the interaction matrix corresponding to $\{c_{i_1 i_k}\}$. A final division by $[V(\Gamma_K(\rho))]^{3/2}$ gives the skewness. Thus, type III approximations can be used fairly easily for obtaining significance levels.

Because small values of $\Gamma_K(\rho_I)$ are of interest, the appropriate normalized measure would be of the form $\Gamma_K^{**}(\rho_I)$. If each sequence consists of the n untied ranks, then $\Gamma_K^{**}(\rho_I)$ is the average, $[1/(K-1)] \sum_{i=1}^{k} r_{1k}$, where r_{1k} is the Spearman correlation between the first sequence and the kth. This expression is denoted by $r_{s,ave}$. Alternatively, if each sequence is standardized to mean zero and variance one, then $\Gamma_K^{**}(\rho_I)$ is the average of the $K - 1$ Pearson correlations, using the usual lower bound of zero on each of the K - 1 sums of squared differences (this latter measure will be denoted by r_B in the sequel).

In terms of the K × n rectangular data matrix defined by the K sequences, the problem of testing for an a priori order leads to a number of interesting observations about aggregation and its effect on the associated significance tests. We will discuss several of these issues, using the data given in Table 3.2.5a, which provide the rates of rape (per 100,000) over a 10-year period (1968 to 1977) for the 15 largest SMSAs whose geographical regions have remained unchanged throughout this period. By inspecting the entries for each SMSA, it would appear that a general increase is present in the rape rate from 1968 to 1977. In fact, because the trend is strong even in the presence of the obvious inconsistencies to a perfect pattern, one may be content with a simple visual inspection and interpretation. In more ambiguous contexts, however, it would be of value to have a formal inference strategy for confirming whether an a priori conjecture of an increase in rate is reasonable, or more generally, for confirming any conjecture based on a source independent of the data themselves.

Given the general problem of pattern confirmation, two more or less obvious evaluation methods could be followed:

Preliminary Data Aggregation. If the data of Table 3.2.5a were aggregated over rows to produce a single number for each year, these summary values could then be compared, say, to an expected pattern of increase. Intuitively, if a general trend exists within each row, it would also be apparent and possibly enhanced in the aggregation. As an illustration of the mechanics of such a strategy, suppose the rates within each row are first ranked from 1 to n according to their size; ties are broken by use of the conservative procedure of assigning ranks contrary to the expected temporal increase (see Table 3.2.5a). The ordering process on column sums produces what may be called a "consensus ranking" (for example, see Kendall, 1970, pp. 101-120); we are essentially using Spearman's rank order correlation coefficient, r_S, to measure the degree of correspondence to our conjecture.

Because this first type of aggregation leads to an LA model of the type discussed in Chapter 1, the significance of the rank order correlation r_S can be assessed in the usual way based on the formulas of Section 1.3.3. Under the independence hypothesis that all n! permutations of the first n integers are equally likely to be the consensus ranking, the expectation of r_S is 0 and its variance is $1/(n - 1)$. Thus, a Z statistic would be defined very simply as $(n - 1)^{1/2} r_S$. In the example, the column sums are 30, 55, 42, 54, 82, 89, 108, 107, 120, 138 (excluding the first sequence), which produces the consensus ranking of 1, 4, 2, 3, 5, 6, 7, 8, 9, 10, and a correlation with the perfect pattern of .952. By assuming the adequacy of a normal approximation, the associated Z statistic of 2.85 would be declared significant at any of the usual levels. (In the case of untied ranks, the distribution of r_S is symmetric, and thus, the skewness parameter is zero. Type III and normal approximations would be the same.)

Correlation Aggregation. Instead of aggregating over cities to obtain a single value for each year, suppose the rows of Table 3.2.5a are considered separately in the same sense discussed earlier in this section. In particular, the entries (that is, ranks) within each row are first compared (that is, correlated) with the expected pattern of increase; the aggregation is then carried out over the K - 1 correlations to provide a single average measure of correspondence. It is apparent that aggregation now forms the last step rather than the first.

As a simple numerical example using the data of Table 3.2.5a, 15 correlations would be generated: .22, .72, .90, .71, .91, .82, .07, .99, .88, .99, .81, .85, .89, .96, .91, giving an average value, $r_{s,ave}$, of .776. If it is assumed that all permutations of the integers within each row are equally likely, the expectation of $r_{s,ave}$ is 0 and its variance is $1/[(K - 1)(n - 1)]$. Thus, the Z statistic, $[(K - 1)(n - 1)]^{1/2}r_{s,ave}$ would be 8.71, which gives a much smaller (that is, better) significance level than the Z statistic generated under preliminary data aggregation. Counterintuitively, however, $r_{s,ave}$ is smaller than r_s, even though the former is more "significant." (Again, because of untied ranks, the distribution of $r_{s,ave}$ is symmetric and the type III and normal approximations would be the same.)

The stage at which aggregation takes place appears to affect dramatically the significance of the final summary statistic as well as its size, and unfortunately, in opposite ways. By extrapolating from the simple example of Table 3.2.5a, preliminary data aggregation will lead to a larger summary correlational measure. However, this larger correlation will generally be less significant when compared with the average of the separate correlations for each of the rows. All this suggests that the aggregation level at which the strongest correlational pattern can be found may be at odds with the level of aggregation that would allow the pattern to be most easily detected statistically. Thus, by pragmatically picking an aggregation level that seems to "smooth out" the data the best, we may also limit the types of relationships that could later be determined as statistically relevant, particularly if researchers have only the secondary summaries at hand. These same relationships are well-known in multifactor analysis of variance. Once data are collapsed (aggregated) over the levels of a factor, the associated sum of squares becomes part of a usually larger error term needed in tests of significance; consequently, these latter tests tend to be less sensitive.

The relationship between the two aggregation schemes can be developed more formally if we assume that the data in each sequence are standardized to z scores. For example, in preliminary data aggregation, let r_A be the correlation between the column sums and the criterion; in correlational aggregation, let r_B be the average correlation between the K - 1 sequences and the criterion. Under the appropriate inference model [n! equally likely orderings of the column sums for r_A and $(n!)^{K-1}$ equally likely realizations

Table 3.2.5a Rape Rates (per 100,000) and Rank Orders of the Rape Rates for the Fifteen Largest SMSAs Whose Geographical Regions Have Remained Unchanged Throughout the Period 1968-1977

SMSA	Year									
	1968	1969	1970	1971	1972	1973	1974	1975	1976	1977
Rape rates										
Chicago	21.4	24.7	25.4	27.4	28.5	30.2	34.8	31.2	24.1	24.6
Los Angeles - Long Beach	45.0	51.8	50.0	51.3	56.0	55.6	54.5	51.2	57.9	64.9
Philadelphia	13.5	16.1	15.2	17.7	19.9	23.5	28.1	26.1	25.6	27.2
San Francisco - Oakland	27.5	46.1	42.9	39.7	45.2	44.9	42.2	48.5	50.4	51.9
Boston	10.2	13.3	14.8	12.8	15.1	19.6	18.5	22.9	20.1	22.6
Pittsburgh	12.7	14.8	14.0	17.7	19.1	16.0	17.8	17.4	21.2	20.9
Baltimore	37.9	41.6	34.9	28.8	31.4	34.4	35.7	36.1	36.3	41.0
Cleveland	11.2	17.9	18.7	24.8	27.0	27.1	27.8	31.6	33.4	32.5
Anaheim - Santa Ana - Garden Grove	16.9	20.6	21.9	27.0	29.2	31.7	30.6	29.2	30.9	36.7
San Diego	14.9	19.1	19.4	20.1	22.1	22.7	27.7	32.1	31.9	38.7
Miami	20.4	17.5	17.0	22.7	21.2	19.4	32.8	31.6	32.4	37.8
Milwaukee	8.6	7.6	8.3	9.6	9.0	16.6	17.3	13.5	15.6	19.4
Seattle - Everett	24.9	30.1	23.8	22.7	31.4	35.4	37.2	38.2	40.1	48.5
Cincinnati	14.2	17.3	16.8	19.5	23.9	23.6	26.8	25.8	30.4	31.1
Buffalo	13.4	14.1	13.7	13.4	16.2	19.0	20.0	20.5	22.8	25.5

Rank orders of rape rates

	1	2	3	4	5	6	7	8	9	10
Perfect temporal order	1	2	3	4	5	6	7	8	9	10
Chicago	1	4	5	6	7	8	10	9	2	3
Los Angeles - Long Beach	1	5	2	4	8	7	6	3	9	10
Philadelphia	1	3	2	4	5	6	10		7	9
San Francisco - Oakland	1	7	4	2	6	5	3	8	9	10
Boston	1	3	4	2	5	7	6	10	8	9
Pittsburgh	1	3	2	6	8	4	7	5	10	9
Baltimore	8	10	4	1	2	3	5	6	7	9
Cleveland	1	2	3	4	5	6	7	8	10	9
Anaheim - Santa Ana - Garden Grove	1	2	3	4	6	9	7	5	8	10
San Diego	1	2	3	4	5	6	7	9	8	10
Miami	4	2	1	6	5	3	9	7	8	10
Milwaukee	3	1	2	5	4	8	9	6	7	10
Seattle - Everett	3	4	2	1	5	6	7	8	9	10
Cincinnati	1	3	2	4	6	5	8	7	9	10
Buffalo	2	4	3	1	5	6	7	8	9	10

of the data table for r_B], $E(r_A) = E(r_B) = 0$; $V(r_A) = 1/(n - 1)$; $V(r_B) = 1/[(K - 1)(n - 1)]$. Some simple algebra shows that $r_A = r_B/\beta$, where $\beta = ([1 + (K - 2)r_{ave}]/(K - 1))^{1/2}$ and r_{ave} is the average of all $\binom{K-1}{2}$ correlations between pairs of rows. As r_{ave} is less than or equal to 1, β must also be less than or equal to 1. Moreover, β can be 0 if and only if $r_{ave} = -(K - 2)^{-1}$, which implies a lack of variance in the column sums (that is, the sums are all zero) and an undefined correlation r_A. Thus, without loss of any essential generality, it is assumed that β is positive and less than or equal to 1, i.e., $0 < \beta \leq 1$.

The implications of the relation between r_A and r_B are somewhat surprising, but are consistent with what is generally known about aggregation phenomena. First of all, $|r_A|$ is always greater than or equal to $|r_B|$, with equality only when $r_{ave} = 1$, that is, when all rows are identical in their standardized scores. Consequently, any nontrivial variability across rows will result in $|r_A|$ being greater than $|r_B|$. Second, as $K \rightarrow \infty$, $\beta \approx r_{ave}^{1/2}$ and $r_A \approx r_B/r_{ave}^{1/2}$. This last relationship is analogous to a correction for attenuation in the psychometrics literature (see Lord and Novick, 1968, pp. 69-74). Here, r_A can be loosely interpreted as a correlation between a "true score" (defined from column data) and an "infallible variate" (defined by the criterion); r_B is the correlation between an imperfect measure of the "true score" (defined by an average over the separate row data) and an infallible variate (again defined by the criterion).

In terms of Z statistics,

$$\frac{Z_B}{Z_A} = [1 + (K - 2)r_{ave}]^{1/2}$$

Thus, if $r_{ave} \geq 0$, $|Z_B| \geq |Z_A|$. In words, if there is some positive degree of correspondence among the rows, that is, $r_{ave} > 0$, the Z statistic for r_A will be less extreme than the Z statistic for r_B. Equality exists only if $r_{ave} = 0$; at the other extreme, if $r_{ave} = 1$, $Z_B = (K - 1)^{1/2}Z_A$. In summary, the greater the internal correspondence as measured by r_{ave}, the greater the discrepancy between the Z statistics and the closer r_A and r_B become. This can be seen in the simple equality:

$$\frac{Z_B}{Z_A} = \frac{(K - 1)^{1/2}r_B}{r_A}$$

suggesting that as r_B and r_A get closer together, Z_B and Z_A get more discrepant.

One possible extension of these ideas might be mentioned that parallels parts of our discussion regarding generalized concordance. Instead of K sequences with the first identified as the one against which the others are to be tested, we could be given K object by attribute data matrices. The first could be assumed to be the criterion and K - 1 assignment matrices defined either by matching over rows or over columns. The general comparison index in equation (3.2.5a) provides one possible raw measure of correspondence between the first object by attribute matrix and the remaining K - 1. In a similar manner, we could relate a single spatially defined variable F_1 to a set of K - 1 additional variables, $F_2, \ldots, F_K$. The K individual assignment matrices would contain the spatial separation measures between those locations that contained the identical ranks on the two variables being considered. The normalized indices for these extensions would parallel the extensions given in expressions (3.2.3c) and (3.2.3d). Based on notation introduced there, for object by attribute data matrices we could use

$$\left(\frac{1}{K-1}\right) \sum_{k=2}^{K} \frac{R_{1k}}{q} \qquad (3.2.5b)$$

and for spatial sequences with the squared euclidean distance measure,

$$\left(\frac{1}{K-1}\right) \sum_{k=2}^{K} \frac{(r_{x_{F_1} x_{F_k}} + r_{y_{F_1} y_{F_k}})}{2} \qquad (3.2.5c)$$

3.2.6 Two-Group Concordance (B)

In Section 3.2.5, a fixed set of scores $x_1^{(1)}, \ldots, x_n^{(1)}$ was assumed to define the hypothesized pattern or criterion. Suppose now that no such static conjecture is available, and instead we are given a split of the K objects into a first set containing K_1 objects and a second set containing K_2 objects, where $K_1 + K_2 = K$. Although each group could be evaluated separately for concordance as in Section 3.2.2, our more immediate concern is with measuring the correspondence between the two given groups. Intuitively, the single set of criterion scores $x_1^{(1)}, \ldots, x_n^{(1)}$ is replaced by a set of rows from one of the two groups. The first group then serves as a target for the second and conversely (see Schucany and Frawley, 1973).

Given the discussion of the last section on evaluating the correspondence to a single sequence, one obvious specification of the K-place assignment score for the two-group context would take the form

$$c_{i_1 i_2 \cdots i_K} = \sum_{k'=K_1+1}^{K} \sum_{k=1}^{K_1} c_{i_k i_{k'}} \qquad (3.2.6a)$$

If the two-place assignment scores are defined, as before, by squared differences, then

$$c_{i_1 i_2 \cdots i_K} = \sum_{k'=K_1+1}^{K} \sum_{k=1}^{K_1} [x_{i_k}^{(k)} - x_{i_{k'}}^{(k')}]^2$$

The observed index

$$\Gamma_K(\rho_I) = \sum_i \left\{ \sum_{k'=K_1+1}^{K} \sum_{k=1}^{K_1} [x_i^{(k)} - x_i^{(k')}]^2 \right\}$$

is based on the sum of all pairwise comparisons between the first set of K_1 sequences and the second set of $K - K_1 = K_2$ sequences.

Assuming the common standardization to mean 0 and variance 1 for each row, the normalized index, $\Gamma_K^{**}(\rho_I)$, would simply be the average of all correlations between the sequences from the two groups (again, using bounds). As notation in the discussion below, this average will be denoted by r_B^*. A parallel discussion could be developed with ranks, but because this is rather redundant with our use of z scores, we merely note that it would follow the same general pattern of Section 3.2.5.[†]

There is a fairly extensive literature on the topic of two-group concordance that the reader may consult, e.g., see Schucany and Frawley (1973), Beckett and Schucany (1979), and Li and Schucany (1975). In addition, various analogs of the two-group concordance idea could be developed in other contexts merely by using the appropriate set of assignment matrices. For example, in the unweighted nominal-scale agreement framework, the two-place assignment scores would be defined by the appropriate zero-one weights that identify identical categories. Similar extensions could be made to assess two-group concordance among a set of object by attribute data matrices or among a sequence of K spatially defined variables. The normalized indices in these latter two cases would be the obvious extensions of the measures given in expressions (3.2.5b) and (3.2.5c).

[†]We might note that in significance testing for r_B^*, a type III approximation would be based on the general formula given in equation (3.2.2b). All that is needed is to let $d_{i_k i_{k'}} = 0$ whenever k and k' belong to the same group.

The aggregation schemes discussed previously have natural analogs in the two-group context. First of all, the data in each group could be aggregated and the two sequences of scores correlated to obtain r_A^*. This would place the two-group task in the context of the LA model of Chapter 1. Alternatively, and as mentioned earlier, each row in group 1 could be correlated with each row in group 2 and the $K_1 K_2$ correlations averaged to give r_B^*. Obvious extensions of the relationship between r_A and r_B given in Section 3.2.5 would generate rather simple connections between r_A^* and r_B^*. For example,

$$r_A^* = \frac{r_B^*}{\beta_1 \beta_2}$$

where

$$\beta_1 = \left[\frac{1 + (K_1 - 1)r_{ave1}}{K_1} \right]^{1/2} \qquad \beta_2 = \left[\frac{1 + (K_2 - 1)r_{ave2}}{K_2} \right]^{1/2}$$

and r_{ave1} and r_{ave2} are the average intercorrelations within each of the two groups. Thus, $|r_A^*| \geq |r_B^*|$.

A loose correspondence to reliability theory is again possible. If $K_1 \longrightarrow \infty$ and $K_2 \longrightarrow \infty$, $r_A^* \approx r_B^*/(r_{ave1} r_{ave2})^{1/2}$, which implies that r_A^* is an analog of a disattenuated correlation. In other words, to generate r_A^*, we merely correct r_B^* for the lack of perfect concordance within each of the two groups separately. By using these same ideas, one obtains

$$\frac{Z_B^*}{Z_A^*} = [1 + (K_1 - 1)r_{ave1}]^{1/2} [1 + (K_2 - 1)r_{ave2}]^{1/2} \qquad (3.2.6b)$$

Thus, if $r_{ave1} \geq 0$ and $r_{ave2} \geq 0$, $|Z_B^*| \geq |Z_A^*|$.

The expression in equation (3.2.6b) can be used to indicate when extreme values of r_B^* will be observed. If the equality is rewritten as

$$Z_B^* = (n - 1)^{1/2} r_A^* [1 + (K_1 - 1)r_{ave1}]^{1/2} [1 + (K_2 - 1)r_{ave2}]^{1/2}$$

then significant values of r_B^* are to be expected if there is concordance within each group and if the aggregate sums are related to some reasonable extent, either positively or negatively. This same type of property for ranks is pointed out by Schucany and Frawley (1973).

Going back to our discussion of object by attribute data tables in Section 2.3.2 from the last chapter, the two schemes of aggregation can be given a slightly different interpretation. In that section we were concerned with two $p \times q$ matrices $\underset{\sim}{B}$ and $\underset{\sim}{C}$ and an index

$$\Lambda(\phi_0) = \sum_{i,j} b_{ij} c_{\phi_0(i)j}$$

where ϕ_0 was (typically) the identity permutation and defined a matching on rows, i.e., the index could be given more simply as $\Sigma_{i,j} b_{ij} c_{ij}$. We essentially have an analog of this problem in the present two-group concordance framework, but the roles of the rows and columns are reversed.

As notation that makes explicit the connection to object by attribute data matrices, suppose group 1 defines an $n \times K_1$ matrix $\underset{\sim}{B}^o = \{b_{ij}^o\}$ and group 2 defines an $n \times K_2$ matrix $\underset{\sim}{C}^o = \{c_{ij}^o\}$, where the columns now define the sequences of z scores. Because there is no one-to-one correspondence between the K_1 columns of $\underset{\sim}{B}^o$ and the K_2 columns of $\underset{\sim}{C}^o$, a raw index that would be an analog of that used for object by attribute matrices would need to compare each column in $\underset{\sim}{B}^o$ to each column in $\underset{\sim}{C}^o$:

$$\sum_{u,v} \sum_i b_{iu}^o c_{iv}^o \qquad\qquad (3.2.6c)$$

A number of points need to be made about the index in (3.2.6c). First of all, and even though the notation has changed, this index is a simple linear transformation of $\Gamma_K(\rho_I)$. The argument parallels our previous observation that products could be used instead of squared differences, when defining the appropriate two-place assignment scores. Thus, if we used an inference model that assumed all permutations of the scores within each column to be equally likely, the normalized index would merely be r_B^*, i.e., the average of all $K_1 K_2$ correlations between the columns of $\underset{\sim}{B}^o$ and $\underset{\sim}{C}^o$. Second, the index in (3.2.6c) can be rewritten as

$$\sum_i b_{i \cdot}^o c_{i \cdot}^o \qquad\qquad (3.2.6d)$$

This last expression is just the raw cross-product statistic between the n row sums of $\underset{\sim}{B}^o$ and $\underset{\sim}{C}^o$. If an inference model were used that merely permuted the row sums of one matrix against those of the second, we would essentially be permuting complete row vectors in either $\underset{\sim}{B}^o$ or $\underset{\sim}{C}^o$, as we did in our discussion of object by attribute data matrices. The normalized index would now be r_A^*.

In summary, the two schemes of aggregation that lead to r_A^* and r_B^* can be interpreted as ways of comparing two object by attribute data matrices where the latter have the same number of rows on which the matching is defined, but have a differing number of columns. Thus, the raw index must be constructed over all possible pairs of columns because there is no obvious one-to-one correspondence between them, as in our original object by attribute comparison task. The use of r_A^* and r_B^* is based on the same raw index in (3.2.6c), but the normalization varies depending on which inference model is assumed.

The two-group results given above can themselves be extended to T groups in an obvious way. Suppose now there are T groups, with K_1, K_2, ..., K_T objects in each. Moreover, let $r_A^{\triangle}$ denote the average of the intercorrelations of the type r_A^* among the T sequences obtained by aggregating within each of the groups separately; let $r_B^{\triangle}$ denote a similar average that uses correlations of the form r_B^*. More specifically, if $r_{A_{uv}}^*$ denotes the correlation (of the form r_A^*) for groups u and v, and $r_{B_{uv}}^*$ denotes the correlation (of the form r_B^*) for groups u and v, then

$$r_A^{\triangle} = \frac{1}{\binom{T}{2}} \sum_{u<v} r_{A_{uv}}^* \qquad r_B^{\triangle} = \frac{1}{\binom{T}{2}} \sum_{u<v} r_{B_{uv}}^*$$

and

$$r_A^{\triangle} = \frac{1}{\binom{T}{2}} \sum_{u<v} \frac{r_{B_{uv}}^*}{\beta_u \beta_v}$$

No immediate simplification is possible, however, as the terms $r_{B_{uv}}^*$ and $\beta_u \beta_v$ depend on the same subscripts. If it is assumed that all component correlations $r_{B_{uv}}^*$ are nonnegative (as they typically would be), then it is easy to show that $r_A^{\triangle} \geqslant r_B^{\triangle} \geqslant 0$. Finally, a related expression holds for the associated Z statistics:

$$\frac{Z_B^{\triangle}}{Z_A^{\triangle}} = \left[\frac{\binom{T}{2}}{\sum_{u<v} K_u K_v} \right]^{1/2} \frac{\sum_{u<v} r_{B_{uv}}^*}{\sum_{u<v} \frac{r_{B_{uv}}^*}{\beta_u \beta_v}}$$

Although we have implicitly assumed that our initially given objects were elemental in some given sense, it should be clear that the T group analysis merely sets up another object by attribute data matrix that could be analyzed as such. In this case, each group would correspond to a single object and the aggregate sums, once standardized, define the observations within a row.

We also might point out another alternative for measuring the concordance among T groups. Schucany and Frawley (1973) considered an index based on the products of the T aggregated sums rather than averaging over the pairwise statistics appropriate for two-group splits. An analysis similar to that given above could be formulated for these alternatives. Finally, we note that in these T-group analyses, the K object by attribute data matrices could replace the K sequences as long as the appropriate assignment matrices and normalized indices are used. A similar comment holds for K spatially defined sequences. These extensions would parallel similar comments made earlier in the section for the two-group application.

3.3 TWO-DEPENDENT SAMPLE PROBLEMS (B)

Our definition of the index $\Gamma_K(\rho)$ was based on the n^K entries in $\underset{\sim}{C}_K$ through the sum

$$\sum_i c_{i\rho_1(i)\rho_2(i)\cdots\rho_{K-1}(i)}$$

Because a separate term was defined for every possible value of the first subscript, the first sequence could be assumed fixed without loss of generality. Suppose now, however, that our interest changes to the individual terms making up the expression in (3.3a), and in particular, to an index based upon the selection of only a single entry from $\underset{\sim}{C}_K$. For instance, one natural optimization analog that depends on the individual entries in $\underset{\sim}{C}_K$ rather than on a sum would use an objective function defined by

$$\underset{i,\rho_1,\ldots,\rho_{K-1}}{\max}\ c_{i\rho_1(i)\cdots\rho_{K-1}(i)}$$

Obviously, because the largest entry in $\underset{\sim}{C}_K$ can be found by inspection, this problem is trivial to solve. As always, however, our interest will center on the distribution of the term(s) making up the objective function and not on the assignment task itself.

Because we are now concerned with one entry from $\underset{\sim}{C}_K$ rather than n such terms as in expression (3.3a), a fixed first sequence cannot be assumed. However, for a random selection of a single term from $\underset{\sim}{C}_K$, say H, the mean and variance can still be obtained very easily as

$$E(H) = \left(\frac{1}{n^K}\right) \sum_{i_1,\dots,i_K} c_{i_1 i_2 \cdots i_K}$$

and

$$V(H) = \left[\left(\frac{1}{n^K}\right) \sum_{i_1,\dots,i_K} c^2_{i_1 i_2 \cdots i_K}\right] - \left[\left(\frac{1}{n^K}\right) \sum_{i_1,\dots,i_K} c_{i_1 \cdots i_K}\right]^2$$

The skewness parameter, $\gamma(H)$, can be given as

$$\gamma(H) = \left\{\left[\left(\frac{1}{n^K}\right)\left(\sum_{i_1,\dots,i_K} c^3_{i_1 i_2 \cdots i_K}\right)\right]\right.$$

$$- \left[3\left(\frac{1}{n^K}\right)^3 \left(\sum_{i_1,\dots,i_K} c_{i_1 \cdots i_K}\right)\left(\sum_{i_1,\dots,i_K} c^2_{i_1 \cdots i_K}\right)\right]$$

$$\left. + \left[2\left(\frac{1}{n^K}\right)^3 \left(\sum_{i_1,\dots,i_K} c_{i_1 \cdots i_K}\right)^3\right]\right\}/V(H)^{3/2}$$

One special case of this random selection task is of particular interest here. Suppose we are given K sequences of observations as in Section 3.2.2 but assume $n = 2$ and define the assignment score to be a simple sum

$$c_{i_1 \cdots i_K} = \sum_k x^{(k)}_{i_k}$$

Furthermore, the two observations in each of the K sequences or pairs sum to 0, and thus $x^{(k)}_1 = -x^{(k)}_2$ for $1 \leq k \leq K$. Under these conditions, the expectation of H is 0, the variance is $\Sigma_k (x^{(k)}_1)^2$, and the distribution is symmetric, i.e., $\gamma(H) = 0$. In general, we will be concerned with the relative size of the single assignment score $c_{1 \cdots 1} = \Sigma_k x^{(k)}_1$ with respect to all 2^K possible assignment scores that could be constructed from the K sequences. As notation for normalized indices that are obvious analogs of $\Gamma^*_K(\rho_I)$ and $\Gamma^{**}_K(\rho_I)$, we define

$$c^*_{1 \cdots 1} = \frac{c_{1 \cdots 1} - E(H)}{\max H - E(H)}$$

and

$$c^{**}_{1\cdots1} = \frac{E(H) - c_{1\cdots1}}{E(H) - \min H}$$

The terms max H and min H are, respectively, the maximum and minimum assignment scores in C_K. We note that since H is symmetric, only the one measure $c^{*}_{1\cdots1}$ has to be used, remembering that $c^{*}_{1\cdots1} = -c^{**}_{1\cdots1}$.

As a starting point, we begin with K pairs of observations over two dependent groups:

Group 1	Group 2
$y^{(1)}_1$	$y^{(1)}_2$
.	.
.	.
.	.
$y^{(K)}_1$	$y^{(K)}_2$

These data will then be used to define the K pairs $(x^{(1)}_1, x^{(1)}_2), \ldots, (x^{(K)}_1, x^{(K)}_2)$. (It might be useful to point out that if a one-sample test were desired, the entries in group 2 could be defined by a single constant, e.g., the mean, median, or another fixed value such as zero. Thus, our discussion includes the usual one-sample tests as well.)

3.3.1 Sign Test

If

$$x^{(k)}_1 = \text{sign}(y^{(k)}_1 - y^{(k)}_2)$$

and

$$x^{(k)}_2 = \text{sign}(y^{(k)}_2 - y^{(k)}_1)$$

then $c_{1\cdots1}$ = the number of times an observation in group 1 is larger than one in group 2 minus the number of times an observation in group 2 is larger than one in group 1. Reducing the earlier formulas, the expectation is again 0 and the variance is K (assuming no ties). If we let c = the number of times an observation in group 1 is larger than one in group 2, and b = number of times an observation in group 2 is larger than one in group 1, then K is equal to c + b, and the assignment score $c_{1...1}$ is c - b. Thus, we could rewrite $c_{1...1}$ as a constant transformation of c, i.e., $c_{1...1} =$ 2c - K, and if we wished, the quantity c itself could be used as our observed measure. In fact, the term c is the typical index used for the sign test with expectation K/2 and variance K/4 based on our earlier formulas and the transformation from $c_{1...1}$. An exact reference distribution could also be used based on the binomial with p = 1/2.

It is interesting to note the equivalence between the sign test and a directional version of McNemar's test discussed in Section 3.2.2. In general, if a pair of the original y observations are tied, we can define the corresponding pair of x scores as zero. Then, the number of untied pairs is b + c, and we may consider the binomial test of p = 1/2 with b + c trials. If b - c is chosen as our test statistic, it has expectation 0 and variance b + c. Thus, a normal approximation would be based on

$$\frac{b - c}{\sqrt{b + c}}$$

whose square defines the usual two-tailed McNemar statistic. In general, the reference distribution is symmetric and the normalized measure, $c^*_{1...1}$, that corresponds to the raw index (e.g., c, b - c, or b) would take a very simple form as the difference between two proportions, [c/(b + c)] - [b/(b + c)]. As mentioned earlier, because $c^*_{1...1} = -c^{**}_{1...1}$, only one index has to be used.

3.3.2 Wilcoxon Test

Let

$$x_1^{(k)} = \text{sign}(y_1^{(k)} - y_2^{(k)}) \text{ times } d_k$$

and

$$x_2^{(k)} = \text{sign}(y_2^{(k)} - y_1^{(k)}) \text{ times } d_k$$

where d_k is the rank of $|y_1^{(k)} - y_2^{(k)}|$ among all K absolute differences. Here, $c_{1...1}$ = sum of positive signed ranks minus the sum of ranks with negative signs. This assignment score is a constant function of the sum of the positively signed ranks, which is the typical test statistic used in this context.

Obviously, other two-dependent sample strategies could be pursued by merely redefining the term d_k that was introduced in the Wilcoxon test. For example, if the observations are commensurable over the K sequences or have been made so by some transformation, d_k could be defined directly as $|y_1^{(k)} - y_2^{(k)}|$. In general, the advantage of concentrating on a single assignment score instead of using the structure of the earlier sections, but specialized for $n = 2$, is that directionality can be introduced if only a single such score is considered. In all cases, tied values within a pair could be eliminated and the analysis carried out on the remaining untied pairs. The normalized index $c_{1\cdots 1}^*$ would be the simple difference between two averages:

$$c_{1\cdots 1}^* = \frac{\sum_k x_1^{(k)}}{\sum_k |y_1^{(k)} - y_2^{(k)}|}$$

$$= \frac{\sum_{\substack{k \\ y_1^{(k)} - y_2^{(k)} > 0}} |y_1^{(k)} - y_2^{(k)}|}{\sum_k |y_1^{(k)} - y_2^{(k)}|} - \frac{\sum_{\substack{k \\ y_1^{(k)} - y_2^{(k)} < 0}} |y_1^{(k)} - y_2^{(k)}|}{\sum_k |y_1^{(k)} - y_2^{(k)}|}$$

As an application of some of these ideas in a slightly different context, we can go back to the task of assessing an a priori order among sequences as discussed in Section 3.2.5. In testing average correlations, such as r_B and r_B^*, the emphasis throughout has been on an inference model that assumes all permutations of the observations within each sequence to be equally likely (the Friedman model). Other options based on less drastic randomization are possible, however. For example, suppose we wish to test the significance of r_B for a perfect temporal trend defined by the ranks from 1 to n in the first row of the usual $K \times n$ matrix of 3.2.5. For each of $K - 1$ remaining rows, we obtain the correlation of each sequence to this pattern and to its exact opposite. These two correlations are equal in absolute value but of opposite sign, that is, if r_i denotes the correlation of the ith row, $2 \leq i \leq K$ to the first, then $-r_i$ is the correlation to the reverse pattern. In short, $K - 1$ pairs of observations are obtained that sum to zero within each pair. If the conjecture of a temporal increase reflected across rows is correct, then the first member of each pair should be positive and the second negative. Thus, the pattern can be confirmed as a special case of the two-sample framework discussed above. The assignment score $c_{1\cdots 1}$

is $(K - 1)r_B$, and if our conjecture is correct, this latter value should be relatively large and positive. Based on the 2^{K-1} equally likely permutations of the K - 1 sequences, we find $E(r_B) = 0$ and $V(r_B) = (\Sigma\ r_i^2)/(K - 1)^2$. As long as $\Sigma\ r_i^2/(K - 1) > (n - 1)^{-1}$, the Z statistic based on the two-dependent sample model will be less than that obtained under the Friedman model, but because the former is based on a weaker randomization condition, it still may be preferable. Finally, an extension of these ideas to the multi-group case is immediate. For instance, for two groups

$$E(r_B^*) = 0 \text{ and } V(r_B^*) = \frac{1}{(K_1 K_2)^2} \sum_{i,j} r_{ij}^2$$

where r_{ij} is the correlation between the ith row in group 1 and the jth row in group 2.

3.3.3 Mielke-Berry Two-Dependent Sample Statistic

As we have noted, the definition of the assignment score $c_{i_1 i_2 \cdots i_K}$ as a simple sum has the advantage of providing a directional statistic in which relatively large or small values of the index may be of particular interest to distinguish. If directionality is not of concern, however, a variety of other two-dependent sample statistics could be developed within this same framework. For example, Mielke and Berry (1982) develop an alternative measure that, in our context, would reduce to the use of an assignment score with an additive form comparable to what was of concern in our discussion of general concordance in Section 3.2.2, i.e.,

$$c_{i_1 \cdots i_K} = \sum_{k<k'} c_{i_k i_{k'}}$$

Specifically, the n = 2 observations in each of the K sequences or pairs would still sum to 0, as before (i.e., $x_1^{(k)} = -x_2^{(k)}$), and

$$c_{i_k i_{k'}} = |x_{i_k} - x_{i_{k'}}|^V$$

where V > 0 (usually, V = 1 or 2). The Mielke-Berry statistic is given by

$$\delta = \frac{1}{\binom{K}{2}} c_{1 \cdots 1}$$

and leads to fairly simple moment formulas, i.e.,

$$E(\delta) = \frac{1}{K(K-1)} \sum_{k<k'} (a_{kk'} + b_{kk'})$$

$$V(\delta) = \left[\frac{1}{K(K-1)}\right]^2 \sum_{k<k'} (a_{kk'} - b_{kk'})$$

$$\gamma(\delta) = \frac{-6 \sum_{k<k'<k''} (a_{kk'} - b_{kk'})(a_{kk''} - b_{kk''})(a_{k'k''} - b_{k'k''})}{\sum_{k<k'} (a_{kk'} - b_{kk'})^2}$$

where

$$a_{kk'} = |x_1^{(k)} + x_1^{(k')}|^V$$

$$b_{kk'} = |x_1^{(k)} - x_1^{(k')}|^V$$

This particular formulation includes a variety of two-dependent sample tests as special cases, but all of these could have been considered as part of the general concordance discussion of Section 3.2.2. Given the way $c_{i_1 \cdots i_K}$ is defined (when n = 2), δ could be represented as

$$\delta = \left[\frac{1}{K(K-1)}\right](c_{1\cdots 1} + c_{2\cdots 2})$$

If V = 2, then δ is a constant, $1/[K(K-1)]$, times the raw concordance index considered explicitly there; that is, $c_{1\cdots 1} + c_{2\cdots 2}$ is the raw index $\Gamma_K(\rho_I)$. For instance, defining $x_1^{(k)}$ and $x_2^{(k)}$ as we did earlier for the directional sign and Wilcoxon test, nondirectional versions of these two tests would be obtained. The reader is referred to the paper by Mielke and Berry (1982) for a discussion of other alternatives based on different definitions for the pairwise assignment score $c_{i_k i_{k'}}$.

3.4 COMPARISONS BETWEEN RELATED LINEAR ASSIGNMENT PROBLEMS (B)

In classical parametric statistics, the use of Fisher's Z transformation in testing for a difference between two independent correlations (i.e.,

correlations based on distinct samples) is a well-known and commonly accepted method. Unfortunately, there is no comparable procedure that enjoys
the same degree of universality for testing the difference between two
dependent correlations, and in fact, a substantial literature has been developed that debates the merits of various approaches (e.g., see Steiger,
1980; Neill and Dunn, 1975). More recently, a simple approach has been
suggested by Wolfe (1976, 1977) that, as will be seen throughout the book,
can be extended to a variety of problems. Wolfe observed that if U, V, and
W denote three random variables with some joint distribution for which
$Var(V) = Var(W)$, then using an obvious notation in a population framework

$$\rho_{u,v-w} = \frac{\rho_{uv} - \rho_{uw}}{\sqrt{2(1 - \rho_{vw})}}, \quad \rho_{vw} \neq 1 \tag{3.4a}$$

Thus, $\rho_{uv} = \rho_{uw}$ if and only if $\rho_{u,v-w} = 0$ and a test of the equality of the
two correlations ρ_{uv} and ρ_{uw} reduces to the test of a zero correlation between U and V - W. The relationship in equation (3.4a) is a somewhat
obvious algebraic identity and brings with it the problem of assuming that
the variances of V and W are equal in the population. It does, however,
suggest that the correlation between U and V - W does contain information
on the relationship between U and V as compared to U and W.

One might consider using the standard parametric procedure based on
the t distribution for testing whether $\rho_{u,v-w} = 0$. However, our interests
are in an alternative randomization strategy, similar to those discussed up
to this point, that is conditional on the obtained univariate data and which
can be stated and given meaning without an explicit population analog for the
sample correlation $r_{u,v-w}$. In addition, because the randomization paradigm is defined only with respect to the observed data, the problem of variance equality can be avoided by an appropriate standardization. Specifically,
suppose a data set is available on three variables U, V, and W that consists
of N trivariate observations, i.e., three sequences of n observations each.
Based on these data, the scores on the V and W variables are both standardized to mean zero and variance one, and n new difference scores of the
form V - W are constructed. In terms of a descriptive analog of equation
(3.4a) using sample correlations, we obtain

$$r_{u,v-w} = \frac{r_{uv} - r_{uw}}{\sqrt{2(1 - r_{vw})}}, \quad r_{vw} \neq 1 \tag{3.4b}$$

The particular hypothesis that is of interest is one of randomness in the
patterning of the observed scores on U and V - W, which falls under our
sequence comparison procedures of Section 1.3.3. Specifically, it is assumed from this baseline conjecture that all n! possible permutations of the

U scores are equally likely compared to the sequence of scores on V - W (or equivalently, fixing the U scores and considering the n! permutations of the scores on V - W as equally likely). As always, we wish to reject the randomness hypothesis that the observed matching could have been reasonably generated by picking one of the n! permutations at random. Operationally, $r_{u,v-w}$ is compared to the complete distribution generated under the randomness assumption, and the number of permutations giving values as extreme as the obtained correlation is reported as the (one-tailed) significance level. Given the algebraic identity in equation (3.4b) and the constancy of r_{vw} under randomization, a sufficiently small significance level would in turn suggest that the observed difference between r_{uv} and r_{uw} is "real" in the sense that it cannot be reasonably attributed to a simple randomization of the obtained data.

The randomization procedure just described may not be the best strategy if we were willing to assume a specific population model within a standard parametric model, e.g., with respect to such a model there may be more powerful procedures. However, the advantage of developing this concept of randomness with respect to the obtained data is that no explicit population model has to be assumed; the randomness conjecture has an operational meaning and interpretation in terms of the original data. More importantly, in other contexts developed later on, the appropriate population models are not at all obvious; thus, randomization may be the only alternative available that has the type of generality desired.

We do note in passing that at least in this simple three variable context, the randomization strategy on $r_{u,v-w}$ could be reinterpreted as a conservative procedure with respect to its significance level if we were willing to assume the usual population model. Specifically, under the null hypothesis that $\rho_{uv} = \rho_{uw} = \rho$, the asymptotic variance of $r_{uv} - r_{uw}$ is $(1 - \rho_{vw}) \times (2 - 3\rho^2 + \rho_{vw}\rho^2)/n$ (Neill and Dunn, 1975). Thus, the asymptotic variance is bounded above by $2(1 - \rho_{vw})/n$, and a conservative test would be based on comparing the statistic

$$\frac{r_{uv} - r_{uw}}{\sqrt{2(1 - r_{vw})/n}} \tag{3.4c}$$

to the standard normal (assuming the usual estimate r_{vw}). Based on randomization and using the procedures of Section 1.3.3, $r_{u,v-w}$ is asymptotically normal with mean 0 and variance $1/(n-1)$ under mild regularity conditions. Thus, using equality (3.4b), the same statistic as in expression (3.4c) could be referred to the standard normal (with n - 1 replacing n).

In short, randomization can be interpreted as a conservative paradigm with respect to the usual population model; consequently, if our sole interest is in this initial three-variable problem, more powerful procedures are available (see Steiger, 1980). In later chapters, however, U, V, and W are replaced by proximity matrices. An analog of the relationship in equation (3.4b) is again of interest, but now the correlations are obtained between the corresponding entries in two different matrices. Comprehensive population models are difficult if not impossible to formulate in this matrix framework, which in turn suggests that randomization of some form may be the only reasonable strategy to follow.

REFERENCES

Beckett, J., and W. R. Schucany. Concordance among categorized groups of judges, J. Educ. Stat. 4 (1979), 125-137.

Campbell, D. T., and D. W. Fiske. Convergent and discriminant validation by the multitrait-multimethod matrix, Psychol. Bull. 56 (1959), 81-105.

Conger, A. J. Integration and generalization of kappas for multiple raters, Psychol. Bull. 88 (1980), 322-328.

David, F. N., and D. E. Barton. Combinatorial Chance, Hafner, New York, 1962.

Dwyer, P. S. The mean and standard deviation of the distribution of group assembly sums, Psychometrika 29 (1964), 397-408.

Hildebrand, D. K., J. D. Laing, and H. Rosenthal. Prediction Analysis of Cross-Classifications, Wiley, New York, 1977.

Hubert, L. J. Kappa revisited, Psychol. Bull. 84 (1977), 289-297.

Hubert, L. J. The comparison of sequences, Psychol. Bull. 86 (1979), 1098-1106.

Hubert, L. J., R. G. Golledge, C. M. Costanzo, and G. D. Richardson Assessing homogeneity in cross-classified proximity data, Geogr. Anal. 13 (1981), 38-50.

Hubert, L. J., R. G. Golledge, T. Kenney, and C. M. Costanzo. Aggregation in data tables: Implications for evaluating criminal justice statistics, Envir. Planning [A] 13 (1981), 185-199.

Kelley, T. L., and A. C. Krey. Tests and Measurements in the Social Sciences, Scribner, New York, 1934.

Kendall, M. G. Rank Correlation Methods, 4th ed., Griffin, London, 1970.

Lehmann, E. L. Nonparametrics: Statistical Methods Based on Ranks, Holden-Day, New York, 1975.

Li, L., and W. R. Schucany. Some properties of a test for concordance of two groups of rankings, Biometrika 62 (1975), 417-423.

Lord, F. M., and M. R. Novick. Statistical Theories of Mental Test Scores, Addison-Wesley, Reading, Mass., 1968.

Mielke, P. W., and K. J. Berry. An extended class of permutation techniques for matched pairs, Commun. Stat.: Theory and Methods A11 (1982), 1197-1207.

Mielke, P. W., and H. K. Iyer. Permutation techniques for analyzing multi-response data from randomized block experiments, Commun. Stat.: Theory Methods A11 (1982), 1427-1437.

Neill, J. J., and O. J. Dunn. Equality of dependent correlation coefficients, Biometrics 31 (1975), 531-543.

Page, E. B. Ordered hypothesis for multiple treatments: A significance test for linear ranks, J. Am. Stat. Assoc. 58 (1963), 216-230.

Pirie, W. R., and M. Hollander. A distribution-free normal scores test for ordered alternatives in the randomized block design, J. Am. Stat. Assoc. 67 (1972), 855-857.

Schmitt, N., B. W. Coyle, and B. B. Saari. A review and critique of analyses of multitrait-multimethod matrices, Multivar. Behav. Res. 12 (1977), 447-478.

Schucany, W. R., and W. H. Frawley. A rank test for two group concordance, Psychometrika 38 (1973), 249-258.

Steiger, J. H. Tests for comparing elements of a correlation matrix, Psychol. Bull. 87 (1980), 245-251.

Wolfe, D. A. On testing equality of related correlation coefficients, Biometrika 63 (1976), 214-215.

Wolfe, D. A. A distribution-free test for related correlation coefficients, Technometrics 19 (1977), 507-509.

PART II

APPLICATIONS AND EXTENSIONS OF THE QUADRATIC ASSIGNMENT MODEL

4

The Quadratic Assignment Model

4.1 INTRODUCTION (A)[†]

This chapter develops a general data analysis strategy, referred to as the quadratic assignment (QA) model, that is very similar in form to the linear assignment (LA) method presented in Part I. As we will show, the notion of quadratic assignment can be used to assess a variety of different data structures that may be reflected in a proximity matrix defined on the objects from some given set. Throughout, the term "proximity" will be interpreted in a very broad sense and refers to any measure of relationship that is specified for object pairs, e.g., correlations, distances, measures of pattern correspondence, flow, and so on.[‡] Obviously, there will be many instances in which the tasks we face in analyzing proximity data will not fall into one of the categories that can be handled routinely by the type of strategy presented in the later sections; nevertheless, the approach taken is flexible enough to give a general point of departure for many of the problems an individual faces in choosing an appropriate methodology, and, more importantly, is inclusive enough to provide a broad organizing principle for an extensive analysis of structure within a proximity matrix.

The discussion of QA will parallel that for the LA model of Chapter 1. Again, an optimization analog is introduced first, before the inference applications of the paradigm are mentioned. Because most of the issues in the

---

[†]The reader is referred to the Preface for an explanation of the use of (A), (B), and (C) in text headings.
[‡]In a more formal sense, we could refer to a proximity function that assigns numerical values to all ordered object pairs constructed from an object set $S = \{O_1, \ldots, O_n\}$. Because all ordered pairs form the cartesian product, $S \times S$, proximity is a mapping or function from $S \times S$ to the real numbers. This terminology will not be used now, but it is convenient in Chapter 5, where we will need proximity functions on higher-order cartesian products, e.g., on $S \times S \times S$ or $S \times S \times S \times S$.

QA framework are essentially the same as they were for LA, our presentation at the beginning can be somewhat abbreviated and only the points of departure have to be emphasized explicitly.

4.2 BACKGROUND (A)

In the literature on data analysis over the past twenty years, a distinction between exploratory and confirmatory procedures has become very popular (Tukey, 1962; Hildebrand, Laing, and Rosenthal, 1977; Kaiser, 1970). Because this same dichotomy is important for us as well, it may be helpful to begin our discussion of QA by noting what this distinction is in rather broad terms. Typically, an exploratory strategy involves the use of an analysis technique on a given data set with the aim of identifying interesting relationships, patterns, and the like. For example, most of the data reduction methods reviewed in the multidimensional scaling context by Carroll and Arabie (1980) would be of this type. A confirmatory approach, on the other hand, requires the test of an a priori conjecture that is generated from a source distinct from the data to be used for the purposes of validation. This latter test in the present chapter (as well as in Part I) is usually correlational, and thus, the term "confirmation" is given a limited meaning that does not imply the absolute correctness of a hypothesis. Because a correlational analysis can never exclude all competing explanations, we argue, when it is justified, that the pattern of data is not unrelated to the conjectured pattern.

It may be obvious that confirmatory analyses would be desirable adjuncts to many of the current exploratory methods used in the study of proximity matrices (such as clustering and multidimensional scaling), but very few techniques have been proposed that could help carry out such a program with any degree of success (for a notable exception, see Ramsay, 1978). Users of the newer data reduction procedures generally lack confirmatory techniques even of a correlational nature and must rely on intuitive arguments based on whatever additional information is available for the objects being studied. Although this practice is commendable given the current state of the art, it is possible to go one step further. Among many other uses, the methods presented in this chapter can be used to incorporate the same information that is relevant to a post hoc explanation, more directly in a confirmatory manner.

Confirmatory methods are emphasized throughout this chapter and the next three, but for background we start with a brief introduction to the QA optimization task. Section 4.2.2 presents the general confirmatory model that will be of concern throughout most of Part II. The latter is also based on the QA notion but not in its optimization sense.

4.2.1 The QA Model in Optimization (A)

The term "quadratic assignment" refers to a general combinatorial optimization task with an extensive literature in the area of operations research and management science (for example, see the reviews by Hanan and Kurtzberg, 1972, and Hubert and Schultz, 1976a). In its most typical multiplicative form, two $n \times n$ matrices, $\underset{\sim}{P} = \{p_{ij}\}$ and $\underset{\sim}{Q} = \{q_{ij}\}$ are given, both with zeros along their main diagonals, i.e., $p_{ii} = q_{ii} = 0$, $1 \leq i \leq n$. We attempt to permute the rows, and simultaneously the columns, of one of the matrices, say $\underset{\sim}{P}$, so as to maximize or minimize

$$\mathcal{A}(\rho) = \sum_{i,j} p_{\rho(i)\rho(j)} q_{ij}$$

where ρ is some permutation of the first n integers. For example, $\{q_{ij}\}$ may be considered a matrix of numerical relationships between n facilities, where the entry q_{ij} reflects the amount of material to be moved from facility i to facility j. In turn, the matrix $\{p_{ij}\}$ could contain distances among n positions or locations.

The permutation ρ represents one possible assignment of facilities to locations, i.e., if $\rho(u) = v$, then facility u is assigned to location v. The QA task in this context is equivalent to finding an assignment that minimizes a total effort or cost represented by the index $\mathcal{A}(\rho)$. It should be apparent that the two matrices P and Q play formally equivalent roles, and thus, without loss of generality, either (or both) could be subjected to row and column permutation. As a notational convenience from now on, however, it will be assumed that ρ acts on the rows and columns of $\underset{\sim}{P}$ alone; the matrix $\underset{\sim}{Q}$ will be considered fixed.

The term "quadratic" comes from a representation of the optimization task using products of indicators and an objective function of the form

$$\sum_{u,v,i,j} p_{uv} q_{ij} x_{iu} x_{jv}$$

which is subject to three constaints:

1.
$$x_{iu} = \begin{cases} 1 & \text{if facility } i \text{ is assigned to location } u \\ 0 & \text{otherwise} \end{cases}$$

and

$$x_{jv} = \begin{cases} 1 & \text{if facility } j \text{ is assigned to location } v \\ 0 & \text{otherwise} \end{cases}$$

2. $\sum_{i=1}^{n} x_{iu} = 1$, implying that each location u has one and only one facility
    assigned to it, $1 \leqslant u \leqslant n$.

3. $\sum_{u=1}^{n} x_{iu} = 1$, implying that each facility i is assigned to one and only
    one location, $1 \leqslant i \leqslant n$.

As in LA, the optimization problem reduces to locating an appropriate set
of indicators either to maximize or minimize the given objective function.
Unfortunately, and in contrast to LA, this task can be very difficult to solve
without an enormous computational effort, unless n is very small and the
complete enumeration of all solutions is a real possibility. For now, and
in analogy to our emphasis on LA in Part I, our interest will center on the
distribution of $\mathcal{A}(\rho)$ over all n! possible permutations.

4.2.2 The QA Model as a Basis for Inference (A)

Given the two matrices, $\underset{\sim}{P}$ and $\underset{\sim}{Q}$, the inference problem is one of comparison and assessing whether the pattern of entries represented by $\underset{\sim}{P}$ is also
present in $\underset{\sim}{Q}$, and conversely. Following the discussion of LA in Part I,
the general approach should seem familiar: If the structures defined by the
matrices are not similar, then the value of $\mathcal{A}(\rho_0)$, where ρ_0 is some identified permutation (and typically the identity, ρ_I),[†] should not be unusually
extreme compared to the distribution we would expect based on the n! pos-
sible labelings of the rows and columns of $\underset{\sim}{P}$ against a fixed $\underset{\sim}{Q}$ (or equivalently, fixing $\underset{\sim}{P}$ and permuting $\underset{\sim}{Q}$). We note, in particular, that because the
rows and columns are permuted together, the internal structure or integrity
of each matrix is preserved under the randomization model. Thus, the
relabeling process defined by each permutation ρ would maintain properties
such as positive definiteness, a tendency toward symmetry, realization of
the proximities as distances in euclidean space, and so on. We return to
this point a little later on.

As always, the use of a randomization distribution for evaluating the
relative size of an index associated with a specific permutation, ρ_0, is
restricted to those assignments that are hypothesized without the aid of the
information provided by $\underset{\sim}{P}$ and $\underset{\sim}{Q}$. It makes no sense to assess the relative
magnitude of $\mathcal{A}(\rho_0)$ through an equally likely labeling assumption, if by
construction that function ρ_0 must already generate an index value at an
extreme percentage point of the reference distribution. Clearly, such a
practice is akin to using a standard t-test on two independent groups pre-
selected for their extreme differences. Although this admonition is obvious

---

[†] As in LA, ρ_0 could always be defined as the identity permutation ρ_I. Given
any permutation, ρ_0, an index $\mathcal{A}(\rho_0)$ comparing $\underset{\sim}{P}$ and $\underset{\sim}{Q}$ is the same as one
of the form $\mathcal{A}(\rho_I)$ that compares $\{p_{\rho_0(i)\rho_0(j)}\}$ and $\underset{\sim}{Q}$.

when stated explicitly, it is not always heeded in the literature, e.g., consult Fillenbaum and Rapoport's (1971) use of Johnson's (1968a) cluster statistic discussed in Section 4.3.3.

The mean and variance of $\mathcal{A}(\rho)$ may be obtained through the indicator function strategy of Section 1.4.1 (Graves and Whinston, 1970a,b, and Mantel, 1967, provide independent derivations; the latter reference; in particular, appears to be the first published discussion of the QA model in the general form used throughout this chapter—also, see Mantel and Bailar, 1970; Mantel and Valand, 1970):

$$E(\mathcal{A}(\rho)) = \left[\frac{1}{n(n-1)}\right] \sum_{i,j} p_{ij} \sum_{i,j} q_{ij}$$

$$V(\mathcal{A}(\rho)) = -\left[\frac{1}{n(n-1)}\right]^2 B_1 + \left[\frac{1}{n(n-1)}\right](B_2 + B_3)$$

$$+ \left[\frac{1}{n(n-1)(n-2)}\right](B_4 + 2B_5 + B_6)$$

$$+ \left[\frac{1}{n(n-1)(n-2)(n-3)}\right]B_7$$

where $B_1, \ldots, B_7$ are given in the Appendix.

Two special cases of these expressions deserve particular mention. First of all, if $\underset{\sim}{P}$ and $\underset{\sim}{Q}$ are symmetric, then the variance simplifies because $B_2 = B_3$, and $B_4 = B_5 = B_6$. Specifically,

$$V(\mathcal{A}(\rho)) = -\left[\frac{1}{n(n-1)}\right]^2 D_1 + \left[\frac{2}{n(n-2)} D_2\right]$$

$$+ \left[\frac{4}{n(n-1)(n-2)}\right]D_3 + \left[\frac{1}{n(n-1)(n-2)(n-3)}\right]D_4$$

where

$$D_1 = \left(\sum_{i,j} p_{ij}\right)^2 \left(\sum_{i,j} q_{ij}\right)^2$$

$$D_2 = \left(\sum_{i,j} p_{ij}^2\right)\left(\sum_{i,j} q_{ij}^2\right)$$

$$D_3 = \left[\sum_i \left(\sum_j p_{ij}\right)^2 - \sum_{i,j} p_{ij}^2\right]\left[\sum_i \left(\sum_j q_{ij}\right)^2 - \sum_{i,j} q_{ij}^2\right]$$

$$D_4 = \left[\left(\sum_{i,j} p_{ij} \right)^2 - 4 \sum_i \left(\sum_j p_{ij} \right)^2 + 2 \sum_{i,j} p_{ij}^2 \right]$$
$$\cdot \times \left[\left(\sum_{i,j} q_{ij} \right)^2 - 4 \sum_i \left(\sum_j q_{ij} \right)^2 + 2 \sum_{i,j} q_{ij}^2 \right]$$

If $\underset{\sim}{P}$ and $\underset{\sim}{Q}$ are skew-symmetric, i.e., $p_{ij} = -p_{ji}$ and $q_{ij} = -q_{ji}$, then $E(\mathcal{A}(\rho)) = 0$, and $V(\mathcal{A}(\rho)) = [2/n(n-1)][D_2 + (2/(n-2))D_3]$, because $B_1 = B_7 = 0$, $B_2 = B_3$, and $B_4 = B_5 = B_6$.

These last two special cases are of particular importance because of a very convenient decomposition of the index $\mathcal{A}(\rho)$ into two conceptually distinct comparisons, one based on symmetric information between $\underset{\sim}{P}$ and $\underset{\sim}{Q}$ and the second based on skew-symmetric information. Specifically, any matrix can be represented as the sum of two others, one symmetric and the second skew-symmetric (see Tobler, 1976). For example, if $\underset{\sim}{T}$ denotes an arbitrary $n \times n$ matrix, then

$$\underset{\sim}{T} = \{ t_{ij} \} = \left\{ \frac{t_{ij} + t_{ji}}{2} \right\} + \left\{ \frac{t_{ij} - t_{ji}}{2} \right\} = \underset{\sim}{T}^+ + \underset{\sim}{T}^-$$

where the symmetric part is denoted by $\underset{\sim}{T}^+ = \{ t_{ij}^+ \}$ and the skew-symmetric part by $\underset{\sim}{T}^- = \{ t_{ij}^- \}$. In terms of the comparison between $\underset{\sim}{P}$ and $\underset{\sim}{Q}$,

$$\mathcal{A}(\rho) = \sum_{i,j} p_{\rho(i)\rho(j)} q_{ij}$$
$$= \sum_{i,j} \left[\frac{p_{\rho(i)\rho(j)} + p_{\rho(j)\rho(i)}}{2} \right] \left[\frac{q_{ij} + q_{ji}}{2} \right]$$
$$+ \sum_{i,j} \left[\frac{p_{\rho(i)\rho(j)} - p_{\rho(j)\rho(i)}}{2} \right] \left[\frac{q_{ij} - q_{ji}}{2} \right]$$
$$= \mathcal{A}^+(\rho) + \mathcal{A}^-(\rho)$$

where $\mathcal{A}^+(\rho)$ and $\mathcal{A}^-(\rho)$ are the cross-product statistics between the symmetric and skew-symmetric parts of $\underset{\sim}{P}$ and $\underset{\sim}{Q}$, respectively. Thus, because $E(\mathcal{A}^-(\rho)) = 0$,

$$E(\mathcal{A}(\rho)) = E(\mathcal{A}^+(\rho)) + E(\mathcal{A}^-(\rho)) = E(\mathcal{A}^+(\rho))$$

Unfortunately, the two components $\mathcal{A}^+(\rho)$ and $\mathcal{A}^-(\rho)$ may not be uncorrelated over the $n!$ permutations, which implies that no similar decomposition of the variance is true in general.

The notion of matrix decomposition will be used in various forms throughout the chapter. At times, we may be able to identify relatively different aspects of pattern correspondence between two given matrices by considering the measures $\mathcal{A}^+(\rho_0)$ and $\mathcal{A}^-(\rho_0)$ separately. As one specification of particular importance, we observe that if one of the matrices is already symmetric, say $\underset{\sim}{P}$, then without loss of generality, Q can be redefined as the symmetric $\underset{\sim}{\text{matrix}}$ $\{(q_{ij} + q_{ji})/2\}$, i.e., $\mathcal{A}^-(\rho)$ is $\underset{\sim}{\text{zero}}$ and and $\mathcal{A}(\rho) = \Sigma_{i,j}\, p_{\rho(i)\rho(j)}q_{ij} = \Sigma_{i,j}\, p_{\rho(i)\rho(j)}\{(q_{ij} + q_{ji})/2\}$. In a similar manner, if $\underset{\sim}{P}$ is skew-symmetric, then without loss of generality, Q can be assumed $\underset{\sim}{\text{skew-symmetric}}$ as well, i.e., $\mathcal{A}^+(\rho)$ is zero and $\underset{\sim}{}$

$$\mathcal{A}(\rho) = \sum_{i,j} p_{\rho(i)\rho(j)}q_{ij} = \sum_{i,j} p_{\rho(i)\rho(j)}\left\{\frac{q_{ij} - q_{ji}}{2}\right\}$$

The comments on significance testing given in Section 1.4.2 are appropriate here as well. The calculation of a Monte Carlo significance level will be our method of choice, typically in conjunction with the routine calculation of a Z statistic. Unfortunately, there are very few general conditions that would justify asymptotic normality for a wide range of applications, and in fact, as we will see in Section 4.3.4, there are important nonpathological cases for which asymptotic normality does not hold. Consequently, the Z statistic will be referred to a type III approximation (see Table 1.4.2a) developed in this context by Mielke, Berry, and Brier (1981). The third moment of $\mathcal{A}(\rho)$, obtained by Mielke (1979), is given in the Appendix to this chapter (also, see Siemiatycki, 1978). Whatever literature is available on explicit conditions for asymptotic normality will be mentioned in our discussion of various interpretations for $\underset{\sim}{P}$ and $\underset{\sim}{Q}$ (for some general work in the area, see Shapiro and Hubert, 1979, and the references they cite; also, the recent paper by Costanzo, Hubert, and Golledge, 1983, gives a detailed comparison of the type III and Monte Carlo approaches for one particular data set).

In certain applications of the index $\mathcal{A}(\rho_0)$, just as for certain uses of the LA statistic $\Gamma(\rho_0)$, the raw index by itself may be interpretable in some absolute sense without any further normalization. This usually occurs when the second matrix $\underset{\sim}{Q}$ contains dichotomous (zero-one) values and functions solely to pick out entries from a proximity matrix $\underset{\sim}{P}$, or possibly, when $\underset{\sim}{Q}$ contains simple constants that serve to weight the entries in $\underset{\sim}{P}$ in some very elementary fashion, e.g., to define an average. If deemed appropriate, however, normalized versions of $\mathcal{A}(\rho_0)$ could be developed as in Section 1.2.4 of Chapter 1. For consistency of notation, these latter indices could be represented as $\mathcal{A}^*(\rho_0)$ and $\mathcal{A}^{**}(\rho_0)$, depending on whether large or small values are of primary concern. Obviously, because $\mathcal{A}(\rho_0)$ is already in a raw cross-product form, the use of bounds as in Section 1.2.4 would lead directly to the simple correlation coefficient between the entries of $\underset{\sim}{P}$ and $\underset{\sim}{Q}$. This measure will be denoted by r_{PQ} and has zero expectation under the randomization assumption. This same parallel to Section 1.2.4 also reminds us that the cross-product measure $\mathcal{A}(\rho)$ could be replaced by the sum of squared differences

$$\sum_{i,j} [p_{\rho(i)\rho(j)} - q_{ij}]^2$$

This latter measure is a constant linear transformation of $\mathcal{A}(\rho)$ and could be substituted for it without any loss of generality.

Even though we may use r_{PQ} as a descriptive measure of correspondence between two matrices, its significance is not assessed merely by permuting at random the values in one matrix against the fixed set from the second. Instead, the permutation model is restricted because the rows and columns of $\underset{\sim}{P}$ are rearranged in exactly the same way. To provide a more detailed comparison, suppose $\underset{\sim}{P}$ and $\underset{\sim}{Q}$ are symmetric and the $\binom{n}{2}$ above diagonal entries in both are considered as two numerical sequences in analogy with Section 1.3.3. Random permutation of one sequence against the second would give an expectation of 0 for $r_{\underset{\sim}{PQ}}$ and a variance of $1 / \left[\binom{n}{2} - 1 \right] \approx$ $1 / \binom{n}{2}$. Although the expectation of $r_{\underset{\sim}{PQ}}$ is again zero in the QA model, the variance of $r_{\underset{\sim}{PQ}}$ would be

$$\left[1 / \binom{n}{2} \right] \left\{ 1 + \frac{2}{(n-2)G_2 H_2} \left[(G_1 - G_2)(H_1 - H_2) + \frac{(2G_1 - G_2)(2H_1 - H_2)}{(n-3)} \right] \right\}$$

where

$$G_1 = \sum_i \left[\sum_j (p_{ij} - \bar{p}) \right]^2$$

$$G_2 = \sum_{i,j} (p_{ij} - \bar{p})^2$$

$$H_1 = \sum_i \left[\sum_j (q_{ij} - \bar{q}) \right]^2$$

$$H_2 = \sum_{i,j} (q_{ij} - \bar{q})^2$$

$$\bar{p} = \left[\frac{1}{\binom{n}{2}} \right] \sum_{i<j} p_{ij}; \quad \bar{q} = \left[\frac{1}{\binom{n}{2}} \right] \sum_{i<j} q_{ij}$$

This expression is obtained from considering $r_{\underset{\sim}{P}\underset{\sim}{Q}}$ as a linear function of the raw cross-product index and using the usual conversion formula for finding the variance. In other words, under QA, the variance of $r_{\underset{\sim}{P}\underset{\sim}{Q}}$ is $[1/\binom{n}{2}][1 + C]$ where C may be positive or negative, depending on the structure of the matrices being compared. The LA model, based on sequences of length $\binom{n}{2}$, destroys the integrity of $\underset{\sim}{P}$ and $\underset{\sim}{Q}$ in generating a reference distribution. This would be reflected by defining C as 0 in the expression for the QA variance (to make this comparison, we ignore the trivial -1 term in the denominator of the variance based on sequences and the LA model).

As another way of pointing out the implications of using the QA model, suppose G_1 and H_1 are both of order n^3 and G_2 and H_2 are both of order n^2. These are reasonable assumptions if the row sums for $\underset{\sim}{P}$ and $\underset{\sim}{Q}$ are not equal for large n; the entries themselves are not a function of n; and the matrices are not too sparse. Under these conditions, the variance of $r_{\underset{\sim}{P}\underset{\sim}{Q}}$ is about $(4/n^3)(G_1H_1/G_2H_2)$ for large n, which is of order $1/n$. In fact, this variance term actually has an upper bound of $1/n$. Because the LA model has a variance of order $1/n^2$, we could expect, in general, that significance would be more difficult to achieve with QA, i.e., as n increases, C remains positive and grows at a very fast rate. We might also note in this context that if the row sums for, say, $\underset{\sim}{P}$ are equal for large n, then $G_1 \longrightarrow 0$ and the variance for $r_{\underset{\sim}{P}\underset{\sim}{Q}}$ is about

$$\frac{2}{n^2}\left(1 - \frac{2}{n}\frac{H_1}{H_2}\right)$$

This latter expression is of order $1/n^2$, which proves important, at least implicitly, in Section 4.3.4. It is part of the reason why asymptotic normality can be particularly inappropriate for certain applications.

As a final clarifying comment, we should point out that the discussion throughout the chapter will emphasize the terminology of proximity matrices, but it could be rephrased equivalently in terms of weighted graphs or networks. The lines or edges of a graph connect the n objects (nodes) in a set S and are weighted by the proximities from a given matrix. The lines could be directed or undirected, depending on whether the proximity matrix is asymmetric or symmetric, respectively. If the proximities are dichotomous (zero-one), then the graph that results is defined by the presence or absence of lines between nodes. The diagonal condition merely implies that an object is not connected with itself. As one short illustration, suppose $q_{ij} = 1$ for $j = i + 1$, and 0 otherwise. Then, $\mathcal{A}(\rho_I) = \Sigma_{i=1}^{n-1} P_{i(i+1)}$ gives the length of a directed path. The significance testing strategy specifies the distribution of the path length statistic $\mathcal{A}(\rho)$ over all n! possible paths that could be constructed through the graph defined by the proximity matrix

P. Some familiar graph theory notions will be reviewed in Section 4.3.10,
but only after a variety of specific applications are first developed from the
general QA model for comparing P and Q.

4.2.3 A Numerical Example (A)

To provide a concrete illustration of the QA inference strategy, we use a
small data set collected by Hare and Bales (1963). The n objects denote
the placement of individuals seated at three sides of a rectangular table
with the unused side facing a one-way mirror. From the observer's position
looking out of the mirror, person 1 was seated at the left side, persons 2,
3, and 4 at the long side, and person 5 at the right side. Experimentally,
12 different groups of five people each were given a topic to discuss, and
the direction of interaction was recorded for a 40-minute session. Table
4.2.3a contains the average number of interactions (reported as integer
numbers) among the five seating positions over the 12 groups. The data of
Table 4.2.3a define the first matrix P.

Although it may be possible to identify structure in the underlying
communication network by simply inspecting the matrix P, the literature
also hypothesizes that certain structures should be present. For instance,
Steinzor (1950) conjectured that group members will talk more to persons
who sit opposite them than to persons next to them. Therefore, the experimentally defined seating arrangement used by Hare and Bales suggests that
certain entries in the data matrix should be relatively large, and conversely,
certain entries be relatively small. In particular, the matrix Q in Table
4.2.3b identifies adjacencies by 0's and nonadjacencies by 1's, and according to Steinzor, the 0's should be reflected by low values in P, and conversely, the 1's by high values.

Table 4.2.3a Average Interaction in 12 Laboratory Groups—Hare and
Bales Data: P

		Person receiving interaction				
		1	2	3	4	5
	1	X	21	29	22	31
	2	25	X	24	20	33
Person initiating interaction	3	35	24	X	20	36
	4	23	19	19	X	21
	5	32	33	34	19	X

Table 4.2.3b Pattern of Interaction Expected According to Steinzor: $\underset{\sim}{Q}$

		Person receiving interaction				
		1	2	3	4	5
	1	X	0	1	1	1
	2	0	X	0	1	1
Person initiating interaction	3	1	0	X	0	1
	4	1	1	0	X	0
	5	1	1	1	0	X

The index $\mathcal{A}(\rho_I)$ is the sum of entries in the data matrix $\underset{\sim}{P}$ that identify communication among nonadjacent seating positions. Thus, if Steinzor's conjecture were reflected in the matrix $\underset{\sim}{P}$, $\mathcal{A}(\rho_I)$ should be relatively large. Moreover, because this data set is small, a complete enumeration of $\mathcal{A}(\rho)$ can be given fairly easily, as in Table 4.2.3c, over the 120 possible permutations. The probability of obtaining an index as large or larger than the observed $\mathcal{A}(\rho_I)$ value of 347 is $2/120 = .017$. Thus, there appears to be support for Steinzor's conjecture in the data of Table 3.2.3a. In fact, as is clear from Table 4.2.3c, there are no other reorganizations of the matrix $\underset{\sim}{P}$ that would lead to a larger index when compared to the same matrix $\underset{\sim}{Q}$. [For completeness, we note that $E(\mathcal{A}(\rho)) = 312.00$, $V(\mathcal{A}(\rho)) = 213.6$, and $\gamma(\mathcal{A}(\rho)) = -.035$, as can be verified from the complete enumeration or from the formulas given earlier and in the Appendix.]

Some Descriptive Considerations. The QA statistic, $\mathcal{A}(\rho_0)$, is a global measure of correspondence; consequently, it may be helpful, at least at a descriptive level, to identify where the agreement between $\underset{\sim}{P}$ and $\underset{\sim}{Q}$ is coming from and where it is not. What we offer is not very profound, but it may help identify anomalies that should be pursued further. A somewhat similar tactic was suggested in using the LA statistic where the interaction matrix provides informal information about why some conjectured assignment, ρ_0, leads to a particular value for the associated index, $\Gamma(\rho_0)$.

Assuming, for convenience, that ρ_0 is the identity, ρ_I, the patterning of entries in the n × n matrix

$$\underset{\sim}{E} = \{(p_{ij} - \bar{p})(q_{ij} - \bar{q})\}$$

Table 4.2.3c Complete Permutation Distribution of
              $\mathcal{A}(\rho)$ for the Hare and Bales Data

Index	Frequency	Index	Frequency
275	2	314	2
281	2	315	2
288	4	316	2
293	2	317	2
294	2	318	2
297	2	319	2
298	2	320	2
299	2	321	2
300	8	322	2
301	2	323	4
302	8	324	4
303	2	325	6
304	4	326	2
305	6	329	4
306	4	330	4
307	4	331	4
309	2	332	2
311	4	333	2
312	2	341	2
313	2	347	2

where $\overline{p}$ and $\overline{q}$ are the means of the off-diagonal entries in P and Q, respectively, may help to isolate what row and column positions are producing
the given index value, $\mathcal{A}(\rho_I)$. If we sum over all the entries in E, we
obtain $\mathcal{A}(\rho_I) - E(\mathcal{A}(\rho))$. Thus, each of the terms in E contributes positively
or negatively to how "far" $\mathcal{A}(\rho_I)$ is away from the expectation under the QA
model. As an example, for the Hare-Bales data we would obtain

		Person				
		1	2	3	4	5
	1	X	+3.0	+1.2	-1.6	+2.0
	2	+ .6	X	+1.2	-2.4	+2.8
Person	3	+3.6	+1.2	X	+3.6	+4.0
	4	-1.2	-2.8	+4.2	X	+3.0
	5	+2.4	+2.8	+3.2	+4.2	X

suggesting that the pattern we conjecture for $\underset{\sim}{P}$ is more or less present throughout <u>except</u> for the fourth seating position and how it interacts with the first and second. The contributions here are negative and contrary to our conjecture (because we are looking for large values of $\mathcal{A}(\rho_I)$). Whether this fact is of any substantive interest really cannot be answered from this simple empirical observation, but it does raise a question that might be of interest to pursue further.

4.3 APPLICATIONS OF THE QA MODEL (B)

There is some degree of latitude as to how various applications of the basic QA model are sequenced and discussed. Almost invariably, each proposed interpretation could just as well be rephrased from a number of perspectives. As a result, some unavoidable redundancy is contained in the discussion to follow, particularly when interrelationships among specific applications are pointed out. It may be true, however, that some redundancy is a virtue which will ultimately help clarify both the richness and generality of the QA model.

Part of any unavoidable redundancy is due, at least indirectly, to the formal equivalence of the roles played by the two matrices $\underset{\sim}{P}$ and $\underset{\sim}{Q}$. They are essentially interchangeable, and only convenience really dictates which of the two should be labeled $\underset{\sim}{P}$ and which should be labeled $\underset{\sim}{Q}$. As in the Hare-Bales example of the previous section, however, we will try to follow the convention that if only one of the two matrices is based on empirically obtained data, it will be labeled as $\underset{\sim}{P}$ and called the "data" matrix. The second matrix, $\underset{\sim}{Q}$, would then typically define some a priori structure to be identified in $\underset{\sim}{P}$, and therefore, it could be referred to as the "structure" matrix. Unfortunately, this dichotomy is tenuous at best. Certain applications of QA will require both matrices to be empirically generated, and any distinction between data and structure disappears. The particular application under discussion will hopefully make it clear whether the data-structure difference is a useful one to consider.

Population Models. In all of the applications we envision for the QA model, the null conjecture will be specified directly in terms of a random relabeling of the rows and columns of one matrix against those of a second. If we wish, however, it may be possible to move one step back and specify a particular hypothesis through a data-generating mechanism in a population that, in turn, would imply the random-relabeling conjecture. Nothing different is done mechanically in obtaining significance levels for an observed statistic, but we would now be indirectly testing a structural assumption for a hypothetical data-generating mechanism. As an example and following similar discussions in Puri and Sen (1971), suppose we have two random vectors, $\underset{\sim}{X}^t = [X_1, \ldots, X_r]$ and $\underset{\sim}{Y}^t = [Y_1, \ldots, Y_s]$, with a joint cumulative distribution $F(x,y)$, and assume that $(\underset{\sim}{X}_1, \underset{\sim}{Y}_1), \ldots, (\underset{\sim}{X}_n, \underset{\sim}{Y}_n)$ represent independent observations on $(\underset{\sim}{X}, \underset{\sim}{Y})$ associated with the n objects in S. The entries in $\underset{\sim}{P}$ and $\underset{\sim}{Q}$, p_{ij} and q_{ij}, are constructed from $\underset{\sim}{X}_i$ and $\underset{\sim}{X}_j$ and from $\underset{\sim}{Y}_i$ and $\underset{\sim}{Y}_j$, respectively. Then, conditional on the two sets of n observations associated with each pair, the independence hypothesis that $F(x,y)$ factors into the product of the marginals, $F_x(\underset{\sim}{x})$ and $F_y(\underset{\sim}{y})$, induces an equally likely distribution over all n! (possibly nondistinct) values of $\mathcal{A}(\rho)$. Given the validity of the general population framework, rejection of this latter equally likely conjecture can then be interpreted as a rejection of the factorization conjecture. This model would be appropriate in the correlational context illustrated by Daniel's generalized coefficient in Section 4.3.2 (for $r = s = 1$).

As another possibility, suppose $\underset{\sim}{P}$ is a given matrix and $\underset{\sim}{Y}_1, \ldots, \underset{\sim}{Y}_n$ represent observations on a random vector $\underset{\sim}{Y}^t = [Y_1, \ldots, Y_s]$, where $\underset{\sim}{Y}$ has the cumulative distribution $F_y(\underset{\sim}{y})$. Conditional on the n vector-valued observations and assuming q_{ij} is constructed from $\underset{\sim}{Y}_i$ and $\underset{\sim}{Y}_j$, the hypothesis that $\underset{\sim}{Y}_1, \ldots, \underset{\sim}{Y}_n$ are independent induces, as before, an equally likely distribution over all n! values of $\mathcal{A}(\rho)$. This latter model would be appropriate for the spatial autocorrelation problem of Section 4.3.9 (s = 1). Of, if independence is assumed, the hypothesis under test could be one of identical distributions as in the analysis-of-variance application of Section 4.3.4.

No matter what population models are developed, there is a question of how sensitive the strategy is to violations of the assumed population preconditions other than the specific conjecture under test, e.g., observations being identically distributed, bivariate pairs being independent, and so on. These are standard robustness concerns in all population-based inference strategies but are beyond our current emphasis. The interest here is merely to suggest that a random relabeling conjecture might be justified from other principles; although we will not develop our discussion from this perspective.

4.3.1 Seriation (B)

The sequencing of objects along a continuum is a basic data analysis task
that has interested applied researchers for years. For example, a political
scientist may wish to place legislators along a liberal-conservative dimension, a psychologist may attempt to seriate subjects along a moral or
developmental continuum, or an archaeologist may try to order artifacts
according to age. In all of these cases, a set of n objects, say $S = \{O_1,$
$\ldots, O_n\}$, is assumed to be sequenced in some linear fashion. The
researcher's objective is to arrange these objects appropriately on the
basis of available data, which we will assume for now is represented by an
$n \times n$ proximity matrix $\underset{\sim}{P}$. Within psychology, much of the literature on
object seriation can be found in Coombs (1964), Torgerson (1958), and
Guttman (1954); in the archaeological and historical sciences, the volume
edited by Hodson et al. (1971) provides a fairly comprehensive source.

Depending on how the proximities in $\underset{\sim}{P}$ are constructed, it may be of
interest to test whether a pattern or gradient, attributable to unidimension-
ality, is reflected by some a priori ordering of the rows and columns of $\underset{\sim}{P}$.
This would correspond to the conjecture that the n objects in S are more or
less scalable along a single dimension, and furthermore, the scale is
defined by some identified permutation ρ_0. Here, $\rho_0(i) = k$ will imply that
object O_k is placed at the ith position, numbering from left to right. In
short, the proximities in $\underset{\sim}{P}$ should reflect the distances between the objects
and their order when the latter are appropriately placed along a continuum
by the permutation ρ_0.

Symmetric Proximities. If $\underset{\sim}{P}$ is symmetric and small proximities denote
similar objects, the test for a unidimensional pattern reduces to evaluating
whether an anti-Robinson form is present in the matrix reordered by ρ_0.
Explicitly, when ρ_0 defines a perfect unidimensional placement, we would
expect the entries in $\{p_{\rho_0(i)\rho_0(j)}\}$ within a row moving to the right or to
the left away from the main diagonal never to decrease (Kendall, 1971;
Shepard, 1978). As one measure of how clear this pattern is, the index
$\mathcal{A}(\rho_0)$ could be defined through a structure matrix $\underset{\sim}{Q}$, where $q_{ij} = |i - j|$.
In other words, we test for a pattern that would be generated through
euclidean distances if the n objects were equally spaced one unit apart
along a continuum (in an optimization sense, this criterion is used by
Szczotka, 1972, and will be referred to by this name). The larger $\mathcal{A}(\rho_0)$ is,
the better the pattern. An obvious normalized statistic in this context, as
well as for most of the variations to be mentioned, would be $r_{\underset{\sim}{P}\underset{\sim}{Q}}$. (Throughout this discussion, if the proximities were keyed so that large values
reflected similar objects, an opposite Robinson pattern would be expected,
but otherwise, the analysis proceeds directly as before. Large negative

values of $r_{\underset{\sim}{P}\underset{\sim}{Q}}$ would be expected if the conjectured ordering were appropriate.)

To some extent, the choice of an index to measure the adequacy of pattern in $\{p_{\rho_0(i)\rho_0(j)}\}$ is arbitrary. For instance, the restriction to an equally spaced rank ordering of the locations could be removed if it were possible to specify the n positions beforehand as, say, $x_1, x_2, \ldots, x_n$ where $x_i \leqslant x_{i+1}$. The term q_{ij} could again be defined by euclidean distance, but the form would now be $q_{ij} = |x_i - x_j|$. In fact, the use of absolute differences is itself somewhat of an arbitrary choice, and other specifications could be considered, e.g., squared differences. As one alternative worth explicit mention (used in an optimization context by Wilkinson, 1971), the matrix $\underset{\sim}{Q}$ could even be redefined as

$$q_{ij} = \begin{cases} 1 & \text{if } |i - j| = 1 \\ 0 & \text{otherwise} \end{cases}$$

Given this variation, the index $\mathcal{A}(\rho_0)$ would be the sum of entries in $\{p_{\rho_0(i)\rho_0(j)}\}$ immediately above and below the main left-to-right diagonal. If the anti-Robinson form holds to some reasonable extent, the proximities between adjacent objects should be relatively small (cf. the discussion of serial correlation in Section 4.3.9). Thus, the smaller $\mathcal{A}(\rho_0)$ is, the better the pattern. Intuitively, however, an index based on, say, $q_{ij} = |i - j|$, would use "more" of the information available in the proximity matrix, and for this reason alone, the Szczotka criterion may be preferred to the Wilkinson alternative. For a further discussion of these two criteria, the reader is referred to Hubert and Schultz (1976b) and to Section 4.5, which discusses the correlation between the two indices. We might also note that Section 5.3.1 presents a number of alternative measures that depend only on the order of the entries within $\underset{\sim}{P}$. This latter topic, to a great extent, resolves the issue of arbitrariness in the definition of $\underset{\sim}{Q}$, but it also brings with it a greater computational effort. In any event, these extensions are still based on the QA indices we develop here; consequently, we delay any further development of these alternatives until the next chapter.

The test for a unidimensional pattern in $\underset{\sim}{P}$ depends on the identification of a permutation ρ_0 and some decision as to how the matrix $\underset{\sim}{Q}$ will be constructed. We have made no explicit substantive assumption about how ρ_0 is to be chosen or why it may be of particular interest to a researcher. Presumably, ρ_0 is based on some source of available information that is distinct from $\underset{\sim}{P}$, possibly from an earlier empirical study or from other ancillary data we may have once it is known what the objects actually signify. In one specific class of applications, discussed separately in Section 4.3.9 under the topic of spatial autocorrelation, the choice of ρ_0 is obvious because it is

based on a given outside variable. Because this same device can be used
more generally, however, it may be helpful at this point to at least mention
the main idea briefly.

As developed in Section 4.3.9, spatial autocorrelation is concerned
with relating a given variable represented by x_1, ..., x_n to a proximity
matrix $\underset{\sim}{P}$. The objects $\{O_1, ..., O_n\}$ are typically geographical locations,
proximity corresponds to spatial separation, and the ith location, O_i, has
a numerical value x_i attached to it. Again, for convenience, the objects
can be assumed subscripted so that $x_i \leq x_{i+1}$. Thus, when q_{ij} is defined
through one of the alternatives mentioned above, $\mathcal{A}(\rho_I)$ measures the degree
to which the values x_1, ..., x_n reflect the proximity data, e.g., the degree
to which similar values on the variable are in spatially close locations.
With this interpretation, the definition for q_{ij} as $|i - j|$ would correspond
to the assignment of the untied ranks from 1 to n to the sequence x_1, ..., x_n

This same device of using some ancillary variable to define ρ_0 is a
generally applicable strategy that finds a variety of uses in topics other
than spatial autocorrelation. For example, suppose the n objects are sites,
areas, communities, and so on, and p_{ij} refers to some measure of correspondence between the sites based on the similarity of organisms present.
If the sites have a natural order (e.g., temporal in terms of the time at
which the site was observed, or spatial in terms of, say, latitude or depth
of the site), then our interests may center on the degree of "gradedness"
for $\underset{\sim}{P}$ as ordered by ρ_0. The notion of "gradedness" is the same as that of
a unidimensional pattern and reflects the gradualness of environmental
change. This topic is discussed in some detail by Pielou (1979); in fact, we
will come back to it for a numerical example later on in this section.

Asymmetric Proximities. When proximities are inherently asymmetric,
the value assigned to the ordered pair of objects (O_i, O_j) will index the
degree of "dominance" of O_i over O_j; by convention, greater degrees of
dominance will be reflected by larger proximities. In this context, we
include various measures of "flow" from O_i to O_j, and in general, any
measure of correspondence from one object to a second that need not be
reciprocal.

For psychologists, the basic interest in the problem of seriation using
asymmetric proximities results from the experimental paradigm commonly
known as the technique of paired comparisons (see Hubert, 1976, and
David, 1963, for comprehensive surveys). In the most standard illustration,
a subject is provided with all pairs of distinct objects from some set S and
is asked to choose in each case one member of the pair; for instance, the
object that is heavier, smarter, more preferable, more liberal, and so on.
If the ordered pair (O_i, O_j) is given a score of 1 whenever O_i is chosen
over O_j and a 0 otherwise, then a simple dichotomous proximity measure
is obtained. More generally, if the same procedure is repeated over subjects or over trials, the final summed proximity measure is asymmetric,

and both the largest and the smallest values are assigned to the object
pairs that contain the most discriminable objects. As a slight variation
that should be mentioned, each subject could be forced to provide a linear
ranking of the n objects immediately; in this task, the set of ordered pairs
that are given a score of 1 is derived directly from the subject's ordering,
e.g., (O_i, O_j) is given a score of 1 if O_i is ranked ahead of O_j. A final
proximity measure could then be obtained by aggregating these dichotomous
scores over subjects. For now, and as a technical convenience, it is
assumed that $p_{ij} + p_{ji} = 1$. Thus, if m_{ij} denotes the raw number of times
O_i was preferred to O_j, a natural measure of dominance could be defined very simply as $p_{ij} = m_{ij}/(m_{ij} + m_{ji})$. Later in this section, we comment on how an arbitrary matrix $\underset{\sim}{P}$ can be handled by the simple decomposition of $\underset{\sim}{P}$ into its symmetric and skew-symmetric components. (Section
6.1.2 discusses the important issue of whether individual data, when available, should be aggregated to the group level before a comparison to an
assumed order is carried out.)

A geometric analogy may help somewhat in clarifying the asymmetric
seriation problem. Suppose the n objects in S are placed along a one-
dimensional continuum with spacing of a single unit; furthermore, the
researcher believes the placement is correct and justifies this contention
by observing that the given proximity function has the following ideal properties:

1. If O_i is placed to the left of O_j, then $p_{ij} \geq p_{ji}$, i.e., whenever O_i is to
   the left of O_j, then O_i has "more" (or at least no less) of the relevant
   quantity that the continuum supposedly represents.
2. If the hypothesized right-to-left ordering is denoted by $(O_1, \ldots, O_n)$,
   then the proximities above the main diagonal have a perfect anti-Robinson
   form (never decreasing moving to the right from the main diagonal
   within a row), and those below the main diagonal have a perfect Robinson
   form (never increasing moving to the left from the main diagonal within
   a row).

Property (1) reflects the basic correctness of the ordering from left to
right induced by the placement, whereas property (2) implies that increasing
distances between two objects along the continuum should be reflected in
increasing proximities above the main diagonal, and, conversely, in decreasing proximities below the main diagonal. The need for the additional
first property, (1), is in contrast to our discussion of symmetric proximities
where only a single gradient condition was considered. Consequently, there
is greater latitude in how the adequacy of a conjectured order is assessed
when the proximities are asymmetric.

If we wish to test the reasonableness of a conjectured seriation defined
by a permutation ρ_0, there are two distinct sources of information that could
be used. First, using property (2), it is expected that the absolute differences

$|p_{ij} - p_{ji}|$, would be directly reflected in the distance between the place-ments for O_i and O_j. No direction of dominance is now implicit, but large absolute differences should imply greater degrees of separation.[†] Moreover, because these absolute differences are symmetric, this pattern could be tested as in the earlier context of symmetric proximities. Second, when appropriately seriated from left to right, the ordering of S should be re-flected in the sign of $p_{ij} - p_{ji}$. If the more dominant objects are convention-ally placed to the left and O_i precedes O_j in this ordering, then the expres-sion $p_{ij} - p_{ji}$ should be nonnegative.

Formally, we can test whether there is an excess of $+1$'s above the diagonal by defining a proximity matrix $\underset{\sim}{P}^O$ as

$$
p_{ij}^o = \begin{cases} +1 & \text{if } p_{ij} - p_{ji} > 0 \\ -1 & \text{if } p_{ij} - p_{ji} < 0 \\ 0 & \text{otherwise} \end{cases}
$$

and $\underset{\sim}{Q}$ as

$$
q_{ij} = \begin{cases} 1 & \text{if } i < j \\ 0 & \text{if } i \geq j \end{cases} \tag{4.3.1a}
$$

The index $\mathcal{A}(\rho_0)$ is the sum of the above-diagonal entries in $\{p^o_{\rho_0(i)\rho_0(j)}\}$, or equivalently, the number of above-diagonal positive differences minus the number of above-diagonal negative differences. Thus, $\mathcal{A}(\rho_0)$ should be large and positive if the conjectured order defined by ρ_0 produces an appro-priate unidimensional pattern in the proximities.

From discussions of similar statistics in the literature (Ager and Brent, 1978), it appears that under the conjecture of randomness and for reason-ably large n, the $\mathcal{A}(\rho)$ statistic based on signs can be considered normal with mean zero and a variance based on our earlier formula of

$$
V(\mathcal{A}(\rho)) = \frac{1}{6}\left[\sum_{i,j} p_{ij}^{o2} + 2 \sum_i \left(\sum_j p_{ij}^o \right)^2 \right]
$$

A formal proof of asymptotic normality could follow the lines of Kendall's discussion (Kendall, 1970, pp. 72-74), which assumes $\sum_i (\sum_j p_{ij}^o)^2$ is of

---

[†]Data of this type have been popularized through the topic of "revealed preference analysis" in which the objects correspond to spatial locations, e.g., see Rushton (1969) and Hubert (1982).

order n^3. As a final descriptive measure of the degree to which the matrix P^0 is reordered appropriately, a measure suggested by Ager and Brent (1978) can be adopted that has the $\overset{\sim}{\mathcal{A}}^*(\rho_0)$ form:

$$\frac{\mathcal{A}(\rho_0)}{\sum_{i<j} p^o_{ij}} \tag{4.3.1b}$$

The term $\Sigma_{i<j}\, p^{o2}_{ij}$ is the number of above-diagonal nonzero entries for any ordering, and at least as a bound, defines the maximum value for the numerator. Alternatively, the correlation $r_{\underset{\sim}{P}\underset{\sim}{Q}}$ could be used, which relies on a denominator of the form

$$\sqrt{\frac{n(n-1)}{2}\,\sum_{i<j} p^{o2}_{ij}}$$

This latter term provides an upper bound on $\Sigma_{i<j}\, p^{o2}_{ij}$, and thus, a smaller normalized index.

Several special cases of these formulas deserve particular mention. First, if there are no off-diagonal zeros in P^0 and no intransitivities or circular triads (defined as a triple (i,j,k) for which $\tilde{p}^o_{ij} = +1$, $p^o_{jk} = +1$, and $p^o_{ik} = -1$), then the inference problem is equivalent to the comparison of two untied rankings based on Kendall's tau statistic (see Section 4.3.2). The numerator of tau is $\mathcal{A}(\rho_0)$, and the variance term reduces to $(1/18)n(n-1) \times (2n+5)$, which is the standard expression used for a significance test (Kendall, 1970). The general variance term given above is also appropriate when there are ties in one of the rankings and this ranking defines the matrix P^0, or even when intransitivities exist in $\tilde{P}^0$. A particularly simple formula results when there are T intransitive triples and no off-diagonal zeros: $(1/18)\{[n(n-1)(2n+5)] - 48T\}$. Obviously, for T = 0, this latter formula reduces to the standard expression for untied rankings. Also, when T is 0, the well-known Goodman-Kruskal gamma statistic is equivalent to our normalized measure in expression (4.3.1b)—see Section 4.3.2.

Although we have phrased the discussion of seriation in the asymmetric case for data of the constrained paired-comparison form, we note that our discussion could be extended very easily to an arbitrary proximity matrix $\underset{\sim}{P}$ in which the entries do not necessarily satisfy the summation condition that $p_{ij} + p_{ji} = 1$. In particular, if we decompose $\underset{\sim}{P}$ into $\underset{\sim}{P}^+$ and $\underset{\sim}{P}^-$, then $\underset{\sim}{P}^+$ provides a symmetric proximity matrix that can be analyzed apart from $\underset{\sim}{P}^-$. The latter matrix, $\underset{\sim}{P}^-$, has an arbitrary entry p^-_{ij} that replaces our earlier difference $p_{ij} - p_{ji}$. In turn, the signs and the absolute values of p^-_{ij} can be analyzed separately just as we can analyze the signs and absolute values of $p_{ij} - p_{ji}$.

<u>Combining Sign and Magnitude Information</u>. Sign and magnitude information provided by differences of the form $p_{ij} - p_{ji}$, or more generally, by $\overline{p_{ij}}$, should lead to consistent results whenever a strong unidimensional scale underlies the original measure p_{ij}. In fact, because the degree to which sign and magnitude data are compatible gives an indication of "scalability," it may be inadvisable in any analysis that has not demonstrated strong scale saliency either to eliminate one source from consideration altogether or automatically to combine both into a single proximity measure prior to showing that they would lead to similar conclusions. Nevertheless, if we wished to have both sources combined to give a single measure of the adequacy of a conjectured ordering, there are several different options that could be considered:

1. Based on $\underset{\sim}{Q}$ in equation (4.3.1a) and $\underset{\sim}{P}$ as the original proximity matrix (or $\underset{\sim}{P}^-$ when $\underset{\sim}{P}$ is not of the paired comparison form), the index $\mathcal{A}(\rho_0)$ is the sum of above-diagonal entries in $\{p_{\rho_0(i)\rho_0(j)}\}$ and should be relatively large if the seriation ρ_0 is being reflected in the proximities.

As another statistically equivalent alternative that will be especially helpful for the discussion of generalized correlation coefficients in the next section, suppose $\underset{\sim}{Q}$ is given as

$$q_{ij} = \begin{cases} +1 & \text{if } i < j \\ 0 & \text{if } i = j \\ -1 & \text{if } i > j \end{cases} \qquad (4.3.1c)$$

Based on this definition, $\mathcal{A}(\rho_0)$ is the above-diagonal sum in $\{p_{\rho_0(i)\rho_0(j)}\}$ minus the below-diagonal sum. Because the entries in $\underset{\sim}{P}$ sum to a constant, say c, twice the index constructed from equation (4.3.1a) minus c leads to the measure constructed from equation (4.3.1c).

2. When a seriation using an asymmetric proximity matrix $\underset{\sim}{P}$ is perfect, we expect an anti-Robinson form above the main diagonal and a Robinson form below. To model this pattern by a placement of objects along the continuum one unit apart, $\underset{\sim}{Q}$ could be defined as

$$q_{ij} = j - i \qquad (4.3.1d)$$

3. Instead of an equal spacing of objects as in (2), we may also test for a pattern that represents unequal spacing. Assuming the n positions are identified by the values $x_1, x_2, \ldots, x_n$ in increasing order along the continuum, $\underset{\sim}{Q}$ could be defined as

$$q_{ij} = x_j - x_i \tag{4.3.1e}$$

Although all three weighting options are presented here as a way of combining sign and magnitude data to test for the adequacy of a given seriation, these same options will be mentioned again in the context of generalized correlation coefficients in the next section.

These three options, in effect, allow a direct test of the ordering implied by a given variable with values $x_1, \ldots, x_n$ against the matrix $\underset{\sim}{P}$. The definition of equation (4.3.1c) depends on the order of the values alone, equation (4.3.1d) depends on ranks, and equation (4.3.1e) depends on actual magnitudes. For example, suppose the objects $\{O_1, \ldots, O_n\}$ correspond to n spatial locations and define the skew-symmetric matrix $\underset{\sim}{P}$ by $p_{ij} = (m_{ij} - m_{ji})d_{ij}$, where d_{ij} is the symmetric distance from O_i to O_j and $m_{ij}(m_{ji})$ denotes the number of people who move from O_i to O_j (O_j to O_i). Thus, the proximity p_{ij} is positive if O_j receives more people from O_i than it loses, and the difference is weighted by the spatial separation between the locations (cf. Tobler, 1979). If the objects are subscripted in such a way that $x_i \leqslant x_{i+1}$, then $\mathcal{A}(\rho_I)$ measures the degree to which the migration pattern in $\underset{\sim}{P}$ is reflected by the sequence $x_1, \ldots, x_n$. The latter may be attractivity values defined by quality of life, unemployment, or other variables that supposedly influence a person's decision to move. It should also be apparent that tied values in the sequence $x_1, x_2, \ldots, x_n$ could be handled by simple variations on the way Q is constructed, e.g., if a definition other than equation (4.3.1e) is used, by letting $q_{ij} = 0$ when $x_i = x_j$ (cf. the definition of the three scoring functions in the next section).

There is an alternative way of representing a QA index based on q_{ij}, defined either as $j - i$ or $x_j - x_i$, as a much simpler LA index (cf. Cliff, 1980). Specifically, if $q_{ij} = x_j - x_i$, then

$$\mathcal{A}(\rho) = \sum_{i,j} p_{\rho(i)\rho(j)} q_{ij} = \sum_{i,j} p_{\rho(i)\rho(j)} (x_j - x_i)$$

$$= \sum_j (p_{\cdot\rho(j)} - p_{\rho(j)\cdot}) x_j$$

Thus, $\mathcal{A}(\rho)$ is equivalent to an index of the form $\Gamma(\rho)$, where the latter is based on an assignment matrix $\underset{\sim}{C} = \{c_{ij}\} = \{(p_{\cdot i} - p_{i\cdot})x_j\}$. Obviously, when $q_{ij} = j - i$, a similar reduction holds if x_j is replaced by j. Furthermore, when $\underset{\sim}{P}$ is itself skew-symmetric, which can be assumed without loss of generality because Q is already skew-symmetric, the matrix $\underset{\sim}{C}$ would take the even simpler form of $\{(2p_{\cdot i})x_j\}$ or $\{(2p_{\cdot i})j\}$, depending on the definition of $\underset{\sim}{Q}$ as equation (4.3.1d) or (4.3.1e), respectively. Unfortunately, the $\pm$ scoring function of equation (4.3.1c) does not lead to such a reduction, and the QA index must be used in its original form.

Circular Placement. A data analysis problem closely related to the sequencing of objects along a continuum involves the ordering of objects around a circle. Circular placements have been discussed by Guttman (1954), Hubert (1974), and others, and complement the earlier emphasis on linear seriation in a very obvious way. Because there is no inherent left-to-right ordering, however, we assume that the proximity matrix is symmetric, or possibly, only symmetric information is of interest, e.g., characterizing proximity by $(p_{ij} + p_{ji})/2$ or $|p_{ij} - p_{ji}|/2$ from an asymmetric matrix $\underset{\sim}{P}$.

For defining $\underset{\sim}{Q}$, each of the alternatives mentioned in the linear sequencing context for symmetric proximities has direct circular parallels. For example, when a cycle is defined by some circular ordering of the objects in S, the sum of the n proximities between adjacent objects is the length of the cycle, and a small total length indicates a potentially adequate ordering. This alternative corresponds to the Wilkinson index in the linear case and is obtained from

$$q_{ij} = \begin{cases} 1 & \text{if } |i - j| = 1 \text{ or } n - 1 \\ 0 & \text{otherwise} \end{cases}$$

where the n locations are numbered in order from 1 to n starting at some arbitrary point. The resulting measure $\mathcal{A}(\rho_0)$ is twice the length of the cycle defined by the permutation ρ_0 (cf. the discussion of cyclical correlation in Section 4.3.9). The Szczotka criterion would be generalized to

$$q_{ij} = \begin{cases} |i - j| & \text{if } |i - j| \leqslant n/2 \\ n - |i - j| & \text{if } |i - j| > n/2 \end{cases}$$

when n is even; or, if n is odd

$$q_{ij} = \begin{cases} |i - j| & \text{if } |i - j| \leqslant (n - 1)/2 \\ n - |i - j| & \text{if } |i - j| > (n - 1)/2 \end{cases}$$

Finally, unequal spacing could be used if q_{ij} were defined by the shortest distance along the circle between the ith and jth positions. In general, the reference distribution for $\mathcal{A}(\rho)$ based on any one of these definitions for q_{ij} and n! possible permutations would reduce to considering (n - 1)! equally likely circular arrangements. This latter reduction is achieved because each possible circular arrangement is generated by n different permutations.

Examples. As an illustration of how the mechanics for one of the comparisons discussed in this section could be carried out, we consider a recent study of figural "goodness" by Glushko (1975), and in particular, his

attempt to verify Garner's (1962) basic hypothesis regarding what makes
one pattern better than another. Each of the 17 patterns used by Glushko,
listed in Figure 4.3.1a, can be characterized by the size of an inferred
equivalence class. The term "equivalence" is used to label the set of patterns that contain a single figure plus all other configurations that result
from reflections or from 90° rigid rotations. As indicated in Figure 4.3.1a,
two of the Glushko patterns construct the same configuration under all of
these operations, eight patterns have four associated figures, and finally,
seven patterns produce eight different members. According to Garner,
the subjective judgment of pattern goodness is a direct function of the size
of a configuration's inferred equivalence class, with the smaller size
classes corresponding to the better patterns.

To test Garner's hypothesis using the 17 patterns of Figure 4.3.1a,
Glushko first obtained a symmetric measure of proximity between each
pair of patterns by using a paired-comparison choice task. All 136 different
pattern combinations were presented to 20 subjects, who were asked to
indicate their preference. These choices were then summed over subjects
and used to define a symmetric proximity matrix based on absolute differ-
ences of the form $|(p_{ij} - p_{ji})/2|$. As notation, the matrix of proximities
constructed in this way is denoted by $|\underset{\sim}{P}^-|$.

Based on $|\underset{\sim}{P}^-|$, given in the lower triangular portion of Table 4.3.1a,
Glushko attempted to represent the structure of the proximity function by
first placing the 17 configurations in a two-dimensional space, using Shepard
and Kruskal's multidimensional scaling routine (see Kruskal and Wish,
1978). Given this geometric representation, Johnson's (1967) diameter
clustering results were then superimposed, producing a representation
similar to that shown in Figure 4.3.1b (here we only indicate the clustering
result defined by three subclasses). Clearly, one strong dimension (the
vertical) can be identified as that of equivalence class size. In addition,
the clusters themselves correspond fairly well to a grouping on the basis of
the same criterion, except for the misplacement of the two configurations
numbered 10 and 11.

The process of verifying Garner's hypothesis through a multidimensional scaling and clustering might be considered rather circuitous, especially because the equivalence class hypothesis implies a definite structure
for the original proximity measure. Although one dimension is very strong
in this example and the clustering and scaling results are clear-cut,
unambiguous outcomes of this type are somewhat rare. In general, when a
strong hypothesis is not reflected as dramatically in the scaling or clustering results, it may be difficult to decide whether the hypothesis is inadequate
or whether the data reduction techniques are at fault. In the typical application, the researcher may be able to identify portions of his or her theory
in a scaling or clustering solution, but may lack a strategy for measuring
in any precise manner the actual degree of confirmation or nonconfirmation.

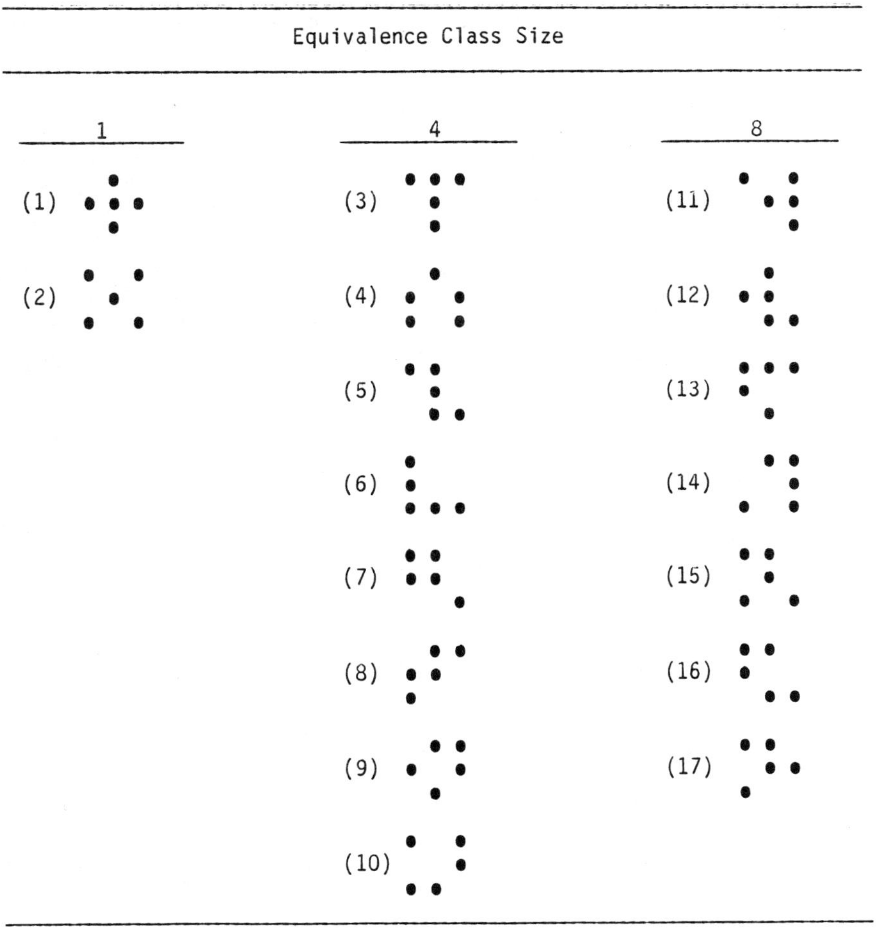

Equivalence Class Size

FIGURE 4.3.1a Patterns used by Glushko in testing Garner's figural good-ness hypothesis.

As an alternative approach, we can test directly whether the pattern goodness hypothesis is reflected in the original proximities. Because each pattern has an associated equivalence class, the 17 figures should demon-strate a unidimensional gradient when the actual size of the class is used to locate a position along a continuum. Specifically, if x_i denotes the equivalence class size for pattern O_i, then the distance between the locations for O_i and O_j will be defined as $|x_i - x_j|$. Obviously, there are many tied

values in the sequence $x_1, \ldots, x_n$ and the choice of an absolute difference
for defining separation along the continuum is somewhat arbitrary. Still,
if equivalence class size is being reflected in the proximities, these distances,
given as the matrix $\underset{\sim}{Q}$ in the upper-triangular portion of Table 4.3.1a,
should reflect the proximities well. (We do note, however, that this arbitrariness can be removed. In Section 5.3.1, a measure is proposed and
illustrated with these same data that depends only on the order of the entries
in $|\underset{\sim}{P}^-|$ and $\underset{\sim}{Q}$.)

The unidimensional pattern in $|\underset{\sim}{P}^-|$ is fairly strong, although obviously
far from perfect, e.g., configurations 10 and 11 produce some clear inconsistencies to the gradient we expect. For example, in the matrix $\underset{\sim}{E} =$
$\{(p_{ij} - \overline{p})(q_{ij} - \overline{q})\}$, the row (and column) sums for objects 10 and 11 are
both negative and rather substantial (no other objects have an aggregate

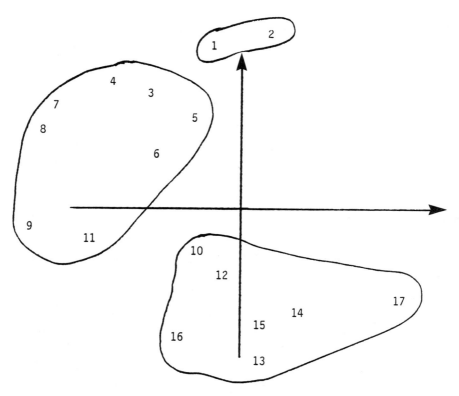

FIGURE 4.3.1b Two-dimensional scaling of the 17 Glushko patterns.

Table 4.3.1a Symmetric Proximity Matrix $|\underset{\sim}{P}^-|$ Obtained by Glushko
for the Patterns of Figure 4.3.1a (Lower Triangle); the
Upper Triangle Represents the $\underset{\sim}{Q}$ Matrix Specifying the
Equivalence Class Conjecture

Pattern	2	3	4	5	6	7	8	9	10	11	12	13	14	15	16	17
1	X	0	3	3	3	3	3	3	3	7	7	7	7	7	7	7
2	1	X	3	3	3	3	3	3	3	7	7	7	7	7	7	7
3	1	2	X	0	0	0	0	0	0	4	4	4	4	4	4	4
4	2	4	0	X	0	0	0	0	0	4	4	4	4	4	4	4
5	3	3	1	1	X	0	0	0	0	4	4	4	4	4	4	4
6	2	4	1	1	1	X	0	0	0	4	4	4	4	4	4	4
7	2	4	3	2	1	2	X	0	0	4	4	4	4	4	4	4
8	3	5	2	1	2	1	0	X	0	4	4	4	4	4	4	4
9	4	4	2	1	5	3	3	4	X	4	4	4	4	4	4	4
10	4	5	4	4	3	3	3	5	4	X	4	4	4	4	4	4
11	5	5	3	4	3	0	2	3	1	1	X	0	0	0	0	0
12	5	6	4	6	4	1	5	5	2	1	3	X	0	0	0	0
13	6	7	7	6	5	4	5	6	5	1	4	1	X	0	0	0
14	7	6	4	4	5	4	6	5	4	2	4	1	1	X	0	0
15	6	7	5	7	4	5	5	4	5	0	3	0	0	1	X	0
16	7	8	5	5	6	4	4	3	4	1	4	2	2	0	1	X
17	7	7	5	5	5	6	5	4	6	3	6	2	3	1	1	1

negative contribution). This last observation clearly indicates a difficulty in our conjecture as it relates to these latter two objects. The
overall observed cross-product index, $\mathcal{A}(\rho_I)$, between $|\underset{\sim}{P}^-|$ and $\underset{\sim}{Q}$ as
given in the lower and upper triangular portions of Table 4.3.1a,
respectively, turns out to be 3350.0, with a normalized value, $r_{|\underset{\sim}{P}^-||\underset{\sim}{Q}}$,
of .64. Table 4.3.1b provides an approximate Monte Carlo reference
distribution based on a sample size of 999, which generates an upper-
tail significance level of .001 for $\mathcal{A}(\rho_I)$. A very small significance

Table 4.3.1b Approximate Monte Carlo Distribution for the
             Comparison of Matrices in Table 4.3.1a
             (Sample Size of 999; Three Significant Digits
             for the Index)

Index	Cumulative frequency
2300	1
2370	10
2410	50
2440	100
2470	200
2500	300
2530	400
2550	500
2570	600
2600	700
2640	800
2690	900
2740	950
2850	990
3030	999

value (less than .001)[†] would be obtained with a type III approximation using
the Z statistic of 7.28 for an expectation of 2557.0, a variance of 11860, and
the calculated skewness parameter of .922.

[†] Because the sample size for a Monte Carlo significance test will be 999
throughout Part II, the smallest significance level that could be reported is
.001. This latter value is achieved whenever the observed index is more
extreme than any value found in the random sample. To maintain compara-
bility, very small significance levels obtained with a type III approximation
will be reported merely as "less than .001" when this statement is appropriate.

A second example deals with the concept of "gradedness" mentioned earlier in this section. Table 4.3.1c presents two proximity matrices based on data, summarized by Pielou (1979), on the sedimentary history of the Labrador Shelf. Matrix $\underset{\sim}{P}_a$ has 12 samples (or objects), matrix $\underset{\sim}{P}_b$ has 11, and each is ordered according to depth. The actual proximities are proportional similarity indices representing organism overlap (see Pielou, 1979). Thus, given the way in which these proximities are "keyed," the pattern we would hope to detect is one of a Robinson form, e.g., large negative values of r_{PQ} would reflect the conjectured structure. Taking each of these matrices in turn as compared to Q for the Szczotka criterion, i.e., $q_{ij} = |i - j|$, the following results would be obtained (the Monte Carlo distributions are given in Table 4.3.1d):

	$\underset{\sim}{P}_a$ versus $\underset{\sim}{Q}$	$\underset{\sim}{P}_b$ versus $\underset{\sim}{Q}$
Observed index	275.2	274.6
Expectation	305.9	275.2
Variance	75.22	43.71
Z	-3.54	-.091
γ	- .796	-.412
Significance level		
Monte Carlo	.005	.476
Type III approximation	.003	.437

Matrix $\underset{\sim}{P}_a$ is more graded than $\underset{\sim}{P}_b$ ($r_{\underset{\sim}{P}_a\underset{\sim}{Q}}$ = -.45 versus $r_{\underset{\sim}{P}_b\underset{\sim}{Q}}$ = -.01). Thus, the former represents a more gradual sedimentation; the latter may suggest the occurrence of some environmental event that disrupted the expected gradual progression. These results are consistent with Pielou's discussion although we rely on a different measure. The actual index used by Pielou is discussed in Chapter 5 but in a more general context.

4.3.2 Generalized Correlation Coefficients (B)

The concept of a generalized correlation coefficient, introduced by Daniels (1944) and used extensively by Kendall (1970) in analyzing contingency tables with ordered attributes, has gained increased prominence in the recent methodological literature of the social sciences, particularly as part of a comprehensive strategy for interpreting ordinal relationships (for example, see Hawkes, 1971). As we will comment later, this topic is really a special case of evaluating a given seriation of n objects based on an asymmetric

Table 4.3.1c Proximity Data from Pielou (1979) Between Samples of
Size 12 and 11 Taken from Two Cores[a]

Matrix $\underset{\sim}{P}_a$

	1	2	3	4	5	6	7	8	9	10	11	12
1	X	.7	.7	.5	.5	.6	.6	.6	.5	.6	.5	.5
2	.7	X	.7	.5	.5	.6	.5	.5	.4	.5	.0	.4
3	.7	.7	X	.6	.6	.6	.7	.6	.5	.6	.4	.6
4	.5	.5	.6	X	.6	.7	.5	.5	.4	.5	.0	.5
5	.5	.5	.6	.6	X	.7	.6	.4	.0	.4	.0	.4
6	.6	.6	.6	.7	.7	X	.6	.5	.4	.5	.0	.5
7	.6	.5	.7	.5	.6	.6	X	.7	.6	.7	.6	.7
8	.6	.5	.6	.5	.4	.5	.7	X	.8	.8	.7	.6
9	.5	.4	.5	.4	.0	.4	.6	.8	X	.7	.8	.7
10	.6	.5	.6	.5	.4	.5	.7	.8	.7	X	.7	.7
11	.5	.0	.4	.0	.0	.0	.6	.7	.8	.7	X	.7
12	.5	.4	.6	.5	.4	.5	.7	.6	.7	.7	.7	X

[a] Each of the matrices is ordered naturally according to sample depth.

Matrix $\underset{\sim}{P}_b$

	1	2	3	4	5	6	7	8	9	10	11
1	X	.7	.7	.5	.5	.5	.6	.6	.5	.6	.6
2	.7	X	.8	.6	.7	.5	.8	.8	.7	.7	.7
3	.7	.8	X	.5	.6	.6	.8	.7	.6	.8	.6
4	.5	.6	.5	X	.7	.0	.6	.7	.7	.5	.6
5	.5	.7	.6	.7	X	.7	.8	.8	.8	.6	.7
6	.5	.5	.6	.0	.7	X	.5	.5	.0	.5	.4
7	.6	.8	.8	.6	.8	.5	X	.8	.7	.7	.6
8	.6	.8	.7	.7	.8	.5	.8	X	.8	.7	.6
9	.5	.7	.6	.7	.8	.0	.7	.8	X	.7	.7
10	.6	.7	.8	.5	.6	.5	.7	.7	.7	X	.7
11	.6	.7	.6	.6	.7	.4	.6	.6	.7	.7	X

Table 4.3.1d Monte Carlo Distributions for the
             Pielou Data in Table 4.3.1c
             (Sample Sizes of 999)

Cumulative frequency	P_a versus Q	P_b verus Q
1	271.	254.
5	276.	257.
10	277.	259.
50	291.	264.
100	296.	267.
200	300.	270.
300	303.	272.
400	305.	274.
500	307.	276.
600	309.	278.
700	311.	279.
800	314.	281.
900	317.	284.
950	319.	285.
990	322.	287.
995	323.	288.
999	325.	291.

proximity matrix. Consequently, some very familiar ideas regarding correlation could have been subsumed under the discussion in Section 4.3.1.

In the typical correlation framework, suppose that n bivariate observations (x_1, y_1), ..., (x_n, y_n) are available and two score functions p_{ij} and q_{ij} are defined on the pairs (x_i, x_j) and (y_i, y_j), respectively. The constraints of skew-symmetry are assumed:

$$p_{ij} = -p_{ji} \text{ and } q_{ij} = -q_{ji}$$

as well as a zero diagonal in the complete matrices $\underset{\sim}{P}$ and $\underset{\sim}{Q}$. If the raw index $\mathcal{A}(\rho_I) = \Sigma_{i,j} \, p_{ij} q_{ij}$ is normalized to form $r_{\underset{\sim}{P}\underset{\sim}{Q}}$, then depending on the definition of the scores, $r_{\underset{\sim}{P}\underset{\sim}{Q}}$ reduces to the three common correlation coefficients. More explicitly, when $p_{ij} = (x_i - x_j)$ and $q_{ij} = (y_j - y_i)$, Pearson's product-moment correlation coefficient is obtained; if p_{ij} = rank x_j - rank x_i and q_{ij} = rank y_j - rank y_i, Spearman's rank correlation statistic ρ_b is defined, and finally, if $p_{ij} = \text{sign}(x_j - x_i)$ and $q_{ij} = \text{sign}(y_j - y_i)$, where

$$
\text{sign(h)} = \begin{cases} +1 & \text{if } h > 0 \\ 0 & \text{if } h = 0 \\ -1 & \text{if } h < 0 \end{cases}
$$

then Kendall's τ_b is obtained.

The permutation mean and variance of $r_{\underset{\sim}{P}\underset{\sim}{Q}}$ can be found easily from the formulas given earlier because the denominator is constant over all permutations. Asymptotic normality is discussed in detail by Kendall (1970). We also note that Pearson's correlation (and, thus, Spearman's index when ranks are used) can also be obtained as a special case of the LA model (Section 1.3.3); therefore, the index $\mathcal{A}(\rho_I)$ must reduce in this case. In particular,

$$
\mathcal{A}(\rho_I) = 2n\{\Gamma(\rho_I) - E(\Gamma(\rho))\}
$$

where the matrix $\underset{\sim}{C}$ is of the form $\{x_i y_j\}$. This equivalence emphasizes the fact that all two-independent sample problems, which are subsumed under LA through a matrix of assignment scores defined by products, could also be rephrased as part of the QA framework.

A rather interesting equivalence exists between the generalized correlation coefficient and the statistics discussed in the seriation context that combine sign and magnitude information in testing for an assumed order. Specifically, the three definitions of Q given in equations (4.3.1c), (4.3.1d), and (4.3.1e) correspond to the skew-symmetric matrices based on the scoring functions for, say, $y_1, \ldots, y_n$; this latter sequence is ordered from smallest to largest. Thus, when proximity is constructed from the corresponding x sequence in exactly the same way and used to define $\underset{\sim}{P}$, the raw generalized correlation coefficient index, $\Sigma_{i,j} \, p_{ij} q_{ij} = \mathcal{A}(\rho_I)$, is identical to the raw index based on ρ_0 and the appropriate definition of Q from equation (4.3.1c), (4.3.1d), or (4.3.1e). The permutation ρ_0 was needed in the last section as a technical device to match the corresponding observations on x and y, whereas it is now "built in" with our use of the generalized correlation coefficient through the definitions of $\underset{\sim}{P}$ and $\underset{\sim}{Q}$ and the matching defined implicitly by the n bivariate pairs. In short, the calculation and testing of one of the usual correlation coefficients is, in effect, a procedure for assessing whether an asymmetric proximity matrix $\underset{\sim}{P}$, obtained from the sequence $x_1, \ldots, x_n$, has the appropriate form, i.e., whether the n objects can be seriated along a continuum by the given permutation that pairs up the x and y observations. Obviously, the x and y sequences play equivalent roles that could be interchanged in this discussion.

Normalization. Although the normalized measure r_{PQ} has been relied on more or less automatically, there are a number of alternatives that could be proposed, using the general forms represented by $\mathcal{A}^*(\rho_I)$ or $\mathcal{A}^{**}(\rho_I)$ and various bounds. Given the LA equivalence for the scoring functions that lead to the Pearson or Spearman correlation coefficients, some of these options have already been discussed in Section 1.3.3. The choice of a normalized index for the sign scoring function, however, cannot be rephrased in the LA paradigm; moreover, a considerable controversy exists in the literature as to what final descriptive statistics are appropriate and should be used.

Because both $\underset{\sim}{P}$ and $\underset{\sim}{Q}$ are skew-symmetric and $E(\mathcal{A}(\rho)) = 0$, our usual normalized indices would have the form

$$\frac{\mathcal{A}(\rho_I)}{\max_{\rho} \mathcal{A}(\rho)} \quad \text{or} \quad \frac{\mathcal{A}(\rho_I)}{\min_{\rho} \mathcal{A}(\rho)}$$

corresponding to $\mathcal{A}^*(\rho_I)$ and $\mathcal{A}^{**}(\rho_I)$, respectively. As in Section 1.2.4, suppose we look for a single bound, B, such that

$$-B \leq \min_{\rho} \mathcal{A}(\rho) \leq \max_{\rho} \mathcal{A}(\rho) \leq B$$

If it is assumed that large index values are of primary interest and should be keyed positively, this bound can be used to define the one index

$$\frac{\mathcal{A}(\rho_I)}{B}$$

which has the general form of $\mathcal{A}^*(\rho_I)$. Then, depending on how B is defined, a variety of different measures of rank correlation can be generated. For example, using the sign scoring function, two bounds are given by the expression

$$n(n-1) \leq \sqrt{\sum_{i,j} p_{ij}^2 \sum_{i,j} q_{ij}^2} \leq \max_{\rho} \mathcal{A}(\rho)$$

Thus, when $B = n(n-1)$, Kendall's τ_a is obtained; and when $\sqrt{\Sigma_{i,j} \, p_{ij}^2 \, \Sigma_{i,j} \, q_{ij}^2}$, we construct Kendall's τ_b, which is based on our original index $\underset{\sim}{r}_{PQ}$. For two coefficients introduced by Somers (1962), we rely on the bounds $\Sigma_{i,j} \, |q_{ij}|$ and $\Sigma_{i,j} \, |p_{ij}|$. The reader is referred to Reynolds (1977) for a more complete discussion of all these indices and to various probabilistic interpretations that may lead to the choice of one over another. For example, each of the $\binom{n}{2}$ pairs, (x_i, y_i) and (x_j, y_j), can be classified as concordant ($p_{ij}q_{ij} = +1$), discordant ($p_{ij}q_{ij} = -1$), tied on x only ($p_{ij} = 0$, $q_{ij} = \pm 1$), tied on y only ($p_{ij} = \pm 1$, $q_{ij} = 0$), or tied on both x and y ($p_{ij} = 0 = q_{ij}$). The Somers indices can be interpreted as the difference between two conditional probabilities: the probability of concordance among pairs not tied on x minus the probability of discordance among pairs not tied on $x(B = \Sigma_{i,j} \, |p_{ij}|)$, or analogously, conditionalizing on pairs not tied on $y(B = \Sigma_{i,j} \, |q_{ij}|)$.

It is interesting to note that the expression in the denominator of Goodman-Kruskal's coefficient gamma, $\Sigma_{i,j} \, |p_{ij}q_{ij}|$, is not necessarily a bound on $\max_\rho \, \mathcal{A}(\rho)$; consequently, this latter measure is not of our usual form because the normalization is specified only with respect to the original orderings of $\underset{\sim}{P}$ and $\underset{\sim}{Q}$. Although we have not discussed this particular type of standardization before, it may be appropriate at this point to develop the basic idea in a little more detail, especially given the popularity of the Goodman-Kruskal coefficient.

We assume for convenience that the raw index is defined with the identity permutation ρ_I, and just as for the earlier measures $\mathcal{A}^*(\rho_I)$ or $\mathcal{A}^{**}(\rho_I)$, consider the deviation of $\mathcal{A}(\rho_I)$ from its expectation (at this point, $\underset{\sim}{P}$ and $\underset{\sim}{Q}$ are arbitrary). Because $\mathcal{A}(\rho_I) = \Sigma_{i,j} \, p_{ij}q_{ij}$ and $E(\mathcal{A}(\rho)) = 1/[n(n-1)] \times \Sigma_{i,j} \, p_{ij} \, \Sigma_{i,j} \, q_{ij}$, this leads immediately to a measure of the form

$$\mathcal{A}(\rho_I) - E(\mathcal{A}(\rho)) = \sum_{i,j} (p_{ij} - \bar{p})(q_{ij} - \bar{q}),$$

where

$$\bar{p} = \frac{1}{n(n-1)} \sum_{i,j} p_{ij} \quad \text{and} \quad \bar{q} = \frac{1}{n(n-1)} \sum_{i,j} q_{ij}$$

If this deviation index is denoted by $\mathcal{A}^\Delta(\rho_I)$, it can be decomposed. In particular, define

$$\mathcal{A}_+^\Delta(\rho_I) = \sum_{\substack{i,j \text{ such that} \\ (p_{ij} - \overline{p})(\overline{q}_{ij} - \overline{q}) > 0}} (p_{ij} - \overline{p})(q_{ij} - \overline{q})$$

$$\mathcal{A}_-^\Delta(\rho_I) = |\sum_{\substack{i,j \text{ such that} \\ (p_{ij} - \overline{p})(q_{ij} - \overline{q}) < 0}} (p_{ij} - \overline{p})(q_{ij} - \overline{q})|$$

Then,

$$\mathcal{A}^\Delta(\rho_I) = \mathcal{A}_+^\Delta(\rho_I) - \mathcal{A}_-^\Delta(\rho_I)$$

Thus, $\mathcal{A}_+^\Delta(\rho_I)$ is the sum of positive products of deviations, and $\mathcal{A}_-^\Delta(\rho_I)$ is the absolute value of the sum of negative products of deviations. Finally, if we desired a normalization of $\mathcal{A}^\Delta(\rho_I)$ with respect to the two matrices as they are given, a natural alternative would be

$$\frac{\mathcal{A}_+^\Delta(\rho_I) - \mathcal{A}_-^\Delta(\rho_I)}{\mathcal{A}_+^\Delta(\rho_I) + \mathcal{A}_-^\Delta(\rho_I)}$$

which could be written as

$$\frac{\sum_{i,j} (p_{ij} - \overline{p})(q_{ij} - \overline{q})}{\sum_{i,j} |(p_{ij} - \overline{p})(q_{ij} - \overline{q})|} \tag{4.3.2a}$$

For example, based on the n bivariate pairs, $(x_1, y_1), \ldots, (x_n, y_n)$, and the sign scoring functions for constructing the two skew-symmetric matrices $\underset{\sim}{P}$ and $\underset{\sim}{Q}$, this last measure in expression (4.3.2a) is Goodman-Kruskal's gamma. Because of the scoring functions in this application, this latter statistic has a nice probabilistic interpretation in terms of picking out concordant versus discordant pairs, i.e., the difference between the probability of concordance and discordance, conditional on pairs untied on either variable (see Reynolds, 1977). Nevertheless, the same general strategy could be used to obtain indices for the other two alternatives, or for that matter, for any of the applications that would use the raw QA cross-product.

The measure in expression (4.3.2a) has one major disadvantage. Because the denominator is not constant over all permutations, a reference distribution based on this normalized measure would not necessarily be a constant linear transformation of one based on the raw index (ρ). As such, the moments of (4.3.2a) cannot be found through our earlier formulas, and Because the normalization may vary from permutation to permutation, this measure, in general, must be interpreted with respect to the very specific way in which the $\underset{\sim}{P}$ and $\underset{\sim}{Q}$ matrices are arranged by the given permutation ρ_I.

Comparing Three Sequences. Given three sequences of matched observations, $\underset{\sim}{x} = \{x_1, \ldots, x_n\}$, $\underset{\sim}{y} = \{y_1, \ldots, y_n\}$, and $\underset{\sim}{z} = \{z_1, \ldots, z_n\}$, Section 3.4 developed a procedure for assessing whether $\underset{\sim}{x}$ is "closer" to $\underset{\sim}{y}$ than to $\underset{\sim}{z}$, based on a comparison of $\underset{\sim}{x}$ to $\underset{\sim}{y} - \underset{\sim}{z}$ by the LA model. Because it was tacitly assumed that y and z are initially commensurable (or are forced to be through a standardization), the difference between $\underset{\sim}{y}$ and $\underset{\sim}{z}$ can be given a legitimate interpretation. As another approach to this same task of evaluating dependent measures of association that demands no such initial commensurability and which fits rather naturally into a discussion of generalized correlation coefficients, suppose we define two skew-symmetric matrices $\underset{\sim}{P}$ and $\underset{\sim}{Q}$ as follows:

$$p_{ij} = \text{sign}(x_j - x_i)$$

$$q_{ij} = \begin{cases} +1 & \text{if } \text{sign}(y_j - y_i) = +1, \ \text{sign}(z_j - z_i) = -1 \\ 0 & \text{if } \text{sign}(y_j - y_i) \times \text{sign}(z_j - z_i) = +1 \text{ or } 0 \\ -1 & \text{if } \text{sign}(y_j - y_i) = -1, \ \text{sign}(z_j - z_i) = +1 \end{cases}$$

In words, q_{ij} is nonzero if and only if the order information provided by the two sequences $\underset{\sim}{y}$ and $\underset{\sim}{z}$ is inconsistent. These latter inconsistencies are then compared to the order information provided in the $\underset{\sim}{x}$ sequence. Large positive values in the cross-product index $\mathcal{A}(\rho_I)$ suggest that $\underset{\sim}{x}$ is closer to $\underset{\sim}{y}$ than to $\underset{\sim}{z}$; conversely, large negative values suggest that $\underset{\sim}{x}$ is closer to $\underset{\sim}{z}$ than to $\underset{\sim}{y}$. The raw index $\mathcal{A}(\rho_I)$ itself can be represented as a - b, where a (or b) refers to the number of ordered pairs for which the order information between $\underset{\sim}{x}$ and $\underset{\sim}{y}$ (or z) is consistent but inconsistent between $\underset{\sim}{x}$ and $\underset{\sim}{z}$ (or y). We proceed in the usual way to obtain a significance test and a normalization of the raw index which will have an operational meaning as the difference between two conditional probabilities.

In the geographical literature, in particular, there appears to be some confusion about the correct procedure for comparing dependent measures of association. As a general rule, it is not legitimate merely to ignore the dependency and treat the difference between two related measures with the same strategies that would be appropriate if they were independent. Unfortunately, this fact is not acknowledged in some current practice, e.g., see Court (1970) and Norcliffe (1977). Both of these authors, for instance, consider the variance of the difference between dependent (medial) correlations to be the sum of the variances for the original components with no corrections for the dependency. Tests of related measures can be constructed, but they require another approach, such as the one suggested above, based on the QA model, or possibly, the strategy of Section 3.4, using LA.

Temporal First Differences. Going back to the beginning of this section, suppose we again have two sequences $\{x_1, \ldots, x_n\}$ and $\{y_1, \ldots, y_n\}$. Now, however, the second sequence will be used to represent n observations along a time continuum ordered by the subscripts. If $p_{ij} = \text{sign}(x_i - x_j)$ as before, various restrictions on the second scoring function based on the sequence $\{y_1, \ldots, y_n\}$ lead to several measures for trend that have been considered in the time series context. For example, suppose $y_i = i$ and the sequence $\{y_1, \ldots, y_n\}$ merely represents the given time order along the continuum. Then,

$$q_{ij} = \begin{cases} \text{sign}(y_j - y_i), & |y_j - y_i| = 1 \\ 0 & \text{otherwise} \end{cases}$$

gives a raw measure $\mathcal{A}(\rho_I)$ defined as a − b, where a (or b) now refers to the number of ordered pairs that are temporally adjacent and $x_i < x_j$ (or $x_i > x_j$). This raw measure is equivalent to one considered by Moore and Wallis (1943). Somewhat more generally, suppose

$$q_{ij} = \begin{cases} \text{sign}(y_j - y_i), & |y_j - y_i| = d \\ 0 & \text{otherwise} \end{cases}$$

for some fixed d between 1 and n − 1. Based on this definition, we obtain a class of statistics equivalent to those considered by Cox and Stuart (1955). The index $\mathcal{A}(\rho_I)$ is again a − b, where a (or b) refers to the number of ordered pairs that are separated by d time intervals and $x_i < x_j$ (or $x_i > x_j$). The values of $d = (1/2)n$ and $(2/3)n$ are the most common in the nonparametric literature.

If the values in the sequence $\{y_1, \ldots, y_n\}$ represent observations ordered through time consistently with the subscripts, then the above definition of q_{ij} for d = 1 leads to a raw index $\mathcal{A}(\rho_I)$ defined by the number of ordered pairs that represent adjacent comovements minus the number of ordered pairs that represent adjacent contramovements between the two series. Here, a comovement occurs when one series decreases and the other increases. This index is similar to those studied by Goodman and Grunfeld (1961) for evaluating the association between two time series, although our inference model based on QA is different.

4.3.3 Simple Cluster Statistics (B)

<u>Compactness and Isolation</u>. In an unpublished paper, Johnson (1968a) developed a statistic for evaluating the distinctness of a single subset or cluster of objects that has since found some acceptance in the substantive literature (e.g., see Fillenbaum and Rapoport, 1971). Johnson's argument is as follows: Suppose, as always, that S is a set of n objects, $\{O_1, \ldots, O_n\}$, and $\underset{\sim}{P}$ represents the matrix of proximities between pairs of elements in S. To each subset or cluster D of S containing K elements, $K \geq 2$, some statistic is defined that measures the "compactness" and/or "isolation" of the subset. Under the assumption that the subset D was drawn at random from all possible K-element subsets of S, Johnson obtained the mean and variance of one particular measure defined as the average of the K(K - 1) proximities within D (which assesses compactness) minus the average of the 2K(n - K) proximities between the objects in D and S-D (which assesses separation or isolation).

The index used by Johnson, as well as a number of other alternatives, can be obtained as special cases of the QA model. For example, if

$$
q_{ij} = \begin{cases}
1/[K(K-1)] & 1 \leq i \neq j \leq K \\
-1/[2K(n-K)] & \begin{array}{l} 1 \leq i \leq K \text{ and } K+1 \leq j \leq n \text{ or} \\ 1 \leq j \leq K \text{ and } K+1 \leq i \leq n \end{array} \\
0 & \text{otherwise}
\end{cases}
$$

then $\mathcal{A}(\rho_0)$ is equal to Johnson's index for the subset D defined by those objects $O_{i_1}, O_{i_2}, \ldots, O_{i_K}$ such that $\rho_0(1) = i_1, \ldots, \rho_0(K) = i_K$. Moreover, the distribution of $\mathcal{A}(\rho)$ over all n! equally likely permutations is the same as that which would be obtained under the hypothesis that all $\binom{n}{K}$ subsets are equally likely a priori. More specifically, for every possible subset that could be constructed from S, K!(n - K)! different permutations would lead to it. Thus, each subset has a probability of $K!(n-K)!/n = 1/\binom{n}{K}$ of occurring when the n! possible permutations are considered equally likely. We note, in particular, that under this assumption, $E(\mathcal{A}(\rho)) = 0$ for Johnson's version of the cluster index.

It should be obvious that other cluster measures could be defined merely by varying the definition of $\underset{\sim}{Q}$. In all cases, they are intended to test the a priori conjecture that a particular subset should display an unusual degree of compactness and/or isolation as reflected by the chosen index. Johnson's statistic combines a measure of the internal cohesion of a subset D through the average within-cluster proximity with a measure of how isolated D is from its complement. The latter isolation component is defined by the average proximity between D and S-D. Thus, Johnson's index can be extreme when D is internally compact and/or not very well connected to its complement, or conversely, when D is well connected to its complement and/or not very internally compact. The former notion is probably the most common interpretation, and is reflected by a negative difference if small proximities denote similar objects.

Because these two concepts of compactness and isolation are conceptually different, they could be evaluated separately. For instance, one obvious alternative for assessing compactness alone, studied by Bloemena (1964), is obtained very simply as the sum of the $K(K - 1)$ within-cluster proximities. Here, $\underset{\sim}{Q}$ is defined as

$$q_{ij} = \begin{cases} 1 & \text{if } 1 \leq i \neq j \leq K \\ 0 & \text{otherwise} \end{cases}$$

Using this specification, Bloemena (1964) has discussed asymptotic normality of $\mathcal{A}(\rho)$ when K is not extreme in size, i.e., when K is either too small or too large. Clearly, if this index is modified slightly by redefining q_{ij} as $1/[K(K - 1)]$ when $1 \leq i \neq j \leq K$, then $\mathcal{A}(\rho_0)$ is the average proximity within the subset generated by ρ_0 and is part of Johnson's original measure. By analogy, the separation of D from its complement could be assessed with the second component of Johnson's statistic. Here, we define q_{ij} as $1/[2K(n - K)]$ for $1 \leq i \leq K$ and $K + 1 \leq j \leq n$ or $1 \leq j \leq K$ and $K + 1 \leq i \leq n$, and 0 otherwise.

Thus far, all of the specifications of $\underset{\sim}{Q}$ have been symmetric; consequently, without loss of generality, $\underset{\sim}{P}$ could be assumed symmetric as well. When the proximities are asymmetric, however, the isolation index itself can be split into two parts: the average proximity from S to S-D and the average proximity from S-D to S. These measures could be constructed, respectively, by letting (i) $q_{ij} = 1/[K(n - K)]$ for $1 \leq i \leq K$ and $K + 1 \leq j \leq n$, and 0 otherwise, or (ii) $q_{ij} = 1/[K(n - K)]$ for $1 \leq j \leq K$ and $K + 1 \leq i \leq n$, and 0 otherwise.

In these latter cases, $\underset{\sim}{Q}$ is not symmetric; therefore, if $\underset{\sim}{P}$ is also not symmetric, it is of some interest to note what questions are being asked in terms of the decomposition of $\mathcal{A}(\rho_0)$ into $\mathcal{A}^+(\rho_0)$ and $\mathcal{A}^-(\rho_0)$. Specifically, the index $\mathcal{A}^+(\rho_0)$ is a measure of subset isolation based on the symmetric matrix $\underset{\sim}{P}^+ = \{(p_{ij} + p_{ji})/2\}$; starting either with (i) or (ii), $\underset{\sim}{Q}^+$ would be exactly the same matrix for obtaining the isolation component of Johnson's statistic. The index $\mathcal{A}^-(\rho_0)$ uses $\underset{\sim}{P}^- = \{(p_{ij} - p_{ji})/2\}$, and, if $\underset{\sim}{Q}$ is given as in (i), then

$$q_{ij}^- = \begin{cases} \dfrac{1}{2K(n - K)} & 1 \leq i \leq K \text{ and } K + 1 \leq j \leq n \\[3mm] \dfrac{-1}{2K(n - K)} & 1 \leq j \leq K \text{ and } K + 1 \leq i \leq n \end{cases}$$

[If Q were given as in (ii), the signs would merely be reversed in this definition.] For an interpretation, suppose p_{ij} represents the amount of "flow" from O_i to O_j. Positive values for $\mathcal{A}^-(\rho_0)$ would indicate an aggregate flow out of the subset D that is larger than the aggregate flow into D; negative values would suggest the opposite. The index $\mathcal{A}^+(\rho_0)$ on the other hand, is concerned with indexing the magnitude of any mutual interaction between D and its complement, irrespective of the direction of flow. Thus, the composite index $\mathcal{A}(\rho_0)$ defined, say, through (i), is a measure of outflow from D to its complement that consists of two parts: the magnitude of mutual flow between D and S-D as given by $\mathcal{A}^+(\rho_0)$, and the degree to which D loses more than it gains as given by $\mathcal{A}^-(\rho_0)$.

In any event, when using cluster statistics descriptively, it may be appropriate to report directly the raw measures without defining normalizations of the form $\mathcal{A}^*(\rho_0)$ or $\mathcal{A}^{**}(\rho_0)$. When the proximities have an inherent meaning and even though normalized indices could be obtained, the use of simple sums, averages, or differences between averages may be sufficient to communicate the notion of compactness and/or isolation by themselves.

<u>Two-Independent Sample Problems</u>. One simple application of this separation notion is to the classical two-independent sample problem encountered in Section 1.3.4. Here, n observations x_1, x_2, ..., x_n are available, where the first K belong to group I and the last n - K belong to group II. If

$$p_{ij} = \begin{cases} 1 & \text{if } x_j < x_i \\ 0 & \text{if } x_j \geq x_i \end{cases}$$

and $q_{ij} = 1$ for $1 \leq i \leq K$ and $K + 1 \leq j \leq n$, then $\mathcal{A}(\rho_I)$ is the well-known Mann-Whitney statistic; furthermore, the QA inference model provides the usual significance test developed in most nonparametric tests, e.g., Gibbons (1971). When $q_{ij} = 1/[K(n - K)]$ for these same index restrictions and one observation is drawn from group I and one from group II at random, $\mathcal{A}(\rho_I)$ is the probability that the group I observation is larger. This is the same measure discussed in Section 1.3.4.

The two-independent sample application can be extended in a variety of ways and to situations that may not simply depend on univariate information. For instance, suppose our data are a set of n vectors $\underset{\sim}{x}_1$, ..., $\underset{\sim}{x}_n$, where $\underset{\sim}{x}_i$ contains p observations on the object O_i. If the proximity p_{ij} is defined in some fashion from $\underset{\sim}{x}_i$ and $\underset{\sim}{x}_j$, then the isolation measure, in effect can be used to obtain a multivariate two-sample test based on this specification. For instance, following Friedman and Rafsky (1979), a euclidean distance measure can first be obtained between every object pair based on their corresponding observation vectors. Then, treating the proximities as if they define a weighted graph, the minimal spanning tree is obtained among the n objects, i.e., the set of n - 1 proximities that generate a connected graph and for which the sum of weights is at a minimum (see Kruskal, 1956, for a very efficient algorithm). Suppose p_{ij} is now defined as 1 if O_i and O_j give an edge in the minimal spanning tree and 0 otherwise, and let q_{ij} be the same as above for the Mann-Whitney example. The index $\mathcal{A}(\rho_I)$ gives the number of edges in the minimal spanning tree between groups I and II. If we add one, this is the statistic considered by Friedman and Rafsky (1979). Obviously, a variety of other alternatives could be generated in essentially the same way, merely by defining proximity in a different manner.

Roll-call Cohesion. It may be helpful at this point to mention one other possible use of the subset compactness notion that is of particular interest in political science. Measures of legislative roll-call cohesion have been widely used in the political science literature to index the degree of solidarity manifested by a definable subgroup of voters (see MacRae, 1970, for a review). Unfortunately, the most popular measures have also been plagued by the lack of inference models that can take into account the degree of cohesion present in the larger unit containing the subgroup. For example, Rice's well-known measure of subgroup cohesion is defined (up to a multiplicative constant) by the absolute difference between the proportions of a positive and a negative vote. Chance cohesion is supposedly indicated by a 50-50 split (i.e., each proportion has a value of $1/2$), and deviations from the 50-50 split are interpreted as evidence for an increased cohesion among the subgroup members. This equally likely baseline is assumed, even though the complete population may not have a perfect 50-50 split. Consequently, it is possible for the larger population to demonstrate more cohesion than the subgroup, even though the latter by itself may appear to be highly cohesive when evaluated against the null assumption of a 50-50-split. This difficulty would be analogous to performing goodness-of-fit tests in contingency tables if marginal information were ignored and expected values were obtained from an equally likely assumption for all the cell probabilities.

By appropriately defining proximity, it is relatively straightforward to develop a suitable inference method for a measure of cohension through the well-known percentage voting agreement measure. We assume that a body of n individuals defines the set S and each member has voted on an issue. The cohesion of a subgroup D of size K is of interest. If there are t voting options, then the percentage voting agreement is defined (Born and Nevison, 1975) as

$$\sum_{j=1}^{t} \frac{m_j(m_j - 1)}{K(K-1)}$$

where m_j is the number of individuals in S using voting alternative j, $1 \leq j \leq t$, and $\sum_{j=1}^{t} m_j = K$. Thus, the percentage voting agreement measure is the ratio of the number of pairs of individuals in D who vote in the same way to the total number of pairs in D. If the proximity p_{ij} is defined as 1 when individuals O_i and O_j vote in the same way and 0 otherwise, and $q_{ij} = 1/[m(m - 1)]$, for $1 \leq i \neq j \leq K$ and 0 otherwise, then $\mathcal{A}(\rho_0)$ defines the percentage voting agreement measure for the subgroup defined by ρ_0. This descriptive index, mentioned earlier, is just the average of the proximities within the subgroup of interest. Furthermore, the QA model compares the degree of cohesion for the subgroup to the degree of cohesion manifested in the entire body of n individuals. Stated in other words, the percentage voting agreement measure can be evaluated directly against what could be expected from the distribution of votes in the entire population. If we reject the conjecture that D was constructed at random, say, by observing a large measure, the obvious substantive implication is that the subgroup has some voting cohesion over and above what could be expected merely by forming a group of the same size at random from the larger population.

Although we have phrased the notion of proximity as a zero-one function in this legislative cohesion application, it should be clear that P could be defined more generally as well. For instance, p_{ij} could denote the number of issues on which O_i and O_j vote the same. Subset cohesion could again be defined in the same way as before through the average of all within-subset proximities (see Hubert and Golledge, 1981a).

<u>2 × t Contingency Tables</u>. The zero-one definition of proximity based on t nominal classes is a general tactic that has applications in areas other than legislative cohesion. The within-subset compactness measures, based on a sum or average, are linear functions of $\sum_{j=1}^{t} m_j^2$, where m_j denotes the number of instances of category j that appear in the chosen subset D, e.g., the sum can be rewritten as $\sum_{j=1}^{t} m_j^2 - K$. Thus the more disproportionate the observed category frequencies are, the larger the absolute size of the chosen compactness measure. In fact, it is conceivable that there are instances where decreased cohesion may be of interest to assess; in the nominal case, this would occur when the category instances are as equal as possible. In any event, the relative size of an observed index would be evaluated with respect to the distribution of category frequencies in the total set S.

A somewhat analogous discussion holds for a separation measure based on the 0-1 proximities from, say, D to S-D. These separation indices are linear functions of $\sum_{j=1}^{t} m_j m_j'$, where m_j and m_j' denote the number of instances of category j that appear in D and S-D, respectively; for instance, the sum of proximities from D to S-D is exactly of this latter form. Consequently, large separation measures indicate that m_j and m_j' tend to be of similar size, and small separation measures indicate the opposite.

The discussion of separation or isolation measures based on 0–1 proximity data can be clarified further by rephrasing the problem in terms of a $2 \times t$ contingency table having the form

	1	2	$\cdots$	t	
I	m_1	m_2	$\cdots$	m_t	K
II	$m_1' = $ $M_1 - m_1$	$m_2' = $ $M_2 - m_2$	$\cdots$	$m_t' = $ $M_t - m_t$	n–K
	M_1	M_2	$\cdots$	M_t	n

Here, I and II refer to membership in D and S–D, respectively, and M_j denotes the number of total instances of category j present in S. In effect, the isolation index, $\Sigma_j \, m_j m_j'$, is a measure of association in this $2 \times t$ contingency table, even though it is not the usual chi-square statistic; the latter can be represented as (Lehmann, 1975):

$$\chi^2 = \left[\frac{n^2}{(n-K)K} \left(\sum_j \frac{m_j^2}{M_j} \right) \right] - \frac{nK}{n-K}$$

However, we can obtain this familiar statistic, at least up to a linear transformation, by one simple variation in how proximity is defined in $\underset{\sim}{P}$. Instead of 0–1 values, suppose we define

$$p_{ij} = \begin{cases} \dfrac{1}{M_j} & \text{if } O_i \text{ and } O_j \text{ are distinct and belong} \\ & \text{to category } j \\ 0 & \text{otherwise} \end{cases}$$

Then, for the subset D defining the first row of the contingency table, the QA index is

$$\sum_j \frac{m_j(M_j - m_j)}{M_j} = K - \sum_j \frac{m_j^2}{M_j}$$

Table 4.3.3a Rothkopf Proximity Matrix Between the Morse Code Symbols

	A	B	C	D	E	F	G	H	I	J	K	L	M	N	O	P	Q	R	S	T	U	V	W	X	Y	Z	1	2	3	4	5	6	7	8	9	0
A	92	04	06	13	03	14	10	13	46	05	22	03	25	34	06	06	09	35	23	06	37	13	17	12	07	03	02	07	05	05	08	06	05	06	02	03
B	05	84	37	31	05	28	17	21	05	19	34	40	06	10	12	22	25	16	18	02	18	34	08	84	30	42	12	17	14	40	32	74	43	17	04	04
C	04	38	87	17	04	29	13	07	11	19	24	35	14	03	09	51	34	24	14	06	17	11	14	32	82	38	13	15	31	14	10	30	28	24	18	12
D	08	62	17	88	07	23	40	36	09	13	81	56	08	07	09	27	09	45	29	06	17	20	27	40	15	33	03	09	06	03	05	02	04	02	03	06
E	06	13	14	06	97	02	04	04	17	01	05	06	04	04	05	01	05	10	07	67	03	03	02	05	06	05	04	03	05	03	05	02	04	02	03	03
F	04	51	33	19	02	90	10	29	05	33	16	50	07	06	10	42	12	35	14	02	21	27	25	19	27	13	08	16	47	25	26	24	21	05	20	05
G	09	18	27	38	01	14	90	06	05	22	33	16	14	13	62	52	23	21	05	03	15	14	32	21	23	39	15	14	05	10	04	10	17	23	03	11
H	03	45	23	25	09	32	08	87	10	10	09	29	05	08	08	14	08	17	37	04	36	59	09	33	14	11	03	09	15	43	70	35	17	04	03	03
I	64	07	07	13	10	08	06	12	93	03	05	16	13	30	07	03	05	19	35	16	10	05	08	02	05	07	02	05	08	09	06	08	05	02	04	05
J	07	09	38	09	02	24	18	05	04	85	22	31	08	03	21	63	47	11	02	07	09	09	09	22	32	28	67	66	33	15	07	11	28	29	26	23
K	05	24	38	73	01	17	25	11	05	27	91	33	10	12	31	14	31	22	02	02	23	17	33	63	16	18	05	09	17	06	08	18	14	13	05	06
L	02	69	43	45	10	24	12	26	09	30	27	86	06	02	09	37	36	28	12	05	06	19	20	31	25	59	12	13	17	15	26	29	36	16	07	03
M	24	12	05	14	07	17	29	08	08	11	23	08	96	62	11	10	15	20	07	09	13	04	21	09	18	08	05	07	06	06	05	07	11	07	10	04
N	31	04	13	30	08	12	10	16	13	03	16	08	59	93	05	09	05	28	12	10	16	04	12	04	06	11	05	02	03	04	04	06	02	02	10	02
O	07	07	20	06	05	09	76	07	02	39	26	10	04	08	86	37	35	10	03	04	11	14	25	35	27	27	19	17	07	07	06	18	14	24	20	12
P	05	22	33	12	05	36	22	12	03	78	14	46	05	06	21	83	43	23	09	04	12	19	19	41	50	30	34	44	24	11	15	17	24	23	25	13
Q	08	20	38	11	04	15	10	05	02	27	23	26	07	06	22	51	91	11	02	03	06	14	12	37	50	63	34	32	17	12	09	27	40	58	37	24

R	13	14	16	23	05	34	26	15	07	12	21	33	14	12	29	08	87	16	02	23	23	62	14	12	13	07	10	13	04	07	12	07	09	01	02
S	17	24	05	30	11	26	05	59	16	03	13	10	05	06	11	04	18	96	09	56	24	12	10	06	07	08	02	02	15	28	09	05	05	05	02
T	13	10	01	05	46	03	06	06	14	06	14	07	06	05	20	04	04	08	96	08	05	04	05	02	06	05	05	03	03	03	08	07	06	14	06
U	14	29	12	32	04	32	11	34	21	07	44	32	11	13	06	11	40	07	06	93	57	34	17	09	11	06	06	16	34	10	09	09	07	04	03
V	05	17	24	16	09	29	06	39	05	11	26	43	04	01	06	20	51	06	32	06	17	92	57	35	17	10	22	14	28	79	44	25	11	01	05
W	09	21	30	22	09	36	25	15	04	25	29	18	15	06	26	17	11	06	04	32	20	17	22	25	10	22	12	28	19	16	05	11	06	03	07
X	07	64	45	19	03	28	11	06	01	35	42	10	06	08	26	10	12	03	19	04	32	17	91	48	26	12	24	20	24	27	16	05	29	17	06
Y	09	23	62	15	04	26	22	17	10	07	30	14	08	24	14	52	05	04	03	09	11	21	44	86	23	26	21	44	40	15	11	26	22	23	16
Z	03	46	45	18	02	22	17	10	07	23	21	51	15	02	15	72	14	04	03	09	09	11	36	42	87	23	16	10	27	09	10	25	66	47	15
1	02	05	10	03	03	05	13	04	02	29	05	14	07	03	07	30	09	04	02	03	12	14	12	19	17	84	22	63	13	08	10	08	19	57	55
2	07	14	22	05	04	20	13	03	25	26	09	14	02	17	28	37	28	06	05	03	06	17	10	30	17	13	62	89	54	20	05	14	20	16	11
3	03	08	21	05	05	32	06	12	02	23	14	13	17	05	19	37	19	07	06	04	16	55	45	25	22	12	18	64	86	31	23	41	17	08	10
4	06	19	19	12	08	25	14	16	07	21	13	19	03	02	29	37	09	03	09	17	55	08	37	24	03	05	26	44	89	42	44	32	10	03	03
5	08	45	15	14	02	45	04	67	07	14	04	41	00	04	13	07	27	02	09	14	45	07	45	10	09	58	10	14	69	90	42	24	10	06	05
6	07	80	30	17	04	23	04	02	07	11	11	27	06	02	07	16	14	03	12	03	30	14	30	03	13	38	15	26	24	17	88	69	85	05	14
7	06	33	22	14	05	25	06	06	02	24	13	32	07	07	36	30	39	12	06	02	36	13	50	13	09	30	18	29	16	15	12	61	70	20	13
8	03	23	40	06	03	15	15	06	02	33	10	14	06	07	14	45	02	06	04	06	14	07	05	07	13	35	24	29	16	16	09	30	60	89	26
9	03	14	23	03	01	06	14	05	02	30	06	07	11	11	31	32	05	07	06	03	06	03	08	06	03	21	57	39	09	12	04	11	42	56	78
0	09	03	11	02	05	07	14	04	04	05	30	08	03	02	25	21	29	02	03	04	05	03	02	12	15	26	09	26	09	11	05	22	17	52	94

This statistic is a constant linear transformation of x^2; moreover, our QA inference model generates a reference distribution for it that is equivalent to the use of the hypergeometric model for constructing the $2 \times t$ contingency table. In short, a simple change in the definition of proximity leads to the familiar chi-square strategy for testing association in a $2 \times t$ contingency table conditional on fixed marginals. An extension of this equivalence to larger tables is given in Section 4.3.7.

Examples. As an illustration of how a subset could be assessed for compactness and isolation, Table 4.3.3a gives a classic confusion matrix collected by Rothkopf (1957) that has been subjected to a number of reanalyses, e.g., see Shepard (1963) and Hubert (1974). The data are a set of proportions obtained from a group of experimental subjects that quantify the confusability of the 36 Morse code signals; the legend is given in Table 4.3.3b. An entry in the asymmetric 36×36 matrix corresponds to the proportion of subjects who stated that the corresponding row and column stimuli were the "same" when the row stimulus was presented first and the column stimulus immediately thereafter. The "self-confusion" proportions reported by Rothkopf are given in Table 4.3.3a but will not be used in our analysis.

To give an explicit example of the results obtainable by using the cluster indices we have discussed, suppose D is the set of eight letters that correspond to Morse code signals with three symbols, i.e.,

$$D = \{D,G,K,O,R,S,U,W\}$$

We might expect that the number of symbols is a very salient cue, and therefore signals having a common number of components would be more confusable with each other and less so with signals that are either longer or shorter.

The results of three different analyses are given below; each is concerned with assessing the compactness and/or isolation of D with respect to the complete 36×36 matrix of Table 4.3.3a. In addition, Table 4.3.3c provides Monte Carlo distributions for each of the measures based on a sample size of 999.

	Johnson's index	Average proximity within D	Average proximity between D and S-D
Observed index	13.77	27.77	14.00
Expectation	.00	17.64	17.64
Variance	10.65	14.17	1.536
Z	4.22	2.69	-2.94
γ	.820	.646	-.695
Significance level			(lower tail)
Monte Carlo	.002	.014	.011
Type III approximation	.001	.012	.008

Table 4.3.3b Morse Code Symbol Correspondence for the
Letters and Digits

Letter/digit	Symbol	Letter/digit	Symbol
A	.−	S	...
B	−...	T	−
C	−.−.	U	..−
D	−..	V	...−
E	.	W	.−−
F	..−.	X	−..−
G	−−.	Y	−.−−
H	....	Z	−−..
I	..	1	.−−−−
J	.−−−	2	..−−−
K	−.−	3	...−−
L	.−..	4	....−
M	−−	5	.....
N	−.	6	−....
O	−−−	7	−−...
P	.−−.	8	−−−..
Q	−−.−	9	−−−−.
R	.−.	0	−−−−−

In all cases, the subset D can be considered special in the sense that it is relatively compact and isolated from its complement, i.e., there is a tendency for confusion to be greater between symbols that contain three elements, and at the same time, these symbols tend to be less confusable with others. Obviously, other subsets could be evaluated in a similar fashion.

4.3.4 Analysis of Variance Procedures Based on Proximity: Comparisons of a Partition to a Proximity Matrix (B)

In certain areas of the behavioral sciences, and particularly in cognitive psychology and related fields, rather complex experimental tasks are now

Table 4.3.3c   Monte Carlo Distributions for the Three Cluster
               Indices Based on a Subset of Size 8 and Table
               4.3.3a (Sample Sizes of 999)

Cumulative frequency	Johnson's index	Average proximity within D	Average proximity between D and S - D
1	-7.3	9.4	12.6
5	-6.3	10.3	13.6
10	-6.0	10.5	14.0
50	-4.6	12.1	15.3
100	-3.7	13.1	15.9
200	-2.7	14.4	16.6
300	-1.9	15.5	17.0
400	-1.0	16.4	17.4
500	- .2	17.3	17.8
600	.5	18.2	18.1
700	1.3	19.4	18.4
800	2.5	20.6	18.7
900	4.7	23.0	19.1
950	6.2	24.5	19.4
990	9.6	28.7	19.9
995	11.5	30.1	20.0
999	14.5	31.6	20.3

being asked of subjects. In fact, researchers may well be collecting data
that cannot be analyzed conveniently with traditional statistical inference
procedures. Analysis of variance and its extensions to more comprehensive
regression models are obviously useful when data come in the form of
numerical variables associated with individual subjects. However, if the
observations being collected are more complex or structured, for example,
as paired comparisons, rankings, item sorts into ordered or unordered
categories, and so on, then the usual statistical methods may no longer be
appropriate. Given this shortcoming and using QA, we can develop a

nonparametric analog to the typical analysis-of-variance procedures for K independent samples that may be used in such situations. No matter what data are available on a subject, simple analysis-of-variance techniques can be applied if some numerical measure of correspondence can be defined between any two subjects.

To provide a somewhat broader context, our discussion will be phrased in terms of generic objects rather than subjects. In effect, we generalize the approach described in the last section for evaluating the compactness of a single cluster of objects to include multiple clusters, and in the process, provide a K-independent sample parallel to the K-dependent sample strategy presented in Section 3.2.3 on the analysis of cross-classified proximity data.

At the outset, it is convenient to assume that each of n objects (e.g., subjects) provides information, in some form or other, about N items; the objects themselves are classified using a one-way categorization defined by K classes containing n_1, n_2, ..., n_K members; thus, $\Sigma_{k=1}^{K} n_k = n$. For instance, if the items are tests, attributes, or so on, then standard parametric inference strategies rely on single numerical variables attached to each item. The a priori classification of the objects defines a between-object factor that is usually tested with a repeated-measures or multivariate analysis-of-variance strategy. Our interest is in evaluating this same type of between-object distinction but in a context that would allow the analysis of more complex relational information among the items which may or may not be reducible to a single vector of N numbers. The approach to be reviewed requires an initial definition of an n × n proximity matrix $\underset{\sim}{P}$ that specifies the correspondence between every pair of objects; the entries in this matrix are based on the data associated with each object over the N items. The between-object factor is then evaluated using the pattern of proximities in $\underset{\sim}{P}$.

In the simplest case we consider, there are n objects categorized into K classes, and only a single numerical variable is available on each, i.e., the number of items, N, is 1. The randomization approach (see Bradley, 1968) for evaluating the salience of the a priori categorization is based on some measure of difference between the K classes (for instance, the standard F ratio) and compares this observed statistic to the distribution generated under the randomness assumption that all $n!/n_1!...n_K!$ ways of allocating objects to the K groups are equally likely. The proportion of such allocations giving values of the measure as extreme or more so than the observed index then defines the level of significance.

The measure chosen to reflect the difference between the K classes is arbitrary to some extent. For example, even if the usual F ratio were selected, any other statistic could be used that was strictly monotone with it over all ways of allocating observations to the K groups. Such a statistic would produce the same significance level. Furthermore, because the total sum of squares is constant over all allocations and can be decomposed

into a between and within sum of squares, these latter two statistics must
be perfectly related in an inverse monotone fashion. Thus, the F ratio
itself is perfectly monotone with respect to each of these statistics, and
either the between or within sum-of-squares could serve as the measure of
difference between the K classes. As a computational convenience, the
within sum of squares will be selected as the measure of group difference;
obviously, this statistic should be small if group differences exist because
a relatively small value for the within sum of squares is equivalent to
either the F ratio or the between sum of squares being relatively large.

There is an alternative way to calculate the within-sum-of-squares
test statistic from an n × n proximity matrix among the n objects, which
will suggest a useful generalization within the QA framework to an arbitrary
proximity matrix. The latter could be constructed from more than the
usual type of univariate data. In the univariate context and between every
pair of objects u_r and v_s, with u_r in group r and v_s in group s, a measure
is defined as $(X_{u_r} - X_{v_s})^2$, where X_{u_r} and X_{v_s} refer to the numerical values
attached to the two objects, respectively.[†] The n × n proximity matrix P
can be represented schematically as in Table 4.3.4a. Based on the entries
in P, or more specifically on the square blocks of entries along the left-to-
right main diagonal of Table 4.3.4a, the within sum of squares can be defined as

$$\sum_{k=1}^{K} c_k \sum_{u_k < v_k} (X_{u_k} - X_{v_k})^2$$

(4.3.4a)

where $c_k = 1/n_k$, i.e., a sum of weighted within-group proximities. Each
simultaneous permutation of the rows and columns of P generates one of
the $n!/n_1! \ldots n_K!$ possible allocations of the n objects to the K groups. In
fact, because each such allocation corresponds to the same number of possible permutations, i.e., $n_1! \ldots n_K!$, a random reordering of the rows and
columns of P induces a random allocation of the n objects among the K cells.

[†] If ranks were used instead of the original scores, we could present this
discussion as part of what is usually called "Kruskal-Wallis analysis of
variance by ranks." In this case, we would use the between sum of squares
rather than the within. If the K categories were ordered, the analysis-of-
variance task would really be part of the discussion of generalized correlation coefficients in Section 4.3.2; for instance, Jonckheere's test compares
the observations in the K classes to a second variable defined from the
ordering imposed on the categories using Kendall's coefficient of rank correlation, tau.

This same procedure can be characterized within the QA model using $\underset{\sim}{P}$ as the (data) proximity matrix and Q as the (structure) matrix of Table 4.3.4b. If $p_{u_r v_s}$ now denotes some arbitrary symmetric proximity between object u_r in group r and v_s in group s, which replaces the entry $(X_{u_r} - X_{v_s})^2$ in Table 4.3.4a, one-half of the QA statistic $\mathcal{A}(\rho_I)$ can be represented in the form

$$\sum_{k=1}^{K} c_k \sum_{u_k < v_k} p_{u_k v_k} \qquad (4.3.4b)$$

This descriptive measure indexes the degree to which the partition structure represented in Q is reflected in $\underset{\sim}{P}$ and is defined by a weighted sum of within-class proximities. (Because $\underset{\sim}{Q}$ is symmetric, $\underset{\sim}{P}$ can also be considered symmetric without loss of generality, as discussed in Section 4.2.2).

The statistic in expression (4.3.4b) weights the sum of within-group proximities by $1/n_k$ and is an obvious measure for assessing an a priori categorization in the univariate one-way framework. Other alternatives could be considered, but for a definition other than $c_k = 1/n_k$, the monotone relationship to the F ratio in the univariate context would be lost. In general, two other natural weightings of the sum of within-group proximities would be: (i) $c_k = 1$, in which the sizes of the classes contribute to the index in direct proportion to the number of objects pairs they contain; (ii) $c_k = 1/\binom{n_k}{2}$, in which each object class contributes equally, irrespective of size.[†] The weighting used in expression (4.3.4a), i.e., $c_k = 1/n_k$, can be considered to be somewhere in between these other two alternatives because the classes contribute in direct proportion to the number of objects they contain [and not in direct proportion to the number of pairs as in (i)]. Irrespective of the weighting used, the QA strategy remains the same except for the specific index chosen to characterize group differences. (When univariate data are being considered, the usual computational form for mean square within is much more efficient than a direct use of expression (4.3.4a) in terms of the number of summing operations required. Consequently, for the simple univariate problem, the usual computational form for mean square within should be used and expression (4.3.4a) bypassed altogether, especially when Monte Carlo significance levels are to be obtained by repeated calculation of the within-sum-of-squares statistic.)

[†] It may be of some interest to note that for this weighting option, the expectation of the index in expression (4.3.4b) is independent of the class sizes $n_1, \ldots, n_K$.

Table 4.3.4a Schematic Form for the Proximity Matrix $\underset{\sim}{P}$ Used to Obtain the Univariate Within–Sum–of–Squares

	Group			
	1	2	$\cdots$	K
Group 1	$-(X_{u_1|}-X_{|v_1})^2-$	$-(X_{u_1|}-X_{|v_2})^2-$	$\cdots$	$-(X_{u_1|}-X_{|v_k})^2-$
2	$-(X_{u_2|}-X_{|v_1})^2-$	$-(X_{u_2|}-X_{|v_2})^2-$	$\cdots$	$-(X_{u_2|}-X_{|v_k})^2-$
$\vdots$	$\cdots$	$\cdots$	$\vdots$	$\cdots$
K	$-(X_{u_k|}-X_{|v_1})^2-$	$-(X_{u_k|}-X_{|v_2})^2-$	$\cdots$	$-(X_{u_k|}-X_{|v_k})^2-$

Whenever the symbol " $-|-$ " is used in a table, we imply that the enclosed entry is to be treated generically and defines a submatrix.

Table 4.3.4b Schematic Form for the Matrix $\underset{\sim}{Q}$ Used to Obtain
the Weighted Sum of Within-Group Proximities

		1	2	$\cdot\ \cdot\ \cdot$	K
	1	$- c_1 -$	$- 0 -$	$\cdot\ \cdot\ \cdot$	$- 0 -$
	2	$- 0 -$	$- c_2 -$	$\cdot\ \cdot\ \cdot$	$- 0 -$
Group	$\cdot$ $\cdot$ $\cdot$	$\cdot$ $\cdot$ $\cdot$	$\cdot$ $\cdot$ $\cdot$		$\cdot$ $\cdot$ $\cdot$
	K	$- 0 -$	$- 0 -$	$\cdot\ \cdot\ \cdot$	$- c_k -$

Within the K on-diagonal subblocks, the main diagonals are still
all zero.

When all the cell sizes $n_1, \ldots, n_K$ are equal, the three weighting
schemes mentioned above lead to indices that are simple multiples of each
other. Thus, the significance levels will be the same, irrespective of the
particular choice of weights. Unfortunately, as noted by Mielke and his
colleagues in a number of publications, the use of asymptotic normality is
not justified when the cell sizes are equal or when the weighting scheme of
$c_k = 1/n_k$ is used (also, see Ascher, 1980; Ascher and Bailar, 1982). Other
options must be used, e.g., Monte Carlo significance testing or approximations based on more than the first two moments (Mielke, 1979). When
all cell sizes are unequal, the use of the weighting options in (i) and (ii), as
compared to $c_k = 1/n_k$, lead to reasonable conditions that justify asymptotic
normality. Although Monte Carlo significance testing or approximations by
more than the first two moments may still be preferable for accuracy, the
use of an asymptotic normal distribution may also prove possible, given
certain regularity properties for the proximities (see O'Reilly and Mielke,
1980). We do note, however, that the use of the weighting scheme of

$c_k = 1/n_k$ is preferred by Mielke and his colleagues because of an asymptotic variance of order $1/n^2$ as compared to $1/n$ for the other two options.[†]

Proximities Between Objects. The strategy just described for evaluating the salience of a K-independent sample distinction requires the initial definition of a matrix $\underset{\sim}{P}$ that contains measures of proximity between every pair of objects. In the randomization approach to univariate one-way analysis of variance, the entries in $\underset{\sim}{P}$ are specified as the squared differences of single numerical values. More generally, however, the same type of randomization strategy can be applied to $\underset{\sim}{P}$ by use of the index in expression (4.3.4b), irrespective of what the proximities actually are. For example, if the objects are subjects and they provide response patterns of some form over a set of N items, the entries in $\underset{\sim}{P}$ could be based on a measure of proximity that indexes the similarity between the patterns. A test of the K-independent sample distinction then assesses whether response patterns within groups are more similar than they are between groups. As one very simple illustration, suppose each subject provides a vector of N numerical responses, and the concern is with whether the patterning of values defining these vectors is more similar for subjects within the same groups. If the measure of proximity is specified very simply as the usual Pearson correlation (which is appropriate for evaluating relative patterning of high and low entries between two vectors), we may conjecture that the subject proximities within groups will be higher than between groups. In other words, a notion of within-group concordance is being assessed as defined by the correlations between subject response vectors. This question is related to our concern in Section 3.2.6 with concordance between groups. Now, however, our interests are in evaluating the between-subject factor itself and an increased concordance within groups. In Section 3.2.6, between-group concordance was of interest, but it was more or less subject to the preexistence of a within-group concordance.

Depending on the research context and on the data collected on each object, there are many different ways of obtaining proximities. In fact, there is a huge literature on this topic developed as part of the current interest in cluster analysis which the reader can consult for a more extensive discussion (see Sneath and Sokal, 1973). As one example from cognitive psychology, which deserves explicit mention because of the generality of the experimental procedure used, suppose a subject is asked to sort N items into a number of groups (the number and type could be specified by the researcher) in such a way that similar items are placed together and

---

[†]Mielke and his coauthors actually use $c_k = 1/(n_k - 1)$ rather than $1/n_k$. This minor change makes the row sums in $\underset{\sim}{Q}$ identical even for small n_1, ..., n_K, which simplifies the moment formulas somewhat.

dissimilar items are kept apart. Based on (free-) sorting data of this kind, one obvious definition of $\underset{\sim}{P}$ would be to count the number of sorting consistencies between two subjects. Formally, for subjects O_i and O_j, p_{ij} would be defined as the number of item pairs that are both placed in one class by each subject, where larger values of p_{ij} would denote similar sorts. Thus, with the use of some weighting scheme for the a priori categorization of subjects, a large value of the index in expression (4.3.4b), based on this cooccurrence definition for p_{ij}, may suggest an increased similarity between subjects that belong to the same class. We might also note that when our interests center on the similarity between subjects within a single class, the various cluster statistics of the previous section are appropriate. The strategy in this section, in effect, tests all clusters simultaneously.

The free-sort experimental paradigm can be used to generate more information than just a proximity matrix between subjects. If our interests center on the objects themselves, the same n free-sorts lead to an $N \times N$ proximity matrix between items. The entry for a particular item pair could be defined as the number of subjects who place those two items within the same class in their respective categorizations. Now, some a priori structure among the N items could be evaluated against this proximity matrix. In any case, given the increasing popularity of the free-sort experimental paradigm, proximity data of this form, between subjects and/ or items, are becoming common in the cognitive psychology literature. The reader is referred to Mandler (1967), Pollard-Gott, McCloskey, and Todres (1979), Anglin (1970), and Mirkin (1979) for possible areas of substantive application.

There is a substantial literature on the partition comparison problem that the reader may wish to consult further (Hubert and Levin, 1976a; Hubert and Baker, 1977a; Hubert, Golledge, and Costanzo, 1982). As one major segment, Mielke and his co-workers (Mielke, 1978; Mielke, Berry, and Johnson, 1976; O'Reilly and Mielke, 1980; Mielke, Berry, Brockwell, and Williams, 1981; Brockwell, Mielke, and Robinson, 1982) have discussed the strategy in detail as a multivariate analysis-of-variance technique. Again, proximity is defined by some measure of similarity between two profiles of observations over multiple variables. For some spatial applications in archaeology, the reader is referred to Berry, Kvamme, and Mielke (1980). We will also return to variations on this same topic in succeeding sections; for example, in Section 4.3.7 the task of comparing two partitions will be rephrased as a problem of nominal-scale agreement. Although this may seem somewhat redundant with the present discussion, it is of clarifying value to come back to this special case once an approach to nominal-scale agreement is developed using what we will call "Scott's model" in the next two sections.

Categorical Clustering in Free Recall. We have commented repeatedly on the generality of the term "proximity," but a simple illustration from the field of psychology may help emphasize this fact again and provide an interesting application of the partition comparison task. We consider the experimental paradigm called "free recall" that is commonly used in research on memory (see Shuell, 1969). A set of words or other stimuli, which are typically assumed to be categorized into mutually exclusive and exhaustive classes, is given to a subject to study in a randomized order; subsequently, the subject is asked to recall as many items as possible from memory. An index of clustering quantifies the amount of correspondence between the subject's protocol and the specific partition of the items hypothesized by the researcher. If clustering in recall occurs according to expectations, then the responses of a subject should be grouped more or less consistently with respect to the a priori categories that theoretically partition the original stimulus list, and in particular, there should be a tendency for related items to be recalled together.

Within the partition comparison framework of this section, we can propose for a single subject both a measure of categorical clustering, as it is called, and a test of its significance. The set of n objects $\{O_1, \ldots, O_n\}$ comprises the stimulus list recalled by the subject, and Q is defined from the hypothesized structure of the list based on the weighting option given earlier in (i). The entry, p_{ij}, in the proximity matrix $\underset{\sim}{P}$ is 1 if objects O_i and O_j were recalled adjacently by the subject and 0 otherwise. If ρ_0 defines the ordering of the n objects produced by the subject, then $\mathcal{A}(\rho_0)$ is twice the number of repetitions, i.e., twice the number of times that adjacent stimuli were from the same category. This measure or some transform of it is the commonly used index of clustering discussed in the literature. In fact, the normalized version $\mathcal{A}^*(\rho_0)$, is one of the more popular measures and is given the name "adjusted ratio of clustering," e.g., see Roenker, Thompson, and Brown (1971). Here, the max $\mathcal{A}(\rho)$ term can be obtained explicitly as $2(n - K)$.

When proximity is defined in terms of simple adjacency information, the raw index $\mathcal{A}(\rho_0)$ can also be understood when the more standard concept of runs of like and unlike objects is used. For example, the partition can be interpreted as a set of K classes of differently colored objects that are placed sequentially along a continuum by the permutation ρ_0. The index $\mathcal{A}(\rho_0)$ is 2(n minus the number of runs).[†]

As an alternative notion of proximity in the free-recall context that uses more than adjacency information from the subject's protocol, p_{ij} could be defined as the absolute difference between the recall positions for

---

[†]In a similar context, the number of runs around a circle could be studied by augmenting q_{ij} to be 1 when $i = 1$, $j = n$ and when $i = n$, $j = 1$. The index $\mathcal{A}(\rho_0)$ is then 2(n minus the number of circular runs)—see David and Barton (1962).

Table 4.3.4c Proximity Matrices for Word Pairs Based on Adult Subjects (Above Diagonal) and Grade 3–4 Subjects (Below Diagonal)

	Words		Nouns						Prepositions				Verbs					Adjectives				
		(a)	(b)	(c)	(d)	(e)	(f)	(g)	(h)	(i)	(j)	(k)	(l)	(m)	(n)	(o)	(p)	(q)	(r)	(s)	(t)
Nouns	(a) Boy	×	98	45	30	8	8	0	0	0	0	5	0	3	10	0	3	5	0	3	0
	(b) Girl	92	×	45	30	8	8	0	0	0	0	5	0	3	10	0	3	5	0	3	0
	(c) Horse	18	20	×	48	33	8	0	0	0	0	0	0	3	3	0	0	3	0	0	5
	(d) Flower	8	8	15	×	35	10	0	0	0	0	3	0	3	13	0	0	3	0	0	8
	(e) Chair	3	3	28	10	×	18	0	0	0	0	3	0	3	3	5	3	0	0	0	8
	(f) Idea	0	0	0	0	3	×	3	0	0	0	13	5	28	15	10	5	3	5	3	5
Prepositions	(g) Above	0	0	3	0	3	10	×	88	65	35	3	0	0	5	8	10	0	0	0	0
	(h) Below	0	0	0	0	0	8	88	×	70	35	0	3	0	0	18	0	5	0	5	0
	(i) Into	0	0	0	0	3	8	38	45	×	48	0	0	0	0	15	0	0	0	0	0
	(j) During	0	0	0	3	0	8	5	5	15	×	0	0	0	0	0	5	0	0	0	0
Verbs	(k) Laugh	3	0	0	0	0	13	3	3	5	3	×	85	60	35	30	5	3	43	3	3
	(l) Cry	0	3	0	3	0	10	3	5	5	3	75	×	53	28	33	0	10	50	8	0
	(m) Listen	0	0	0	0	0	40	3	5	3	5	30	25	×	43	43	3	3	20	5	3
	(n) Grow	0	0	3	30	0	3	5	5	5	8	10	10	10	×	45	10	5	3	18	3
	(o) Fall	0	0	0	0	5	8	20	23	23	5	5	5	5	15	×	3	5	10	10	3
Adjectives	(p) Rich	3	0	0	0	0	3	5	5	3	3	8	3	3	3	5	×	75	20	35	30
	(q) Poor	3	0	0	0	0	5	5	5	3	5	3	5	3	3	5	83	×	25	35	25
	(r) Angry	3	0	0	0	0	10	3	3	3	3	45	55	25	13	5	5	5	×	28	18
	(s) Dead	0	0	0	3	0	3	3	3	3	5	3	3	5	25	8	10	13	3	×	30
	(t) White	0	0	5	13	8	0	0	0	0	0	0	0	0	3	0	0	0	0	3	×

O_i and O_j. Now, small values of $\mathcal{A}(\rho_0)$ would indicate clustering, and a normalized measure $\mathcal{A}^{**}(\rho_0)$ could be based on the explicit minimum for $\mathcal{A}(\rho)$ of $\Sigma_{k=1}^{K} [n_k(n_k^2 - 1)]/3$. In effect, if we turn around the roles of $\underset{\sim}{P}$ and $\underset{\sim}{Q}$, we would be testing for a unidimensional pattern in the symmetric proximity matrix represented by $\underset{\sim}{Q}$, as discussed in Section 4.3.1, using Szczotka's criterion. In an analogous fashion, the use of the repetition measure can be interpreted as a test for a unidimensional pattern in the proximity structure represented by $\underset{\sim}{Q}$, using Wilkinson's criterion.

The generality of interpretation allowed by reversing the data/structure roles of $\underset{\sim}{P}$ and $\underset{\sim}{Q}$ also suggests how structures other than a simple categorical representation of the stimulus set may be studied. For instance, the matrix $\underset{\sim}{P}$ could again represent the protocol information for a subject, and $\underset{\sim}{Q}$ could contain (numerical) inter-item associative norms from some published source. In fact, if $\underset{\sim}{Q}$ were asymmetric, the protocol information could be coded in a directional manner, e.g., to represent directional adjacency, we could let $p_{ij} = 1$ if O_i were recalled immediately before O_j, and 0 otherwise. Or, we could compare the rank order in which the items were presented with the rank order in which they are produced, either using directional information from each sequence, as in Section 4.3.2, or relying only on the absolute magnitudes of separation, as in our discussion of seriation for symmetric proximities in Section 4.3.1. Moreover, if we have the times at which each stimulus was recalled, these latter values could replace the rank-order information by defining an unequally spaced set of locations along a hypothetical continuum. For a discussion of some of these applications, the reader is referred to Hubert and Levin (1976b, 1977) and Pellegrino and Hubert (1982).

Example. As an illustration of how a partition can be compared against a proximity matrix, Table 4.3.4c presents proximities for 20 words taken from Anglin (1970). The above-diagonal entries were obtained from the free-sort responses provided by 40 adults; the below-diagonal entries were similarly obtained from 40 children in grades 3 and 4. In each case, the values in the table are the proportions of times the words were sorted together by that group of subjects (decimals are omitted).

The conjecture of interest to Anglin may be stated in terms of a "parts of speech" partitioning of the words: The first six are nouns, the second group of four are prepositions, the third group of five are verbs, and the last group of five are adjectives. Based on the index in (4.3.4b), which is one-half of the complete QA statistic, the results of comparing this one partition against the two data sets can be summarized as follows (the Monte Carlo distributions are given in Table 4.3.4d):

Weighting options

	(a): $c_k = 1$		(b): $c_k = 1/\binom{n_k}{2}$		(c): $c_k = 1/n_k$	
	Adult	Child	Adult	Child	Adult	Child
Observed index	1549.	719.0	163.2	78.01	312.4	146.7
Expectation	447.	296.9	43.60	28.96	37.21	57.93
Variance	11600.	6922.	126.4	75.29	457.6	271.7
Z	10.23	5.07	10.63	5.65	10.53	5.38
γ	.641	.613	.693	.733	.644	.641
Significance level						
Monte Carlo	.001	.002	.001	.001	.001	.002
Type III	.001	.001	.001	.001	.001	.001

Obviously, the parts-of-speech partition is reflected in both the adult and child data, although it may be stronger for the adults, judging by the size of the Z statistics. As we will see in Section 4.4, the evidence for such a conclusion can be made more precise.

The parts-of-speech hypothesis may be the most obvious experimental conjecture and was of primary interest to Anglin, but other hypotheses that are posited on different grounds could be evaluated in a similar manner. For instance, it may be seen that within the noun category, all but the last item, "idea," are relatively concrete in the Paivio, Yuille, and Madigan (1968) sense. At the same time, the items in the verb category could be alternatively conceptualized as processes associated with human beings, suggesting that "idea" could be added as well as "angry" from the fourth category. This alternative partition would have the form

Concrete nouns: Boy, Girl, Horse, Flower, Chair
Prepositions: Above, Below, Into, During
Human processes: Idea, Laugh, Cry, Listen, Grow,
    Fall, Angry
Adjectives: Rich, Poor, Dead, White

Paralleling the previous analyses, we would obtain the following results (the Monte Carlo distributions are given in Table 4.3.4e):

Weighting options

	(a): $c_k = 1$		(b): $c_k = 1/\binom{n}{2}k$		(c): $c_k = 1/n_k$	
	Adult	Child	Adult	Child	Adult	Child
Observed index	1608.	927.	164.40	91.16	312.57	176.80
Expectation	468.8	311.4	43.60	28.96	87.21	57.93
Variance	11690.	7060.	142.4	85.15	452.5	268.7
Z	10.54	7.33	10.12	6.74	10.59	7.25
γ	.648	.599	.731	.814	.658	.673
Significance level						
Monte Carlo	.001	.001	.001	.001	.001	.001
Type III	.001	.001	.001	.001	.001	.001

A comparison of these last results to those given previously has to be somewhat tentative, but given the size of the Z statistics, it could be conjectured that this second decomposition of the word set may be more reasonable for the grade 3-4 sample than the alternative parts-of-speech decomposition. The evidence for this statement will be made more precise in Section 4.4.

4.3.5 Nominal-Scale Agreement: Scott's Model (B)

The discussion of Section 1.3.1 on nominal-scale agreement included the unweighted kappa index as a special case. Here, the number of rows and columns are equal to a common value, say T; the rows and columns define the same categories, a_1, a_2, ..., a_T, and n_{uv} denotes the number of observations in row u and column v, $1 \leq u, v \leq T$. Furthermore, because the cell weights are 1 along the main diagonal of the square contingency table and 0 otherwise, the expected proportion of agreement (P_e) under the LA model reduces to

$$P_e = \sum_{u=1}^{T} p_{u.} p_{.u} = \frac{\sum_{u=1}^{T} n_{u.} n_{.u}}{n^2} \qquad (4.3.5a)$$

At least in psychology, this latter expectation (4.3.5a) is the term commonly used for defining unweighted kappa:

Table 4.3.4d Monte Carlo Distributions for the "Parts of Speech"
Partition Based on the Data of Table 4.3.4c (Sample
Size of 999)

Cumulative frequency	Weighting option					
	(a): $c_k = 1$		(b): $c = 1/\binom{n_k}{2}$		(c): $c_k = 1/n_k$	
	Adult	Child	Adult	Child	Adult	Child
1	203.	132.	17.6	12.7	38.1	25.6
5	231.	147.	21.6	14.0	45.2	28.7
10	248.	149.	22.7	14.4	47.3	29.2
50	294.	176.	27.2	17.0	56.6	34.5
100	318.	197.	30.3	18.8	61.7	38.4
200	355.	223.	33.7	21.3	68.8	43.3
300	384.	245.	36.9	23.6	74.7	47.6
400	407.	268.	39.6	25.6	79.7	51.9
500	434.	289.	42.4	27.9	85.5	56.0
600	461.	314.	45.4	30.3	90.6	61.2
700	496.	339.	48.5	32.9	97.4	65.9
800	540.	364.	53.4	35.9	105.	71.2
900	604.	402.	59.4	40.1	118.	79.0
950	652.	433.	65.5	44.0	130.	85.9
990	743.	510.	74.4	51.4	144.	101.
995	769.	559.	77.2	57.2	155.	108.
999	817.	752.	85.8	76.0	162.	148.

$$\kappa = \frac{P_o - P_e}{1 - P_e}$$

where $P_e = \Sigma_{u=1}^{T} p_{uu}$ is the observed proportion of agreement. The field
of sociology, however, generally relies on an alternative expression for
P_e suggested by Scott (1955), although the same general form for kappa

Table 4.3.4e Monte Carlo Distributions for the Alternative Partition
(as Given in the Text) Based on the Data of Table 4.3.4c
(Sample Size of 999)

	Weighting option					
	(a): $c_k = 1$		(b): $c_k = 1/\binom{n_k}{2}$		(c): $c_k = 1/n_k$	
Cumulative frequency	Adult	Child	Adult	Child	Adult	Child
1	216.	130.	18.1	11.9	39.1	24.6
5	254.	145.	19.4	13.2	44.6	26.8
10	260.	161.	21.2	14.0	46.8	29.8
50	309.	184.	25.5	16.3	55.3	33.9
100	339.	206.	29.0	17.9	61.4	37.4
200	375.	239.	33.3	20.6	68.2	43.3
300	404.	263.	36.4	23.5	74.9	48.3
400	431.	284.	39.7	25.7	79.9	53.1
500	460.	310.	42.7	28.2	85.7	57.1
600	485.	330.	45.6	31.0	90.5	61.3
700	519.	352.	48.3	33.2	97.5	66.0
800	561.	380.	53.2	36.4	105.	71.8
900	618.	421.	59.1	41.5	115.	79.6
950	663.	455.	65.2	45.5	126.	86.6
990	766.	525.	75.7	54.3	145.	103.
995	792.	559.	77.8	61.2	148.	111.
999	871.	619.	82.3	64.9	164.	114.

is still used. Specifically, Scott (1955), Krippendorff (1970), and others
define

$$P_e = \frac{1}{4} \sum_{u=1}^{T} (p_{u.} + p_{.u})^2 = \frac{\sum_{u=1}^{T} (n_{u.} + n_{.u})^2}{4n^2} \qquad (4.3.5b)$$

which is based on an assumption of rater independence and a fixed sum for each <u>pair</u> of marginals that refer to the <u>same</u> category (e.g., see Levene, 1949, for an early discussion of a matching model on which this expectation can be based).

Both of these agreement notions have been discussed in some detail by Hubert (1977a), who suggested (following an example presented by Light, 1971) that the expectation given in equation (4.3.5b) may be particularly appropriate when consistency between subject pairs is measured, for example, in a task requiring the categorization of 2n subjects who respond individually to a single object. In particular, the $T \times T$ contingency table now represents the responses of the 2n subjects rating the single object, and where the subjects themselves can be divided into separate groups, e.g., brothers and sisters paired within the same family. Thus, in this type of illustration, each diagonal entry n_{uu} represents the number of matched pairs (e.g., brother-sister combinations) that place the single object in the uth category, $1 \leq u \leq T$. As we will mention later, if the observed split that defines the row and column objects produces a similarity in the marginal frequencies, $n_{u.}$ and $n_{.u}$, then this latter consistency typically becomes part of an elevated index of agreement when assessed under Scott's model.

Although both expectations in equations (4.3.5a) and (4.3.5b) are used more or less interchangeably in the literature (see the discussion in Krippendorff, 1970), there are basic differences in the random processes being assumed in the definitions for P_e and the corresponding significance tests. In general, the expectation in equation (4.3.5a) can be justified through LA, as we have seen in Section 1.3.1. The alternative given in equation (4.3.5b) can be obtained through QA (for large n), and consequently, this interpretation will be the major emphasis below.

In the LA interpretation for weighted nominal-scale agreement (specialized here to a common number of categories, T), we, in effect, carried out a process of duplication and matching. First, the n objects for rater 1 were categorized according to T classes that defined the rows of the $n \times n$ assignment matrix $\underset{\sim}{C}$. These same n objects were "duplicated" and recategorized according to these same T classes used by rater 2 to define the columns of $\underset{\sim}{C}$ (or for whatever attributes specify the classification used on rows and columns). Second, the n objects from rater 1 were matched to the same n objects from rater 2 to generate the observed weighted proportion of agreement, $\Sigma_{u,v} w_{uv} n_{uv}$. Specifically, the number of instances were counted in which rater 1 placed an object in the uth category and rater 2 placed the same object in the vth category; these frequencies were then weighted by the appropriate assignment scores w_{uv} that make up the body of the assignment matrix $\underset{\sim}{C}$. Thus, we have 2n objects in total, where n are from rater 1 and n are from rater 2 and the matching is constructed from one set to the other. Each possible realization of the contingency table with the fixed marginals, including the observed table, can be defined by such a matching or permutation. Moreover, when all such permutations are considered equally likely, the usual hypergeometric model for a contingency table with fixed marginals is generated.

The strategy of duplicating objects so that separate sets correspond to each of the two raters serves as a key to a discussion of the Scott model that leads to the expectation in equation (4.3.5b). Previously, it was apparent how the 2n objects were to be matched, because n objects (call this set A) were associated with rater 1, n duplicated objects (call this set B) were associated with rater 2, and only those objects between the fixed sets A and B were to be paired. Suppose now, however, 2n objects are again given, possibly from the type of duplication used above or from the type of illustrative example presented earlier that dealt with n brothers and n sisters. Furthermore, suppose n objects are selected from this pool of 2n elements and called set A; the remaining n elements define set B. Although the matching is carried out as before, the Scott model assumes that an additional random process decides, first of all, the total number of objects of a particular kind for each rater, or for whatever attributes are used to form the categories for the rows and columns. The numbers in the T categories are held constant, but the actual row and column marginals may vary within these specified limits.

To formalize this matching notion within the QA framework, suppose a 2n × 2n matrix $\underset{\sim}{P}$ is defined, having a zero main diagonal but otherwise of the form

$$
\underset{\sim}{P} = \quad
\begin{array}{c}
n_{.1} + n_{1.} \left\{ \vphantom{\begin{array}{c}a\\b\end{array}} \right. \\[2ex]
n_{.2} + n_{2.} \left\{ \vphantom{\begin{array}{c}a\\b\end{array}} \right. \\[2ex]
\vdots \\[2ex]
n_{.T} + n_{T.} \left\{ \vphantom{\begin{array}{c}a\\b\end{array}} \right.
\end{array}
\begin{array}{c}
\overbrace{n_{.1} + n_{1.}} \qquad \overbrace{n_{.2} + n_{2.}} \qquad \cdots \qquad \overbrace{n_{.T} + n_{T.}} \\
\left[
\begin{array}{cccc}
-w_{11}- & -w_{12}- & \cdots & -w_{1T}- \\[2ex]
-w_{21}- & -w_{22}- & \cdots & -w_{2T}- \\[2ex]
\cdot & \cdot & \cdots & \cdot \\
\cdot & \cdot & \cdots & \cdot \\
\cdot & \cdot & \cdots & \cdot \\[2ex]
-w_{T1}- & -w_{T2}- & \cdots & -w_{TT}-
\end{array}
\right]
\end{array}
$$

and also a second 2n × 2n matrix $\underset{\sim}{Q}$:

$$\underset{\sim}{Q} = \begin{array}{|c|c|} \hline & \begin{matrix} 1 & & 0 \\ & \ddots & \underset{\sim}{} \\ \underset{\sim}{0} & & \\ \underset{\sim}{} & \ddots & \\ & & 1 \end{matrix} \\ \underset{\sim}{0} & \\ \hline \underset{\sim}{0} & \underset{\sim}{0} \\ \hline \end{array}$$

where the four square submatrices are each of the size $n \times n$. The matrix $\underset{\sim}{Q}$ essentially matches the first n row objects in $\underset{\sim}{P}$ to the second set of n column objects in $\underset{\sim}{P}$. Each possible outcome of the contingency table (including the observed distribution of entries in the original table) is defined by some permutation ρ on the first 2n integers that is applied to the rows and columns of $\underset{\sim}{P}$. The QA index $\mathcal{A}(\rho)$ produces a raw weighted agreement index for that particular realization of the contingency table defined by ρ. Stated in other words, any function ρ simultaneously reorganizes the rows and columns of $\underset{\sim}{P}$ and has the effect of first choosing the sets A and B at random, and then of inducing a random mapping or pairing from A to B. A permutation (which is not unique) that produces the agreement index for the observed table will again be denoted by ρ_0.

Obviously, the expectation and variance of the raw agreement statistic under the QA model can be obtained as special cases of our earlier formulas. Although the variance term for the QA model is much more complicated than for the LA model (see Hubert, 1980), the expectation for the raw agreement statistic $\mathcal{A}(\rho)$ is reasonably straightforward:

$$\frac{1}{2(2n-1)}\left[\sum_{u,v} w_{uv}(n_{u\cdot} + n_{\cdot u})(n_{v\cdot} + n_{\cdot v}) - \sum_u w_{uu}(n_{u\cdot} + n_{\cdot u})\right]$$

Based on this expression in the unweighted case and for large n, the raw proportion of agreement, $P_o = \Sigma_u n_{uu}/n = \Sigma_u p_{uu}$, has expectation

$$\frac{1}{4} \sum_u (p_{u\cdot} + p_{\cdot u})^2$$

which is the same formula for P_e used by Scott.

In the typical case of weighted nominal-scale agreement with a common number of row and column categories, both the LA and QA approaches can be used to assess significance for the weighted proportion of agreement. Thus, at this level, we have two solutions to exactly the same problem. However, each of the models leads to generalizations that cannot be

accommodated by the other. For example, using LA, we can extend the
notion of nominal-scale agreement, as in Section 1.3.1, to arbitrary R × C
contingency tables where R and C are not necessarily equal. On the other
hand, using QA, we can define nominal-scale agreement when the row and
column objects cannot be identified unambiguously. For example, instead
of brother-sister combinations, suppose we have n pairs of siblings, e.g.,
n pairs of brothers. If the cell weights are assumed symmetric, the Scott
model can still be used to assess sibling consistency. As long as each of
the n families is represented in the set defining the rows (and similarly for
the columns) of the T × T contingency table, the weighted proportion of
agreement and the associated reference distribution would remain the
same. This same idea comes up again in the context of the intraclass
correlation coefficient mentioned in the next section.

As we have noted, the Scott model based on QA essentially combines
(or confounds) two separate sources of information about the patterning of
entries in the T × T contingency table. Besides incorporating a process
that matches row and column objects given the observed marginals, an
initial selection strategy is also carried out that produces the marginal
frequencies subject to certain constraints. The LA model, on the other
hand, is concerned solely with the matching process and not with how the
margins are constructed. Thus, if we wish to consider the matching
process by itself, the LA strategy should be followed rather than one based
on QA.

We could assess separately the process of forming the marginal frequencies, using some of the ideas from Section 4.3.3 on cluster statistics
and how they relate to contingency tables. For example, suppose the 2n ×
2n matrix $\underset{\sim}{P}$ is defined as before, but let $\underset{\sim}{Q}$ take the form

$$
\underset{\sim}{Q} =
\begin{bmatrix}
\underset{\sim}{0} &
\begin{matrix}
\frac{1}{n^2} & \cdots & \frac{1}{n^2} \\
\vdots & & \vdots \\
\frac{1}{n^2} & & \frac{1}{n^2}
\end{matrix} \\
\underset{\sim}{0} & \underset{\sim}{0}
\end{bmatrix}
$$

We, in effect, are constructing an isolation measure as in Section 4.3.3, defined here by the average proximity from D to S-D, in which the row objects from the observed contingency table provide the identified cluster D and the column objects define the complement. If a permutation ρ_0 has been applied to $\underset{\sim}{P}$ to place the identified n row objects from the contingency table into the first n rows of $\underset{\sim}{P}$ and the n column objects from the contingency table in the last n columns of $\underset{\sim}{P}$, then the index $\mathcal{A}(\rho_0)$ is equal to $\Sigma_{u,v} \, w_{uv} \times$ $(n_{u} . n_{.v}/n^2)$. In other words, the distribution of $\mathcal{A}(\rho)$ is actually the distribution over the expected values for each of the individual LA tasks if the marginals are assumed fixed. Thus, an assessment of whether the objects defining the rows and columns of the contingency table could be considered a random split of the total set of 2n objects might be approached through the relative size of the expected value for the associated matching task in the LA model. Some of these same issues of subset selection will be discussed in a broader context again in Section 4.3.11.

This last discussion of evaluating the pattern of row and column frequencies suggests some rather interesting practical concerns in assessing nominal-scale agreement that may be of value to note at least in a rather informal way. Suppose, for example, that a given agreement statistic is evaluated as being nonsignificant based on LA. However, the particular distribution of marginal frequencies leads to an expected value much larger than what usually would be obtained under a random construction of the row and column object sets; for example, in the unweighted case this would occur when the row and column marginal frequencies referring to the same category are as equal as possible. It could then turn out that the agreement statistic would be evaluated as relatively large if Scott's model were used; the reference distribution for the latter would be shifted down, compared to that for LA. In this way we take advantage of the consistency in marginal frequencies in assessing the overall degree of nominal-scale agreement. Obviously, a number of other such variations might be possible as well, e.g., significance under LA but nonsignificance under Scott's model, and so on.

4.3.6 Generalizations of the Scott Matching Model (B)

The Scott matching model has been presented in the specific context of nominal-scale agreement, but the approach is more widely applicable than this. In summary, if $\underset{\sim}{P}$ is an arbitrary 2n × 2n matrix and $\underset{\sim}{Q}$ has the structure as before, the QA inference model involves two stages: (1) the construction of an n × n matrix $\underset{\sim}{C}$ from the larger matrix $\underset{\sim}{P}$, and (2) the generation of all possible assignments within this chosen matrix. In stage (1), the 2n objects are first split randomly into two groups, A and B, each of size n. That part of $\underset{\sim}{P}$ containing the proximities from the objects in A to those in B is then used to define an n × n matrix of assignment scores just as in the LA context. In stage (2), all n! possible assignments based on the latter matrix are considered equally likely.

As might be expected, this particular two-stage generalization has several other applications that parallel uses of the simpler LA model. For most of the applications envisioned, however, P will be symmetric; consequently, it is convenient to make this assumption immediately at the outset. Thus, without loss of generality, Q can also be considered symmetric and of the form

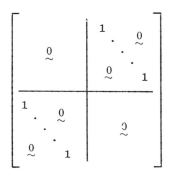

Because we omit the constant divisor of $1/2$ as a convenience, this version of Q produces an index twice that of the original definition from the last section.

Correlational Relationships. Suppose n bivariate observations, (x_i, y_i) for $1 \le i \le n$, are given, and the task is to test the association between the x's and the y's. The usual permutation test can be obtained through the LA model if we define $C = \{x_i y_j\}$ and consider the distribution of $\Gamma(\rho)$ over all n! possible permutations. Scott's model, on the other hand, would assume that a single set of 2n observations is given, $u_1, u_2, \ldots, u_{2n}$, and the problem is to index and test the strength of some pairing within this set. The off-diagonal weights in P are now of the form $p_{ij} = u_i u_j$, and the index $\mathcal{A}(\rho_0)$ is an unnormalized intraclass correlation coefficient where ρ_0 matches the n bivariate pairs.

If the 2n observations are z scores, then $(1/2n)\mathcal{A}(\rho_0)$ is in reality a Pearson product-moment correlation. Because there is no "first" and "second" set, however, each of the 2n z scores is used to form a first member and also a second. The given pairing then matches up the observations to generate 2n pairs. For example, if the 2n observations are heights of brothers paired within the same family, then our concern is with assessing the similarity in familial stature. We note that the expected value of the intraclass correlation is not zero, but rather $-1/(2n - 1)$. Furthermore, other proximities could be used in place of products, e.g., absolute differences as in Spearman's footrule. For a discussion of this type of intraclass correlation coefficient when the 2n observations are the untied ranks from 1 to 2n, the reader is referred to the recent paper by Shirahata (1981).

Extensions of Scott's Model to Multiway Contingency Tables. Although the "selection" matrix Q is given in the form

$$
\begin{bmatrix}
\begin{matrix} & & \\ & 0 & \\ & & \end{matrix} &
\begin{matrix} 1 & & \\ & \cdot & \; 0 \\ & & \cdot \\ 0 & & \cdot \\ & & 1 \end{matrix} \\
\begin{matrix} 1 & & \\ & \cdot & \; 0 \\ & & \cdot \\ 0 & & \\ & 1 & \end{matrix} &
\begin{matrix} & & \\ & 0 & \\ & & \end{matrix}
\end{bmatrix}
$$

by an appropriate reorganization of the rows and (simultaneously) the columns, an equivalent representation could be written as

$$
\begin{bmatrix}
\begin{matrix} 0 & 1 \\ 1 & 0 \end{matrix} & \cdots & \begin{matrix} 0 \\ \end{matrix} \\
\vdots & \ddots & \vdots \\
0 & \cdots & \begin{matrix} 0 & 1 \\ 1 & 0 \end{matrix}
\end{bmatrix}
$$

With this reorganization it may be observed that our inference problem is one of comparing a specific partition of the 2n objects (into subsets with 2 members each) against the original 2n × 2n weight matrix P. In this sense, we have a partition comparison problem of exactly the same type discussed in Section 4.3.4. The notion of proximity is now defined in terms of the symmetric weights that define blocks of entries in P, but otherwise the structure of the task is identical.

Continuing in this way, suppose that 3n objects are given (e.g., n brothers, n sisters, and n mothers), and a measure of consistency is again desired. The appropriate weight matrix is of order 3n × 3n and is categorized according to the sums of the marginals (e.g., $n_{u..} + n_{.u.} + n_{..u}$ for $u = 1, \ldots, T$). Moreover, any particular grouping into 3's can be represented by some permutation of the rows and the columns of a matrix having the form

$$
\begin{bmatrix}
\underset{\sim}{0} &
\begin{matrix} 1 & & \underset{\sim}{0} \\ & \cdot & \\ \underset{\sim}{0} & & 1 \end{matrix} &
\begin{matrix} 1 & & \underset{\sim}{0} \\ & \cdot & \\ \underset{\sim}{0} & & 1 \end{matrix} \\[2em]
\begin{matrix} 1 & & \underset{\sim}{0} \\ & \cdot & \\ \underset{\sim}{0} & & 1 \end{matrix} &
\underset{\sim}{0} &
\begin{matrix} 1 & & \underset{\sim}{0} \\ & \cdot & \\ \underset{\sim}{0} & & 1 \end{matrix} \\[2em]
\begin{matrix} 1 & & \underset{\sim}{0} \\ & \cdot & \\ \underset{\sim}{0} & & 1 \end{matrix} &
\begin{matrix} 1 & & \underset{\sim}{0} \\ & \cdot & \\ \underset{\sim}{0} & & 1 \end{matrix} &
\underset{\sim}{0}
\end{bmatrix}
$$

or alternatively, as

$$
\begin{bmatrix}
\begin{matrix} 0 & 1 & 1 \\ 1 & 0 & 1 \\ 1 & 1 & 0 \end{matrix} & \cdots & \underset{\sim}{0} \\[2em]
\vdots & & \vdots \\[2em]
\underset{\sim}{0} & \cdots & \begin{matrix} 0 & 1 & 1 \\ 1 & 0 & 1 \\ 1 & 1 & 0 \end{matrix}
\end{bmatrix}
$$

This matrix induces a sum of the weights within each group of three and then a final addition over those partial sums (called, in general, the "pairwise" definition of agreement in Section 3.2.1). In fact, any partition whatsoever can be so represented, and there is no inherent restriction to equal size groupings.

Extensions of the Experimental Matching Paradigm. As mentioned in Section 1.3.1, the model that was developed using the index $\Gamma(\rho)$ in the LA context is directly relevant to the experimental paradigm referred to as matching (see Gilbert, 1956). For example, suppose we have 10 samples of handwriting, and a subject is told that 2 were written by doctors, 4 by lawyers, and 4 by teachers. If the subject is required to match a handwriting sample with a profession, then the index $\Gamma(\rho_0)$ represents the subject's performance, i.e., the number of weighted matches where ρ_0 defines the assignment produced by the subject. Here, the weights could be developed, say, from norming information obtained from groups of subjects required to perform the same experimental task. In any event, using our previous terminology, the set A would now refer to the "true" professions, and the set B to the "guesses."

Scott's model, in a similar way, seems particularly appropriate for a simple variation in the above experimental procedure. As an illustration, suppose we are given 20 handwriting samples, and the subject is told to pair the samples in such a way that the same professions are matched. In this instance, the sets A and B are chosen by the subject, and possibly this initial decision should be reflected in the inference strategy as well, i.e., through the use of Scott's model.

4.3.7 Nominal-Scale Agreement Based on Object Pairs: Comparing Two Partitions (B)

The approach to nominal-scale agreement in Section 4.3.5 based on Scott's model was limited to one special case: The row and column categories of the contingency table had to be in one-to-one correspondence. By appropriately redefining the two matrices, P and Q, however, the QA model can also be used to construct a measure of nominal-scale agreement between two "raters," based on the consistency of categorization for object pairs when the classes used by the raters are not assumed to correspond or even be the same in number (see Brennan and Light, 1974; Rand, 1971; Hartigan, 1975; Hubert, 1977b). Such a measure has the one distinct advantage that it can be defined without the use of arbitrary weights. This is in direct contrast to Cohen's index of Section 1.3.1, where the rows and columns of the contingency table are not in one-to-one correspondence.

As one possible approach, suppose the row attribute in the original $R \times C$ contingency table corresponds to the R classes used by rater 1 in classifying the n objects $\{O_1, \ldots, O_n\}$, and the column attribute to the

C classes used by rater 2. If we define the entries on the main diagonals of $\underset{\sim}{P}$ and $\underset{\sim}{Q}$ as zero and the off-diagonals as

$$
p_{ij} = \begin{cases} +1 & \text{when } O_i \text{ and } O_j \text{ are placed in the same category by rater 1,} \\ & O_i \neq O_j \\ -1 & \text{otherwise, } O_i \neq O_j \end{cases}
$$

and

$$
q_{ij} = \begin{cases} +1 & \text{when } O_i \text{ and } O_j \text{ are placed in the same category by rater 2,} \\ & O_i \neq O_j \\ -1 & \text{otherwise, } O_i \neq O_j \end{cases}
$$

then

$$
\mathcal{A}(\rho_I) = 2(e - f)
$$

where

   e = the number of object pairs placed in the same category by rater 1 and in the same category by rater 2 plus the number of object pairs placed in different categories by rater 1 and in different categories by rater 2;

   f = the number of pairs other than those counted in e, i.e., $f = \binom{n}{2} - e.$

Thus, e represents agreements in categorizing object pairs and f represents disagreements. In terms of a formula,

$$
e = \binom{n}{2} + \sum_{v=1}^{C} \sum_{u=1}^{R} n_{uv}^2 - \frac{1}{2}\left(\sum_{u=1}^{R} n_{u\cdot}^2 + \sum_{v=1}^{C} n_{\cdot v}^2 \right)
$$

Operationally, it is assumed that larger values of $\mathcal{A}(\rho_I)$ correspond to greater degrees of nominal-scale agreement. Thus, one obvious normalized measure of the form $\mathcal{A}^*(\rho_0)$ would be

$$
\frac{2(e - f) - E(\mathcal{A}(\rho))}{n(n - 1) - E(\mathcal{A}(\rho))},
$$

using $n(n - 1)$ as an upper bound on $\mathcal{A}(\rho)$ and where

$$E(\mathcal{A}(\rho)) = \left[\frac{1}{n(n-1)}\right]\left[2\sum_{u=1}^{R} n_{u\cdot}^2 - (n+1)n\right]\left[2\sum_{v=1}^{C} n_{\cdot v}^2 - (n+1)n\right]$$

Or alternatively, r_{PQ} could be obtained between $\underset{\sim}{P}$ and $\underset{\sim}{Q}$. Finally, given the $R \times C$ contingency table, the distribution of $\mathcal{A}(\rho)$ over all n! permutations under the QA model is equivalent to what would be obtained from the well-known hypergeometric distribution, i.e., the standard distributional assumption for a contingency table with fixed marginals.

Because the task of measuring nominal-scale agreement is really equivalent to the task of comparing two different partitions of n objects, the approach discussed in Section 4.3.4 on comparing a partition to a proximity matrix really subsumes this discussion. Here, $\underset{\sim}{P}$ could correspond to the row categorization and could be defined by the usual 0-1 structure; the second matrix $\underset{\sim}{Q}$, which is already in this form, would represent the column categorization. The index $\mathcal{A}(\rho_I)$ would be twice the number of object pairs, denoted by a, that were placed in the same category by rater 1 and in the same category by rater 2. In short, this latter measure is a constant transformation of the one obtained by the ±1 scoring function for the entries in $\underset{\sim}{P}$ and $\underset{\sim}{Q}$, and obviously, could be used instead.

It is also of some interest to note that by defining the matrices $\underset{\sim}{P}$ and $\underset{\sim}{Q}$ differently, a number of familiar measures for $R \times C$ contingency tables can be developed as part of this same framework. For instance, generalizing the discussion of the chi-square measure of association for $2 \times t$ contingency tables in Section 4.3.3, suppose $\underset{\sim}{P}$ and $\underset{\sim}{Q}$ are defined as follows:

$$p_{ij} = \begin{cases} \dfrac{1}{n_{u\cdot}} & \text{when } O_i \text{ and } O_j \text{ are distinct and are placed in the uth} \\ & \text{category by rater 1, } 1 \leqslant u \leqslant R \\ 0 & \text{otherwise,} \end{cases}$$

$$q_{ij} = \begin{cases} \dfrac{1}{n_{\cdot v}} & \text{when } O_i \text{ belongs to category v, but } O_j \text{ belongs to a different category for rater 2, } 1 \leqslant v \leqslant C \\ 0 & \text{otherwise} \end{cases}$$

Then, the QA index $\mathcal{A}(\rho_I)$ can be expressed as

$$C - \sum_u \sum_v \frac{n_{uv}^2}{n_{u\cdot} n_{\cdot v}}$$

which is a constant linear transformation of the usual chi-square measure for assessing association in an $R \times C$ table. Specifically, because

$$\chi^2 = n\left(\sum_u \sum_v \frac{n_{uv}^2}{n_{u\cdot}\, n_{\cdot v}} - 1\right)$$

we have

$$\chi^2 = -n\mathcal{A}(\rho_I) + n(C - 1)$$

As a slight variation, suppose $\underset{\sim}{P}$ is given in the same way as above but $\underset{\sim}{Q}$ is defined as

$$q_{ij} = \begin{cases} 1 & \text{when } O_i \text{ and } O_j \text{ are distinct and belong to different categories} \\ & \text{for rater 2} \\ 0 & \text{otherwise} \end{cases}$$

then $\mathcal{A}(\rho_I) = n - \Sigma_u \Sigma_v\, n_{uv}^2/n_{u\cdot}$. Again this quantity is a constant transformation of a familiar measure of association, usually referred to as Goodman and Kruskal's τ_b (not to be confused with Kendall's τ_b) (see Light and Margolin, 1971; Margolin and Light, 1974; Goodman and Kruskal, 1972; Hildebrand, Laing, and Rosenthal, 1977). In particular, because

$$\tau_b = \frac{n \sum_{u,v} \frac{n_{uv}^2}{n_{u\cdot}} - \sum_v n_{\cdot v}^2}{n^2 - \sum_v n_{\cdot v}^2}$$

we know that

$$\tau_b = \left(\frac{-n}{n^2 - \sum_v n_{\cdot v}^2}\right)\mathcal{A}(\rho_I) + 1$$

In both of these instances, the usual permutation assumption generates a hypergeometric distribution for the entries in the contingency table. Thus, by using the transformations given above, the distributions of χ^2 and τ_b can be found under the usual fixed-margins assumption.

No matter what specific measure is used, the task of comparing two partitions has a variety of substantive applications. Besides being a way to assess nominal-scale agreement between two raters, we may compare two different roll-call votes for a body of n legislators (cf. Section 4.3.3) or develop variations on the free-sort experimental paradigm mentioned in

Section 4.3.4. In the free-sort context, for example, a subject is required to categorize a list of words or other stimuli into an unspecified number of groups, usually on the basis of perceived similarity. Thus, we could compare this observed sort or partition to a subject's previous sorting of the same stimuli or to a categorization that is hypothesized by the researcher. In fact, the same approach can be followed for constrained sorting tasks in which a subject must provide a certain number of classes or a certain number of stimuli within each class. The latter are typically called Q sorts in the psychological literature.

Both of the terms e and f introduced earlier in the framework of nominal-scale agreement have been suggested in the literature as raw indices of partition correspondence, usually with a division by $\binom{n}{2}$ to scale the measure between 0 and 1 (see Arabie and Boorman, 1973, for a variety of different measures for comparing two partitions, including some of the alternatives we discuss). For example, Rand (1971) has suggested using $e/\binom{n}{2}$; Johnson (1968b) and Mirkin (1979) have adopted $f/\binom{n}{2}$. All of these index variations based on e and f, or various linear combinations are equivalent statistically because constant linear transformations relate one to the other. We also note that once some normalized index is obtained, it may function as a dependent measure for a traditional statistical method. For example, suppose each subject from one of K groups sorts a set of items. The chosen index between each sort and an experimenter's hypothesis could be obtained, and location differences among the K groups detected by an analysis of variance on the obtained indices. Obviously, this same strategy could be applied in other instances and when other statistical methods are used. In all of these cases, the measures based on LA or QA merely serve as first-stage data that can be subjected to other analyses, which, possibly, use specific interpretations of the LA and QA models again.

Finally, for completeness we should mention the rather obvious applications of the generalized correlation coefficients of Section 4.3.2 to contingency tables with ordered classes, and thus, to ordinal-scale agreement. Certain observations are tied to each of the variables, but the coefficients still can be used to assess the correspondence between the ordered row and column attributes. In fact, given the equivalence between the comparison of two partitions and the comparison of two attributes in a contingency table, the generalized correlation coefficients provide a means for comparing data collected through certain variations in the free-sort experimental paradigm. For example, besides requiring a subject to group a set of items, suppose he or she must also impose an order on the classes so defined. In this case, the various measures of association for contingency tables with ordered categories would provide the basis of an appropriate inference strategy for comparing the resulting sorts or partitions. (We note that the comparison of ordered and unordered partitions has already been discussed in Section 4.3.4 through our analog for one-way analysis of variance.)

4.3.8 Evaluating the Symmetry of a Proximity Matrix (B)

Many of the data-reduction techniques developed in the behavioral sciences, such as clustering and the more popular versions of multidimensional scaling, require a symmetric measure of relationship or proximity between each pair of objects chosen from some larger class S. For instance, when achievement tests are to be clustered or scaled, correlations may serve as the basic data, and by definition, a correlation coefficient is a symmetric measure. In certain instances, however, the original measures of relationship collected by a researcher are defined in terms of a possibly asymmetric measure. Consequently, for these latter cases a secondary averaging process is often used to obtain a suitable symmetric data set required for the analysis. As an example, if the objects are stimuli, then typically we construct the estimated conditional probabilities of responding with stimulus i, given the presentation of stimulus j, and conversely. Because these two probabilities are generally different, they are typically averaged in some way to generate a single symmetric measure for each stimulus pair.

   Although a particular symmetrizing operation may be justified for the purposes of a final analysis, these summing or averaging operations do lose some information. In fact, it is conceivable that the type and degree of symmetry in a measure of proximity could be an important consideration in giving a meaningful substantive interpretation to a particular data set, possibly through the use of an index of the form $\mathcal{A}^-(\rho_0)$ discussed in Section 4.2.2 Unfortunately, very little methodological literature is available in the behavioral sciences that deals with symmetry in any depth by proposing, say, a formal measure of its strength, and secondly, an inference procedure for assessing the relative size of an obtained measure.

   The symmetry of a proximity matrix $\underset{\sim}{P}$ can be characterized by the equality of two matrices, $\underset{\sim}{P}$ and its transpose $\underset{\sim}{P}^t$; consequently, it is reasonable to assess the degree of asymmetry by some function of the differences between $\underset{\sim}{P}$ and $\underset{\sim}{P}^t$. Obviously, there are numerous ways of indexing a discrepancy between $\underset{\sim}{P}$ and $\underset{\sim}{P}^t$, but one natural alternative depends on the quantity used in the QA model:

$$\mathcal{A}(\rho_I) = \sum_{i,j} p_{ij} p_{ji}$$

In terms of an associated normalized index, $r_{\underset{\sim}{P}\underset{\sim}{P}^t}$, the bound of +1 can be achieved when perfect symmetry occurs; -1 is achieved for perfect skew-symmetry, or more generally, when the corresponding across-diagonal terms all sum to the same constant value. Obviously, when Q is interpreted as $\underset{\sim}{P}^t$, a significance test for the $\mathcal{A}(\rho_I)$ index (and, thus, for $\underset{P \underset{\sim}{P}^t}{\underset{\sim}{r}}$) can be generated under the usual QA strategy. The reader should consult Hubert and Baker (1979) for a further discussion.

In a much more specific context, the problem of measuring symmetry is well-known in sociometry, where typically the function p_{ij} assigns a value of 1 to the ordered pair (O_i, O_j) if O_i chooses O_j, and a zero otherwise (see Katz and Wilson, 1956). The raw index $\mathcal{A}(\rho_I)$ is then twice the number of mutual choices and forms a statistic of basic concern in analyzing a sociometric matrix. The inference model used by Katz and Wilson capitalizes on the 0-1 structure of the choices through a discrete probability model and is different from the one proposed above; hence, different moments will be obtained. Unfortunately, the Katz and Wilson strategy does not have any simple generalization to choice data based on strengths of preference that are measured, for instance, on an ordinal scale. The use of the correlation index, $\underset{P \underset{\sim}{P}^t}{r}$, is suited for more complete choice information, and, for this reason, our approach may be the more parsimonious even when the standard sociometric context is being considered.

Example. As a simple illustration of how symmetry may be assessed, Table 4.3.8a presents two well-known confusion matrices collected by Miller and Nicely (1955). Matrix I was obtained under a very severe noise condition (S/N = -18db), and Matrix II under a somewhat less serious degradation (S/N = -12db). Based on the Miller and Nicely original data, the two matrices of Table 4.3.8a present the estimated conditional probabilities of responding with a particular column consonant, given the presentation of a particular row consonant. It is expected that the more severe noise condition would induce a greater degree of random responding (or guessing), and thus, obscure the expected symmetry in the conditional probability matrix associated with the greater error levels. The correlation between the off-diagonal elements of each matrix and its transpose do reflect in a dramatic way the differences in error level, because for Matrix I, $\underset{P \underset{\sim}{P}^t}{r}$ is .15, whereas for Matrix II we obtain a much larger value of .88.

For the two conditional probability matrices of Table 4.3.8a, the Monte Carlo distributions appropriate for each are given in Table 4.3.8b. When the results summarized below are used, some significant degree of symmetry is evident in both matrices although the severe noise condition greatly obscures the structure expected for Matrix I, as measured by the disparity between the normalized indices obtained for the two matrices.

Table 4.3.8a Miller and Nicely Conditional Probability Confusion Matrices
(Matrix I: S/N = –18db; Matrix II: S/N = –12db)

Matrix I

	p	t	k	f	θ	s	ʃ	b	d	g	v	ð	z	ʒ	m	n
p	×	.102	.083	.087	.095	.083	.053	.057	.061	.027	.064	.042	.045	.042	.061	.045
t	.073	×	.092	.095	.068	.082	.064	.032	.045	.027	.077	.041	.059	.050	.041	.059
k	.083	.092	×	.063	.058	.121	.050	.017	.046	.038	.050	.042	.067	.046	.071	.058
f	.101	.082	.101	×	.049	.045	.037	.071	.075	.052	.060	.060	.056	.011	.049	.067
θ	.071	.075	.075	.054	×	.088	.050	.058	.083	.058	.096	.025	.058	.038	.050	.058
s	.071	.067	.091	.044	.071	×	.067	.044	.095	.060	.060	.063	.044	.052	.067	.020
ʃ	.060	.075	.101	.063	.049	.138	×	.037	.078	.026	.075	.067	.034	.030	.060	.056
b	.045	.041	.090	.056	.071	.056	.045	×	.075	.071	.090	.045	.056	.041	.067	.063
d	.054	.081	.061	.044	.051	.051	.047	.074	×	.071	.084	.057	.061	.044	.051	.084
g	.036	.066	.095	.030	.059	.059	.049	.086	.099	×	.059	.046	.053	.066	.079	.072
v	.040	.076	.080	.049	.031	.054	.040	.112	.063	.058	×	.067	.085	.049	.054	.076
ð	.074	.051	.046	.032	.028	.065	.046	.093	.079	.083	.069	×	.079	.056	.083	.083
z	.074	.074	.061	.037	.053	.078	.029	.090	.057	.037	.086	.049	×	.041	.090	.049
ʒ	.036	.073	.077	.064	.055	.068	.032	.100	.082	.036	.068	.050	.068	×	.082	.059
m	.079	.100	.063	.058	.058	.058	.033	.058	.063	.050	.054	.033	.046	.025	×	.117
n	.047	.076	.085	.025	.038	.076	.038	.059	.059	.055	.038	.034	.042	.051	.140	×

Matrix II

	p	t	k	f	θ	s	ʃ	b	d	g	v	ð	z	ʒ	m	n
p	×	.207	.254	.086	.074	.023	.043	.008	.000	.008	.012	.012	.004	.020	.031	.020
t	.219	×	.253	.068	.082	.075	.048	.007	.010	.003	.003	.007	.003	.003	.017	.003
k	.212	.178	×	.093	.076	.068	.047	.017	.017	.004	.004	.008	.000	.000	.017	.008
f	.121	.086	.109	×	.113	.059	.043	.012	.020	.000	.031	.031	.012	.000	.012	.000
θ	.096	.081	.092	.232	×	.099	.044	.022	.033	.011	.040	.033	.011	.007	.026	.007
s	.069	.065	.069	.142	.103	×	.207	.013	.022	.026	.013	.004	.026	.009	.000	.004
ʃ	.127	.177	.111	.078	.150	.139	×	.006	.022	.028	.017	.000	.033	.017	.022	.011
b	.016	.008	.008	.070	.027	.027	.004	×	.070	.070	.172	.098	.055	.023	.078	.039
d	.013	.000	.004	.017	.030	.017	.047	.078	×	.151	.069	.103	.112	.060	.039	.052
g	.013	.004	.004	.004	.017	.021	.029	.083	.158	×	.067	.121	.121	.158	.042	.038
v	.000	.004	.004	.051	.021	.017	.021	.157	.085	.097	×	.068	.059	.017	.059	.038
ð	.000	.004	.015	.063	.007	.011	.007	.198	.116	.093	.187	×	.086	.019	.049	.022
z	.025	.004	.008	.008	.025	.059	.034	.097	.123	.114	.102	.081	×	.110	.013	.025
ʒ	.013	.009	.009	.004	.000	.026	.030	.030	.129	.099	.039	.030	.168	×	.022	.060
m	.000	.009	.000	.000	.009	.009	.000	.097	.026	.053	.070	.097	.000	.009	×	.527
n	.008	.000	.000	.008	.000	.008	.000	.015	.015	.046	.054	.008	.008	.070	.649	×

	Matrix I	Matrix II
Observed index	.922	1.835
Expectation	.906	.689
Variance	.0000322	.0141
Z	2.79	9.66
γ	.094	1.64
Significance level		
Monte Carlo	.002	.001
Type III	.004	.001

Symmetry Within Subsets. The comparison of P and P^t is a global assessment of symmetry for a given proximity matrix and is not directed at any
specific part of P. At times, however, it may be of interest to test the
conjecture that a subset, D, of the complete set S should, for whatever
reasons, display an unusual degree of asymmetry or symmetry. This latter
test can be carried out fairly readily, by use of the subset compactness
notion of Section 4.3.3 and the skew-symmetric information present in P.
Explicitly, suppose we define a "new" proximity matrix by some nonnegative
function of the differences $p_{ij} - p_{ji}$; for example, $|p_{ij} - p_{ji}|$. When $q_{ij} =$
$1/[K(K - 1)]$ if $1 \leqslant i \neq j \leqslant K$, and 0 otherwise, $\mathcal{A}(\rho_0)$ is the average absolute
difference within the K-element subset, D, defined by ρ_0. Thus, $\mathcal{A}(\rho_0)$
should be relatively large or small, depending on whether the proximities
within D display an unusual degree of asymmetry or symmetry, respectively.
In short, the degree of asymmetry in D can be evaluated by assessing the
compactness of D based on a special measure of proximity.

4.3.9 QA Procedures for Assessing Spatial Autocorrelation:
       Relating a Variable to a Proximity Matrix (B)

Given a set S containing n geographical units, spatial autocorrelation refers
to the relationship between some arbitrary variable observed in each of the
n localities and a measure of geographical proximity defined for all n(n - 1)
pairs constructed from S. As we note in this section, many of the techniques for analyzing the effects of geographical proximity can be rephrased
within a QA framework (Hubert, Golledge, and Costanzo, 1981). Because
the literature on spatial autocorrelation is extensive, however, a complete
review of the area is well beyond our intent, and we emphasize only a few
aspects of the topic that are particularly pertinent to the QA interpretation.
The reader interested in pursuing the field more thoroughly should refer to
the seminal monographs by Cliff and Ord (1973, 1981).

Table 4.3.8b   Monte Carlo Distributions for Assessing
               Symmetry for the Miller-Nicely Matrices
               of Table 4.3.8a (Sample Sizes of 999)

Cumulative frequency	Matrix I	Matrix II
1	.888	.467
5	.891	.491
10	.893	.498
50	.896	.537
100	.899	.563
200	.901	.593
300	.902	.618
400	.904	.642
500	.906	.664
600	.907	.691
700	.909	.726
800	.910	.772
900	.913	.840
950	.915	.902
990	.918	1.25
995	.920	1.33
999	.922	1.42

We might also note by way of introduction that our approach to the
assessment of spatial autocorrelation could really be subsumed under the
earlier discussion of seriation for symmetric proximity matrices given in
Section 4.3.1. Nevertheless, this redundancy is excusable because the
topic of spatial autocorrelation is very important in its own right, and a
number of variations on the idea will be specifically mentioned here.
Throughout this section, the interpretations given to $\underset{\sim}{P}$ and $\underset{\sim}{Q}$ will result
in at least one of these matrices being symmetric, and thus, without loss
of generality, both could be assumed to have this property. The comparison
of the skew-symmetric information provided by a single variable in comparison to a skew-symmetric matrix has already been developed in detail in
Section 4.3.1 and will not be referred to again.

To be precise about notation, the object set $S = \{O_1, \ldots, O_n\}$ contains n localities, and the values on some variable measured on each of the members in S are denoted by $x_1, \ldots, x_n$. The notion of geographical proximity is expressed through an $n \times n$ proximity matrix $\underset{\sim}{P}$, where for convenience, larger weights are usually assigned to pairs that are "more related"; for example, p_{ij} could be 1 if O_i and O_j were adjacent localities and 0 otherwise, or more generally, p_{ij} could specify the inverse of the actual physical distance from O_i to O_j. Given this information, the presence of spatial autocorrelation implies that the pattern of weights in $\underset{\sim}{P}$ is related to the variate values $x_1, x_2, \ldots, x_n$. Traditionally, two measures of this relationship have been used—Moran's I statistic and Geary's c coefficient:

$$I = \left(\frac{n}{\sum\limits_{i,j} p_{ij}} \right) \frac{\sum\limits_{i,j} p_{ij}(x_i - \bar{x})(x_j - \bar{x})}{\sum\limits_{i} (x_i - \bar{x})^2}$$

$$c = \left(\frac{n-1}{2 \sum\limits_{i,j} p_{ij}} \right) \frac{\sum\limits_{i,j} p_{ij}(x_i - x_j)^2}{\sum\limits_{i} (x_i - \bar{x})^2}$$

where $\bar{x} = (1/n)\Sigma_i x_i$. Both measures are normalized QA cross-product coefficients in which each weight p_{ij} is compared to q_{ij} defined either as $(x_i - \bar{x})(x_j - \bar{x})$ or as $(x_i - x_j)^2$. Based on expectations that can be derived from the usual QA formula, positive spatial autocorrelation is obtained when I is greater than $-[1/(n-1)]$ or when c is less than 1; negative spatial autocorrelation is obtained when I is less than $-[1/(n-1)]$ or when c is greater than 1. The QA inference model, in effect, merely permutes the n observations at random among the n locations.

In addition to the traditional normalizations given above, others could be used; for instance, we could rely on r_{PQ} or various alternatives based on the general form of $\mathcal{A}^*(\rho_0)$ or $\mathcal{A}^{**}(\rho_0)$. Also, other functions of the difference between x_i and x_j could be considered, e.g., q_{ij} could be defined as the absolute difference $|x_i - x_j|$ mentioned in Section 4.3.1, or possibly, q_{ij} could be a more comprehensive measure based on, say, vectors of observations for O_i and O_j. The latter might include Mahalanobis or euclidean distance, various measures of correlation, and so on. The absolute difference measure along with the QA inference model, in particular, has been important in the biometric literature and with applications of the QA model that deal with disease contagion (see Mantel, 1967). Here, the n localities correspond to places of disease occurrence, and x_i specifies the time of disease onset. Evidence for contagion or what can be called space-time clustering exists when relatively small differences in onset time are associated with spatially close cases of a disease. A related application that also uses absolute differences is given by Royalty, Astrachan, and Sokal (1975) in which $x_1, \ldots, x_n$ are ranks and p_{ij} is a zero-one variable specifying geographical contiguity. The QA model includes this latter work as a special case (see Hubert, 1978; Sokal, 1979).

As traditionally defined, spatial autocorrelation procedures are concerned with relating a given variable represented by $x_1, x_2, \ldots, x_n$ to an $n \times n$ proximity matrix $\underset{\sim}{P}$ through some (normalized) cross-product index. If small or large differences between x_i and x_j consistently correspond to small or large proximities p_{ij} and p_{ji}, the variable is assumed to reflect the information in $\underset{\sim}{P}$, either positively or negatively, depending on how the proximities are keyed. When this occurs, the relational data provided by $\underset{\sim}{P}$ can be partially explained in a correlational sense by the values $x_1, \ldots, x_n$. Although we have emphasized the geographical interpretation, $\underset{\sim}{P}$ could also contain distances or measures of relationship that are reconstructed, say, from some data-reduction strategy such as multidimensional scaling, hierarchical clustering, or additive tree analysis (Sattath and Tversky, 1977). The keying of the proximities may change in these latter applications to imply that small values in $\underset{\sim}{P}$ denote similar objects, but otherwise, the comparison strategy is carried out and interpreted as before.

To be somewhat more specific, a hierarchical clustering method produces a sequence of K partitions from level 0 to level $K - 1$, where (i) level $k + 1$ is obtained from level k by uniting one or more subsets in the latter; (ii) level 0 contains all n objects in separate classes; and (iii) level $K - 1$ has all objects united into one all-inclusive class. The proximity p_{ij} can be defined as the minimum level at which the two objects O_i and O_j appear together in a single subset, producing a set of values satisfying the ultrametric property: $p_{ij} \leq \max\{p_{ik}, p_{kj}\}$ for any triple O_i, O_j, and O_k. (In general, a proximity function could be reconstructed in this same way, even though the hierarchically arranged sequence of subsets have some overlap at each level.) An additive tree analysis, on the other hand, would generate a proximity p_{ij} as the sum of the branch lengths in the shortest path connecting the two objects. Here, a four-object condition is satisfied by the proximities:

$$p_{ij} + p_{kh} \leq \max \begin{cases} p_{ik} + p_{jh} \\ \\ p_{ih} + p_{kj} \end{cases}$$

for any quadruple O_i, O_j, O_k, and O_h. Finally, multidimensional scaling would obtain p_{ij} as the interpoint distance in a T-dimensional space, and where the proximities would now satisfy the triangle inequality: $p_{ij} \leq p_{ik} + p_{kj}$ for any triple O_i, O_j, and O_k. In this last area of multidimensional

scaling, a regression strategy is typically used to carry out a similar type
of evaluation with respect to a given variable, where the coordinates for
each point are treated as predictors of it (Chang and Carroll, 1968; Kruskal
and Wish, 1978). This strategy depends on an explicit coordinate representation, whereas the QA approach does not, and therefore the latter
strategy is more generally applicable.

Given applications of this type and the inherent generality of the task
of comparing a given variable to a proximity matrix, we should not be surprised to find uses for it in a variety of substantive areas and in contexts
that may or may not be explicitly spatial. For example, in sociology,
Winsborough, Quarantelli, and Yutzky (1963) discuss a problem of relating
a variable of productivity to the presence or absence of social interaction
for a group of n individuals, or, stated in another way, do individuals who
are similar in output also interact socially? Here, $\underset{\sim}{P}$ could represent a
zero-one matrix in which a 1 corresponds to a pair of individuals who dis-
play a certain type of social interaction, and a 0 if they do not; $\underset{\sim}{Q}$ could contain absolute values of the differences between output values for each pair
of individuals. Thus, $\mathcal{A}(\rho_I)$ is a raw measure of the degree to which people
who interact socially are similar in productivity as well (see Baker and
Hubert, 1981).

As a second application in sociology that is explicitly spatial, Campbell,
Kruskal, and Wallace (1966) developed an index of seating aggregation
(segregation) and an associated significance testing strategy for determining
whether the observed black-white seating adjacencies within a classroom
might be considered random. As stated, their strategy is a special case of
the QA model. The spatial framework is defined by the occupied seats
within a classroom, or more specifically, by the spatial locations of the
students in a two-dimensional plane. The outside variable of interest is
dichotomous, i.e., black or white, and the inference task is one of determining whether the spatial positioning of blacks and whites indicates
aggregation, e.g., whether blacks sit with blacks and whites sit with whites
(for a further discussion, see Freeman, 1978). Formally, $\underset{\sim}{P}$ contains
measures of spatial distance obtained from the observed seating pattern,
which Campbell et al. define as a very simple dichotomous variable:

$$p_{ij} = \begin{cases} 1 & \text{if } O_i \text{ and } O_j \text{ are seated adjacently within a single row} \\ 0 & \text{otherwise} \end{cases}$$

The second matrix $\underset{\sim}{Q}$ is also dichotomous and is based on race:

$$q_{ij} = \begin{cases} 1 & \text{if } O_i \text{ and } O_j \text{ are both black or both white} \\ 0 & \text{otherwise} \end{cases}$$

Consequently, the index $\mathcal{A}(\rho_I)$ is the number of same-race adjacencies observed in the given seating pattern, and evidence for aggregation is indicated by a large value. Obviously, these same strategies would be of value in other areas of social psychology as well, e.g., when the objects are census tracts and the task is to relate an outside variable to contiguity of the tracts defined geographically.

Serial and Cyclical Correlation. There are several special cases of the spatial autocorrelation idea that have a very long history in statistics. One particular application is to serial correlation where the n geographical (or time) locations are n positions, in order, along a continuum, and $q_{ij} = x_i x_j$, when $i \neq j$. If p_{ij} specifies simple adjacency, i.e., $p_{ij} = 1$ when $|i - j| = 1$ and 0 otherwise, then $\mathcal{A}(\rho_I)$ is twice the common serial correlation statistic of lag 1 discussed by Wald and Wolfowitz (1943) and others. Lag k statistics require $p_{ij} = 1$ when $|i - j| = k$ and 0 otherwise. In a similar way, cyclical correlation measures of lag k can be developed by redefining $p_{ij} = 1$ when $|i - j| = k$ or $n - k$ and 0 otherwise. This last specification essentially orders the n possible positions around a circle. Obviously, both cyclical and noncyclical versions of these correlation ideas could be generalized by using more than zero-one information between the given positions for the n objects.

The basic notion of serial correlation can be extended to a variety of substantive contexts and also to a variety of different measures of similarity between two observations. For example, because directionality is irrelevant along the continuum, our observations $x_1, x_2, \ldots, x_n$ could be obtained from an unoriented ecological transect (see Knight, 1974). Or, if our observations merely represent nominal class and

$$q_{ij} = \begin{cases} 1 & \text{if } x_i \text{ and } x_j \text{ refer to different classes} \\ 0 & \text{otherwise} \end{cases}$$

then a lag 1 index based on simple adjacency is merely twice the number of runs minus one. In a similar manner, the idea of serial correlation for nominal data could be easily extended to lags greater than one.

Association Between Spatially Defined Variables. The discussion in Section 2.3.1 generalized Tjøstheim's original index of association between two spatially defined variables to one of the form

$$\sum_{i,j} c_{ij} (f_i - g_j)^2 \qquad (4.3.9a)$$

where $\underset{\sim}{C} = \{c_{ij}\}$ defines a symmetric $n \times n$ matrix of spatial separation measures between the n locations, $O_1, \ldots, O_n$. The sequences $f_1, \ldots, f_n$

and $g_1, \ldots, g_n$ denote observations on two variables, F and G, in each of these n locations, respectively. Commensurability for the observations on F and G is assured by using, say, z scores or ranks (we refer to the variables as F and G to maintain consistency with the notation in Section 2.3.1).

When specialized to square matrices, the strategy for comparing two rectangular matrices discussed in Chapter 2 does not take advantage of the fact that the row and column objects are the same. Failure to do so, as noted in Section 2.4.1, results in a reference distribution that is partially contaminated with association that is aspatial in design. There is a rather simple way to resolve this difficulty, however, as long as the F and G variables are not identical. The index in (4.3.9a) is really of the QA form if $\underset{\sim}{P}$ is treated as the spatial separation matrix (with a zero main diagonal) and $\underset{\sim}{Q}$ is given by

$$q_{ij} = \begin{cases} (f_i - g_j)^2 & \text{if } i \neq j \\ 0 & \text{if } i = j \end{cases}$$

Or, because $\underset{\sim}{C}$ will be assumed symmetric, $\underset{\sim}{Q}$ could just as well be constructed from

$$q_{ij} = \begin{cases} \dfrac{(f_i - g_j)^2 + (f_j - g_i)^2}{2} & \text{if } i \neq j \\ 0 & \text{if } i = j \end{cases} \tag{4.3.9b}$$

Because the main diagonal of $\underset{\sim}{P}$ is assumed to be zero in the QA framework, the spatial separation of an object with itself will no longer affect the index or its distribution. This is in direct contrast to the Tjøstheim approach based on LA and the statistic in expression (4.3.9a). Furthermore, the QA inference model permutes rows and columns together, which, in effect, keeps each of the original observation pairs, e.g., (f_i, g_i), together in one of the n locations. Thus, all permutations used in constructing the reference distribution lead to exactly the same value for one of the usual aspatial measures of association, such as the Pearson correlation between F and G. In short, what may appear to be a minor change in the usual measure of spatial autocorrelation for a single variable, leads to a strategy for assessing the association between two variables in a purely spatial sense. We evaluate the degree to which the association between the F and G variables can be "explained" by their spatial realizations through the use of a reference distribution that is unrelated to one of the common characterizations of aspatial association between the two variables.

As mentioned in the discussion of Tjøstheim's index in Section 2.3.1, these same ideas could be used to assess the degree of association manifested by F and G as reflected through an arbitrary proximity matrix $\underset{\sim}{P}$, e.g., $\underset{\sim}{P}$ could be obtained from a hierarchical clustering, a multidimensional scaling, and so on. Such an application would parallel our earlier discussion in this section on comparing a single variable to a matrix. Furthermore, we could generalize the index in (4.3.9a) to other than squared functions of the difference between f_i and g_j. For example, $(f_i - g_j)^2$ could be replaced by absolute differences. Or possibly, if F and G consist of the n untied ranks, a zero-one function could be used, based on Tjøstheim's original idea of matching identical ranks only, except that now the spatial separations of an object to itself would not be relevant; i.e., the main diagonal of P is assumed to be zero and does not affect the index or its distribution.

Besides these rather obvious extensions to absolute values or zero-one scoring functions, it may also be appropriate to discuss how the task of assessing association between spatially defined variables can be rephrased in a way that suggests a number of less apparent alternatives to the use of q_{ij} in equation (4.3.9b). This latter definition merely produces a particular index of proximity based on how similar the values on F and G are for the two given locations. Thus, it should be possible to use other measures of proximity in exactly the same way to assess whether the values on these two variables covary with one another in relation to the magnitude of the spatial separation between the corresponding locations.

To be more explicit, suppose for the moment we ignore the spatial placement of each of the n bivariate pairs and consider one of the usual statistics of association between the two sequences $f_1, \ldots, f_n$ and $g_1, \ldots, g_n$, e.g., the coefficients of Pearson, Kendall, and Spearman. At least up to a multiplicative constant (i.e., a division), all of these can be put into a very specific form:

$$\sum_{i,j} q_{ij}$$

where q_{ij} is a symmetric measure between (f_i, g_i) and (f_j, g_j):

1. $q_{ij} = (f_j - f_i)(g_j - g_i)$ (Pearson)
2. $q_{ij} = \text{sign}(f_j - f_i)\text{sign}(g_j - g_i)$ (Kendall)
3. $q_{ij} = (\text{rank } f_j - \text{rank } f_i)(\text{rank } g_j - \text{rank } g_i)$ (Spearman)

where the ranks in (3) are defined within the sets $\{f_1, \ldots, f_n\}$ and $\{g_1, \ldots, g_n\}$. In all of these cases, the fact that the variables F and G are realized in particular spatial locations is of no consequence. Any permutation of the bivariate pairs among the n locations would lead to exactly the same value for the association index.

Because each of the traditional measures of association between the two sequences is additive with respect to the values of $\{q_{ij}\}$, an obvious further question might be to relate these latter entries to the n locations containing $(f_1, g_1), \ldots, (f_n, g_n)$. In other words, if the value q_{ij} is considered a measure of proximity between (f_i, g_i) and (f_j, g_j), and if p_{ij} codifies spatial separation information between locations O_i and O_j, an index $\mathcal{A}(\rho_I)$ comparing $\underset{\sim}{P}$ and $\underset{\sim}{Q}$ is an attempt to assess whether spatial association exists between the two sequences, i.e., the degree to which the pattern of entries in $\underset{\sim}{Q}$ is similar to the pattern of entries in $\underset{\sim}{P}$. Moreover, because $\Sigma_{i,j}\, q_{ij}$ remains constant under an inference model that permutes the observation pairs among the n locations, the idea of spatial association as measured by $\mathcal{A}(\rho_I)$ is distinct from that characterized by the original association measure $\Sigma_{i,j}\, q_{ij}$ [for a more complete discussion of the spatial association notion, along with several numerical examples, the reader is referred to Hubert, Golledge, Costanzo, and Gale (1985a,b)].

The spatial association notion can be interpreted in terms of comovement and contramovements between two sequences or series. For us, a comovement occurs between two locations whenever both series increase or decrease; a contramovement occurs when one series increases and the second decreases. As an illustration, suppose our context is unidimensional (e.g., temporal), and $\{p_{ij}\}$ defines the lag 1 adjacency criterion mentioned in our discussion of serial correlation. Using the Kendall specification of q_{ij} as $\text{sign}(f_j - f_i)\text{sign}(g_j - g_i)$, $\mathcal{A}(\rho_I)$ is the number of comovements minus the number of contramovements among adjacent positions along the continuum. This statistic is very similar to those discussed by Goodman and Grunfeld (1961) and Stuart (1952), although the inference model assumed here is different. Obviously, the specifications of Pearson and Spearman would weight these comovements and contramovements in particular ways. In any case, extensions to a multidimensional spatial context follow immediately.

The use of q_{ij} as defined through one of the usual association measures may be an obvious choice, but a variety of other alternatives could be pursued in the same way. We have already used the alternative in equation (4.3.9b) earlier, but we might note here that $\Sigma_{i,j}\, q_{ij}$ again gives a familiar measure of association because this term is a constant linear transformation of the Pearson product-moment coefficient or Spearman's index, depending on whether the observations are z scores or ranks, respectively. As another type of application, suppose the sequences $f_1, \ldots, f_n$ and $g_1, \ldots, g_n$ represent nominal classes and define the off-diagonal entries in $\underset{\sim}{Q}$ as:

$$\begin{cases} +1 & \text{if } f_i = f_j \text{ and } g_i = g_j \\ & \text{or } f_i \neq f_j \text{ and } g_i \neq g_j \\ \\ -1 & \text{if } f_i = f_j \text{ and } g_i \neq g_j \\ & \text{or } f_i \neq f_j \text{ and } g_i = g_j \end{cases}$$

Then, $\Sigma_{i,j} q_{ij}$ is the number of location pairs containing the same or different labels for the two sequences minus the number of location pairs containing the same labels in one sequence but different labels in the second. Using q_{ij} as a measure of relationship between (f_i, g_i) and (f_j, g_j), a notion of spatial association for nominal sequences could be developed in parallel to what can be done when the measurement level is higher.

4.3.10 The Comparison of Two Graphs (B)

An $n \times n$ matrix containing zero-one entries represents an unweighted graph defined explicitly as a set S of n nodes, $\{O_1, O_2, \ldots, O_n\}$, and a set of e edges joining certain pairs of these nodes. An entry of 1 denotes the presence of an edge, and a 0 denotes the absence. When the matrix is symmetric, the graph is said to be undirected and represents a symmetric relation on the object or node set. Alternatively, when the matrix is asymmetric, the corresponding graph is directed and the edges must be specified from one node to another. In either case, when the two matrices $\underset{\sim}{P}$ and $\underset{\sim}{Q}$ used in the QA paradigm are dichotomous, the graphs they represent will be denoted by G_P and G_Q.

Barton and David (1966), in developing a statistical approach to the study of contagious diseases, define a measure of correspondence between G_P and G_Q by the number of object pairs joined by edges in both G_P and G_Q. Under the hypothesis that the node labels in G_Q are fixed but the $n!$ possible labelings of the nodes in G_P are equally likely, Barton and David obtained the mean and variance of this statistic, which are special cases of the general moment formulas based on arbitrary $\underset{\sim}{P}$ and $\underset{\sim}{Q}$. When the graphs are undirected, the statistic $\mathcal{A}(\rho_I)$ is twice the number of distinct unordered object pairs that are joined in the both G_P and G_Q. If the graphs are directed, then $\mathcal{A}(\rho_I)$ is the number of common directed edges between G_P and G_Q. In the Barton and David application, the n nodes refer to outbreaks of some disease, G_P represents those cases within a critical distance from each other, and $\tilde{G}_Q$ defines those cases that had "close" times of onset. Thus, $\mathcal{A}(\rho_I)$ is an index of disease contagion.

In the undirected case, Barton and David (1966) and Abe (1969) discussed the asymptotic distribution of $\mathcal{A}(\rho)$ and noted that when G_P and G_Q are sparse (i.e., containing few edges), $\mathcal{A}(\rho)$ has a limiting Poisson distribution. Abe (1969) went on to demonstrate asymptotic normality when the number of edges satisfy certain regularity or growth conditions. In particular, if $n^{1/2} < d_P d_Q \longrightarrow \infty$, but $(d_P d_Q / n)^r \longrightarrow 0$ for $r > 2$ as $n \longrightarrow \infty$, then $[\mathcal{A}(\rho_I) - E(\mathcal{A}(\rho))] / \sqrt{V(\mathcal{A}(\rho))}$ has an asymptotic normal distribution with mean zero and variance 1. Here, d_P and d_Q are the average degrees in G_P and G_Q, where average degree is defined as twice the number of edges/n. When $d_P d_Q < n^{1/2}$, an asymptotic Poisson distribution is appropriate.

The problem of comparing two (directed) graphs is a classical one in sociometry and is referred to as evaluating the conformity of sociometric measurements (see Katz and Powell, 1953; Hubert and Baker, 1978a). As a common type of illustration in this context, suppose the set S contains n individuals, and define two (zero-one) sociometric choice matrices P and Q based on two tasks that we ask the individuals to perform: The first task requires each person to choose those members of S that he/she likes the most; the second requires the subject to select those members of S that he/she believes likes him/her the most. The matrices P and Q correspond to choices received and choices perceived, respectively:

$$p_{ij} = \begin{cases} 1 & \text{if subject j chooses subject i} \\ 0 & \text{otherwise} \end{cases}$$

$$q_{ij} = \begin{cases} 1 & \text{if subject i perceives a choice from subject j} \\ 0 & \text{otherwise} \end{cases}$$

Thus, $\mathcal{A}(\rho_I)$ denotes the number of consistencies between the two graphs G_P and G_Q, i.e., the number of cases where a choice was expected and also given.

Given two directed graphs, such as in the sociometric context, the correlation r_{PQ} takes on a fairly simple form:

$$r_{PQ} = \frac{n(n-1)n_{PQ} - n_P n_Q}{(n_P n_Q \bar{n}_P \bar{n}_Q)^{1/2}}$$

where n_{PQ} is the number of entries in P and Q both defined by a 1, $n_P(n_Q)$ is the number of 1's in $P(Q)$, and $\bar{n}_P(\bar{n}_Q)$ the number of off-diagonal 0's in $P(Q)$. This measure is suggested by Katz and Powell (1953), but their inference strategy essentially treats the P and Q matrices as sequences within the LA model of Part I, and does not recognize the possible "linked" character of the entries within each matrix. In particular, the strategy they rely on is the 2×2 Fisher exact test of Section 1.3.4 based on $n(n-1)$ "objects" categorized into two levels (0 and 1) from the data provided by P and Q. The QA procedure, on the other hand, maintains any internal structure present in the original sociometric matrices that is independent of a particular labeling of the subjects, e.g., a general tendency toward mutual choice (reciprocity), or more specifically, the presence of some type of clustering or clique structure. Finally, we note that varying strengths of choice, represented numerically, could be permitted by treating the socio-metric conformity problem within the general QA model for arbitrary matrices.

Many of the applications we have discussed in earlier sections could, if we wished, be rephrased in a graph theory terminology. For example, partitions of an object set S construct very specific types of undirected graphs in which an edge exists between two nodes if and only if the associated object pair belongs to the same class. Thus, the comparison between two partitions is, in effect, a comparison between two such graphs. More generally, if overlap is allowed between the classes of, say, a free-sort, then these structures could be compared through their corresponding graphs, where again an edge exists if and only if an object pair is placed in the same class. In an analogous way, the aggregation index of Campbell, Kruskal, and Wallace (1966) discussed in Section 4.3.9 could be restated as a comparison between two undirected graphs, and so on.

Because the language of graphs (or networks) is currently very popular in the social sciences (e.g., see Holland and Leinhardt, 1979) and provides a convenient way of discussing certain structural concepts,[†] it may be of interest to note a few graph theory notions in greater detail and how they might be incorporated within the QA framework. To provide a somewhat broader perspective, however, it will be assumed that P is now an arbitrary proximity matrix that defines a weighted graph, which is undirected or directed, depending on whether P is symmetric or not, respectively. When P is asymmetric, each ordered pair of nodes (O_i, O_j) in G_P defines an edge that is weighted by p_{ij}; if P is symmetric, then the unordered pair of nodes $\{O_i, O_j\}$ is weighted by p_{ij}. For now, the matrix Q will still be assumed dichotomous.

A (directed) Hamiltonian path is identified by a series of contiguous (consistently directed) edges passing through all nodes once and only once, and its length is defined by the sum of the weights on the $n - 1$ edges. If Q is defined by

$$q_{ij} = \begin{cases} 1 & \text{if } i = j - 1 \\ 0 & \text{otherwise} \end{cases}$$

[†] In applications to social network problems, the objects are typically in-dividuals, and the graphs reflect what are called "ties." The latter refer to a relationship that may or may not exist among the n individuals. For example, we can compare a given tie, represented as a dichotomous matrix P, to another tie or variable represented through Q (see Kawabata, Mul-herin, and Sonquist, 1981).

then $\mathcal{A}(\rho_0)$ defines the length of the path for the node sequence O_{i_1}, O_{i_2}, ..., O_{i_n}, where $\rho_0(i_k) = k$ (when $\underset{\sim}{P}$ is symmetric, this is one-half of Wilkinson's criterion discussed in the seriation context of Section 4.3.1). Thus, the reference distribution based on $\mathcal{A}(\rho)$ is for all n! possible (directed) Hamiltonian paths that could be constructed in G_P.

A simple modification of $\underset{\sim}{Q}$ to

$$q_{ij} = \begin{cases} 1 & \text{if } i = j - 1, \text{ or } i = n \text{ and } j = 1 \\ 0 & \text{otherwise} \end{cases}$$

produces (directed) Hamiltonian cycles rather than paths (see the discussion of circular placement in Section 4.3.1).

Instead of using the length of a directed Hamiltonian path as a measure, suppose we consider the directed node sequence $O_{i_1} \longrightarrow O_{i_2} \longrightarrow \cdots \longrightarrow O_{i_n}$ defined by $\rho_0(i_k) = k$ and delete all edges in G_P that are not consistent with this order, i.e., an edge $(O_{i_k}, O_{i_{k'}})$ where $O_{i_{k'}}$ comes before O_{i_k} in the above sequence. When all such edges are deleted, an acyclic directed graph is obtained (i.e., one without cycles); moreover, if

$$q_{ij} = \begin{cases} 0 & i < j \\ 1 & \text{otherwise} \end{cases}$$

then $\mathcal{A}(\rho_0)$ is the sum of the weights on the deleted edges [cf. (4.3.1a) which defines the complement sum of edge weights that remain]. Thus, the distribution for $\mathcal{A}(\rho)$ gives a distribution over the sums of weights for all possible sets of directed edges that when deleted would produce an acyclic graph.

A cut in an undirected graph G_P is defined as the set of edges that connect two complementary subsets of nodes D and S − D. The weight of a cut is the sum of the weights attached to the edges in the cut and measures the degree to which the subsets are "related" to one another. If the weights are 0 and 1, then this sum is merely the number of edges in the cut. In any case, if there are K nodes in D, and thus, n − K in S − D, and

$$q_{ij} = \begin{cases} 1 & \text{if } 1 \leqslant i \leqslant K \text{ and } K + 1 \leqslant j \leqslant n \\ 0 & \text{otherwise} \end{cases}$$

then the reference distribution for $\mathcal{A}(\rho)$ gives the distribution of weight over all cuts of this given size. Obviously, asymmetric generalizations are

possible and would correspond to our discussion of cluster statistics that
define isolation or separation in Section 4.3.3.

Finally, we note that the discussion of Scott's model in Section 4.3.5
could be presented in the context of matching in a graph with 2n nodes. The
term "matching" refers to a subset of edges such that no two meet at the
same node; thus, Scott's model provides the distribution of total weight
over all possible matchings in the graph. For a further development of the
matching problem in an optimization context, consult Edmonds (1965).

4.3.11 Subset Selection (B)

In the discussion of cluster statistics in Section 4.3.3, our concern was
with the distribution of an index of compactness and/or isolation for a subset of a given size K. Moreover, because of the way in which $\underset{\sim}{Q}$ was defined,
the n! equally likely permutations of $\underset{\sim}{P}$, in effect, produced only $\binom{n}{K}$ equally
likely subsets. Thus, the reference distribution for a cluster index was
reduced to a consideration of its value over this latter set. From a somewhat broader point of view, the reduction from equally likely permutations
to equally likely subsets will occur whenever $\underset{\sim}{Q}$ has the form

$$\underset{\sim}{Q} = \left[\begin{array}{c|c} \underset{\sim}{Q}_{11} & \underset{\sim}{Q}_{12} \\ \hline \underset{\sim}{Q}_{21} & \underset{\sim}{Q}_{22} \end{array} \right] \tag{4.3.11a}$$

where $\underset{\sim}{Q}_{11}$ and $\underset{\sim}{Q}_{22}$ are square and of size $K \times K$ and $n\text{-}K \times n\text{-}K$, respectively, and the entries within each of the submatrices are constant, except
possibly for the zero main diagonals in $\underset{\sim}{Q}_{11}$ and $\underset{\sim}{Q}_{22}$.

Even though the entries in $\underset{\sim}{Q}_{12}$, $\underset{\sim}{Q}_{21}$, and $\underset{\sim}{Q}_{22}$ are constant, as soon
as the off-diagonal values in $\underset{\sim}{Q}_{11}$ are allowed to vary, the notion of what
must be viewed as equally likely changes. Depending on how $\underset{\sim}{Q}_{11}$ is constructed, as many as $\binom{n}{K}K!$ different "structures" could result from the
original set of n! equally likely permutations of $\underset{\sim}{P}$, i.e., all subsets of size
K plus all K! orderings of the objects within each. In short, subset selection is again at issue, but now the reduction is to all $\binom{n}{K}K!$ possible
ordered subsets of size K.

This notion of subset selection arises in a number of contexts that
have been mentioned in earlier sections of this chapter. For example, in
the graph theory context of Section 4.3.10, when $\underset{\sim}{Q}_{11}$ is based on the same
definition used to generate Hamiltonian paths and $\underset{\sim}{Q}$ itself has the form

$$Q = \left[\begin{array}{c|c} \underset{\sim}{Q}_{11} & \underset{\sim}{0} \\ \hline \underset{\sim}{0} & \underset{\sim}{0} \end{array} \right] \qquad (4.3.11b)$$

then the distribution of $\mathcal{A}(\rho)$ is actually a distribution of length over all possible (directed) paths that pass through K nodes once and only once (see Carroll, Romney, Farmer, and Delvac, 1976). In a similar way, the Hamiltonian cycle definition generates a distribution of length over all (directed) cycles on K nodes. In fact, many of the structures that have been discussed for proximity matrices, e.g., partitions, seriations, and the like, could be used to define $\underset{\sim}{Q}_{11}$ in (4.3.11b). If the second matrix $\underset{\sim}{P}$ is partitioned in the same manner, i.e.,

$$P = \left[\begin{array}{c|c} \underset{\sim}{P}_{11} & \underset{\sim}{P}_{12} \\ \hline \underset{\sim}{P}_{21} & \underset{\sim}{P}_{22} \end{array} \right] \qquad (4.3.11c)$$

then $\mathcal{A}(\rho_I)$ is actually a cross-product index based on $\underset{\sim}{Q}_{11}$ and $\underset{\sim}{P}_{11}$ alone. However, the relative size of $\mathcal{A}(\rho_I)$ is not compared only to the K! permutations that would result if our attention were restricted to $\underset{\sim}{Q}_{11}$ and $\underset{\sim}{P}_{11}$. Instead, $\binom{n}{K} K!$ "structures" must be considered in generating a reference distribution against which $\mathcal{A}(\rho_I)$ can be compared. What we are doing, in effect, is to conjecture that part of the proximity matrix $\underset{\sim}{P}$, as given by $\underset{\sim}{P}_{11}$, will be patterned in a particular way, and furthermore, the specific pattern can be characterized by $\underset{\sim}{Q}_{11}$. Thus, the inference model based on (4.3.11b) is two-stage: First, it considers the selection of that part of $\underset{\sim}{P}$ we have identified (by assuming that all $\binom{n}{K}$ subsets are equally likely); and second, within the given subset, the proximities from $\underset{\sim}{P}$ should be reflected by the pattern represented by $\underset{\sim}{Q}_{11}$ (by assuming that all K! orderings within a given matrix are equally likely).

The evaluation of $\mathcal{A}(\rho_I)$ in this two-stage context depends intimately on how the matrix $\underset{\sim}{P}_{11}$ relates to the complete matrix $\underset{\sim}{P}$. For instance, as in one of the illustrations used above, suppose we wish to test the relative length for a path that is defined on K nodes. It could occur that the length is very small when $\underset{\sim}{P}_{11}$ and $\underset{\sim}{Q}_{11}$ are considered separately with the usual QA paradigm; but when they are imbedded in the matrix implied by (4.3.11b) and (4.3.11c), the path length could be relatively large if all the proximities in $\underset{\sim}{P}_{11}$ are themselves relatively large when considered as a subset of $\underset{\sim}{P}$. Or, the converse could happen.

The effects of a two-stage process can be seen fairly clearly in the free-recall context discussed in Section 4.3.4. When a subject is given a list of n words to study but recalls only K in some order, there are two ways of taking this initial selection process into consideration. First, it can be ignored, and an analysis based only on the K words actually recalled. This is essentially what was done in Section 4.3.4, without explicit mention, when it was assumed that all n words were recalled. In other words, the clustering index is defined and tested conditionally on what was actually produced. Second, a selection process could be incorporated into the inference model through a Q matrix of the form given in (4.3.11b). The observed index remains the same, but the baseline model would include the idea that the subject first chooses the subset of recalled objects at random and then provides a random order within the subset. These same comments apply whether we are using Q_{11} to pick up simple adjacency information or absolute differences in recall rank or reaction time.

A substantial debate exists in the free-recall literature as to which approach should be used, i.e., the conditional strategy or what may be called the unconditional strategy based on Q in (4.3.11b). As argued by Hubert and Levin (1977), the unconditional paradigm possibly confounds two rather important and separate phenomena, i.e., the degree of "pure" clustering based on what is actually recalled and the salience of the items chosen for recall. As developed there, the notion of clustering per se should probably be approached conditionally. However, a separate assessment should be carried out for the subset of items a subject chooses to recall. For example, the usual categorical structure of the total word list defines a zero-one matrix P that represents the experimenter-defined partition. A Q matrix of the form in (4.3.11b), where the off-diagonal entries in Q_{11} are equal to 1, provides a measure of compactness for the recall set produced by the subject. In this categorical case, such an assessment is formally equivalent to our discussion of legislative subgroup cohesion in Section 4.3.3. In a similar way, the salience or compactness of the recall set could be evaluated against other inter-item information, e.g., inter-item association norms, presentation order, and so on.

The distinction between a conditional versus an unconditional model is particularly relevant to at least one other application we have discussed— Scott's matching model for nominal-scale agreement introduced in Section 4.3.5. Here, Q is of size 2n × 2n and has the form

$$Q = \begin{bmatrix} 0 & Q_{12} \\ \hline 0 & 0 \end{bmatrix}$$

where Q_{12} is square and of size $n \times n$ with 1's along its main diagonal and
0's elsewhere. In developing a reference distribution for evaluating the
relative size of $\mathcal{A}(\rho_I)$, a first-stage random process picks out from P a
square off-diagonal matrix of the same size as Q_{12}. The rows and columns
of this chosen submatrix are then randomly permuted separately, producing
$(n!)^2$ different row and column orderings. In actuality, however, because
of the structure of Q_{12}, only n! such orderings need to be considered, just
as in the LA model of Part I.

From a slightly different perspective, suppose we consider the definition of Q in (4.3.11b). Because we are, in effect, sampling from the larger
matrix defined by P when a subset of size K is picked from a pool of size n,
what is present in the subset should also tell us something about the larger
matrix P in a classical inference sense. These graph sampling ideas have
been developed in some detail by Frank (e.g., see Frank, 1971). For
example, suppose our aim is to estimate the total sum of all proximities
in P, which is assumed unobservable, from the sum of proximities in a
randomly chosen subset of size K by defining all the off-diagonal entries of
Q_{11} to be 1. The expectation, $E(\mathcal{A}(\rho_0))$, where ρ_0 defines the randomly
selected subset, has the form

$$\frac{K(K-1)}{n(n-1)} \sum_{i,j} p_{ij}$$

Thus, when we "reverse" the equality, the expression

$$\frac{n(n-1)}{K(K-1)} \mathcal{A}(\rho_0)$$

is an unbiased estimate of $\Sigma_{i,j} p_{ij}$. Frank (1971) goes on to discuss how an
unbiased estimator of the variance of this latter quantity can be obtained
starting with the variance for $\mathcal{A}(\rho)$.

The notion of estimation we have just described can also be used to
demonstrate that Kendall's tau coefficient in the untied case is an unbiased
estimate of a population counterpart (see Kendall, 1970). To be explicit,
suppose a finite population is defined by n bivariate pairs, (x_i, y_i), $1 \leq i \leq n$,
and assume the subscript indexing is such that $y_i < y_{i+1}$. If we let

$$p_{ij} = \begin{cases} 1 & \text{if } x_i < x_j \\ 0 & \text{if } x_i > x_j \end{cases}$$

then $\Sigma_{i,j} p_{ij}$ is twice the number of pairs of observations for which the order on the x variable is consistent with the ordering on the y variable. The
population value, τ_{pop}, is a simple linear transformation of $\Sigma_{i,j} p_{ij}$:

$$\tau_{pop} = \frac{2 \sum\limits_{i,j} p_{ij}}{n(n-1)} - 1$$

If we sample K objects at random from the population of size n, then $\mathcal{A}(\rho_0)$ is twice the number of pairs in the sampled subset for which the order on the x variable is consistent with the order on y. The sample value of tau for these K objects, τ_{sam}, is based on a simple linear transformation of $\mathcal{A}(\rho_0)$:

$$\tau_{sam} = \frac{2\mathcal{A}(\rho_0)}{K(K-1)} - 1$$

Thus,

$$E(\tau_{sam}) = \frac{2E(\mathcal{A}(\rho))}{K(K-1)} - 1 = \frac{2}{K(K-1)} \left[\frac{K(K-1)}{n(n-1)} \sum_{i,j} p_{ij} \right] - 1 = \tau_{pop}$$

A similar problem of estimation for Spearman's rank order correlation is mentioned in Section 5.3.

4.3.12 Directional Data (B)

The task of evaluating association between two sets of matched directions was discussed earlier in Section 2.3.3, using the LA model and a simple measure of proximity between the appropriate unit vectors. On the basis of this same notion of proximity, many of the analysis problems encountered in the previous sections of this chapter will have obvious and direct parallels whenever the data set consists of directional observations. To give an illustration in the same association context as in Section 2.3.3 (but using a slightly different notation that is more consistent with the current chapter), suppose two sets of n directions are given: $X = \{x_1, \ldots, x_n\}$ and $Y = \{y_1, \ldots, y_n\}$, where x_i and y_i are p-component vectors of unit length that are matched in some a priori manner. If (off-diagonal) proximity is defined as (cf. Section 2.3.3)

$$p_{ij} = x_i^t x_j$$

$$q_{ij} = y_i^t y_j$$

then the task of evaluating the association between these two matched sets of directions can be reduced to a comparison of P and Q through the QA model.

The most obvious application of this correlational paradigm may be
to matched sets of physically determined directions. As an alternative
source of such data, however, suppose the n vectors in a single set $X =$
$\{x_1, \ldots, x_n\}$ are observed at times $\{t_1, \ldots, t_n\}$, where the times are
assumed important only in a periodic sense. For instance, on a daily
basis, $t_1, \ldots, t_n$ could be observed on the usual 24-hour cycle; thus, observations t_i and t_j are similar if the times of occurrence on the clock are
close, irrespective of the particular day, month, year, and so on. The
set of times can be converted to a set of directions around the clock to
define $Y = \{y_1, \ldots, y_n\}$. Thus, an evaluation of association between X and
Y is concerned with the degree to which similar directions in X correspond
to similar times on the clock. Obviously, the same conversion can be
made for any particular period of interest, e.g., daily, weekly, monthly,
and so on, and the second set of directions, Y, generated accordingly.

Besides the obvious application to assessing association between two
matched sets, the QA model is appropriate in a variety of specialized contexts, assuming that only one vector of directional observations is available.
The first proximity matrix P is generated from X in the usual way, but Q
is constructed by the researcher to obtain the specific analyses of interest.
To give one illustration, suppose the vectors in $X = \{x_1, \ldots, x_n\}$ have been
collected successively in some manner, e.g., through time or along a
unidimensional spatial continuum such as an ecological transect. The task
of assessing serial correlation is essentially one of evaluating the extent
to which observations that are relatively close together along the continuum
are also similar in a directional sense. Depending on how we wish to
measure this relationship, a variety of indices could be proposed by specifying q_{ij} appropriately. As one possibility that is a special instance of our
discussion of serial correlation in Section 4.3.9, Watson and Beran (1967)
and Epp, Tukey, and Watson (1971) depend on simple adjacency that can be
obtained by defining

$$q_{ij} = \begin{cases} 1 & \text{if } |i - j| = 1 \\ 0 & \text{otherwise} \end{cases}$$

Thus, the QA paradigm includes the work of Watson and Beran (1967) and
Epp et al. (1971) as a particular case.

Generalizing the simple adjacency structure given above, the definition
of Q could be changed to include detailed information on spatial separation.
In this way, an explicit strategy can be obtained for evaluating spatial autocorrelation for the directions in X. In fact, once it is recognized that a
set of directions can be used to provide a proximity matrix P, a variety of
data analysis strategies are immediately provided. Thus, the discussion
of spatial autocorrelation in Section 4.3.9 shows how similarity among
directions can be compared to the similarity among the values on a univariate variable. In fact, the values on this univariate variable may

represent the magnitudes of the n vectors. Thus, the comparison task is concerned with the degree to which direction and magnitude are related. Or, based on the extensions of this spatial correlation idea, the association between two spatially defined variables can be related to a similarity in the directions associated with the variables. In addition, some of our earlier discussion in this chapter gives obvious two-sample and/or analysis-of-variance procedures for directional data based on the tests for discrete structures such as subsets or partitions (see Hubert, Golledge, Costanzo, Gale, and Halperin, 1984, for a more complete discussion of these possible uses).

All of these applications indicate that the QA model gives an extremely flexible approach to the study of directional data. Even though the short presentation here points out how a variety of different analysis problems can be handled merely by redefining the matrix Q, there are several other variations that can be adopted to provide even greater degrees of generality. As one example, axial data come in the form of directional observations, but the sign of the usual proximity measure is irrelevant. All that is required to incorporate axial data into our approach is to replace the standard proximity measure by its corresponding absolute value. Clearly, other proximity measures could be used in the same way to index proximity between directional observations with no other changes required in the basic method of analysis.

Besides periodic phenomena or physically determined directional data, there are a number of other situations in which directional observations arise. At least one important application should be mentioned to multi-dimensional scaling by the individual differences model. Here, each subject who supplies part of the original data also provides a direction that defines the relative weights on the recovered dimensions that he/she uses to evaluate the set of scaled stimuli or objects. Consequently, the methods that have been presented can also be used to provide a broad class of data analysis strategies for evaluating the results of a study based on an individual-differences scaling model. The reader is referred to Schiffman, Reynolds, and Young (1981) for a discussion of the importance of these directional considerations in this specific scaling context.

4.3.13 Some Final Comments on Applications (B)

As should be obvious from all the special cases discussed thus far, the QA strategy is a very inclusive method that does not depend on any particular specification of the term "proximity." Judging from the preceding sections, the list of possible applications is very great; in fact, we are limited only by our ability to recast a specific analysis problem as an explicit comparison of two square matrices. We might also mention in this context the papers by Lerman (1980), Sokal (1979), Glick (1979), and Hubert (1983) that have also suggested, in various ways, the potential for the approach we have tried to document in this chapter.

Some of the more obvious uses of matrix comparison based on QA are in the psychometric field, where the n objects would typically be tests or items and the proximities (e.g., correlations) obtained on the same sample of subjects at two time points or from two independent samples (e.g., Schultz and Hubert, 1976). Or possibly, $\underset{\sim}{P}$ may be an empirically generated proximity matrix and Q obtained from the analysis of another data source, for example, by reconstructing a proximity matrix from a pattern found within a factor analytic framework, an ultrametric defined by some hierarchical clustering, the reconstructed euclidean distances from a multidimensional scaling, the metric obtained from an additive tree, and so on (e.g., see Hubert and Subkoviak, 1979). As always, our concern is with evaluating the degree to which the structure represented by Q is present in $\underset{\sim}{P}$. Consequently, if $\underset{\sim}{P}$ itself represents the results of some analysis based on a source distinct from that used in obtaining Q, the QA model provides one way of comparing the two different analyses on the same set of objects, e.g., two different classification schemes or two different multidimensional scalings. For a more complete discussion of this approach in the clustering area, the reader is referred to Hubert and Baker (1977b).

In addition to all the various general applications mentioned above as well as in the earlier sections, there are a wide variety of different substantive uses of a matrix comparison notion encountered in the literature that are directed toward rather specific problems. Although we obviously cannot mention all of them here, a few short illustrations may help emphasize the wide range of possibilities that do exist.

1. Suppose that our n objects are birds, and we have spatial locations for these n birds before and after migration. If P and Q contain before and after interpoint distances among nesting locations, respectively, then $\mathcal{A}(\rho_I)$ is a measure of locational pattern transference (see Besag and Diggle, 1977).

2. Recalling our discussion of multitrait-multimethod matrices of Section 3.2.4, the matrices $\underset{\sim}{P}$ and Q could now correspond to two different methods, and the n objects to traits. The concern here is with a consistency in correlational pattern among the traits across methods. In a similar manner, other correlational patterns could be tested within the other subblocks of a multitrait-multimethod table (see Hubert and Baker, 1978b).

3. The set S could contain a set of n units in a curriculum subject to certain prerequisite relationships that dictate when they should be taught. The matrix Q could be dichotomous and specify the assumed instructional hierarchy in terms of an acyclic directed graph; P could be obtained from the pass-fail proportions for item pairs that supposedly assess performance on the given units. A comparison of $\underset{\sim}{P}$ and Q should reflect the degree to which the assumed hierarchy on which the instruction is based is actually mirrored in the test data (see Baker and Hubert, 1977, for a more complete discussion).

4. If the n objects, $\{O_1, O_2, \ldots, O_n\}$, refer to a sequence of n behavioral instances that fall into one of K categories, the proximity p_{ij} could be 1 if behavior O_i appears immediately before behavior O_j. The matrix Q could then represent the given categorization and be used to obtain an index of the number of transitions from one category to another or some combination of possible transitions. Thus, our interests would be in evaluating whether an observed pattern of category transitions in the behavioral sequence could or could not have reasonably arisen by chance (e.g., see Wampold and Margolin, 1982). To provide a specific example, we might conjecture that the presence of a certain category k in the sequence would facilitate the immediate occurrence of a behavior from a certain other category k'. Thus, if Q is defined by $q_{ij} = 1$ if O_i belongs to category k and O_j belongs to category k', and 0 otherwise, then $\mathcal{A}(\rho_0)$ is the observed order of the n behavioral instances. Thus, if our conjecture were correct, $\mathcal{A}(\rho_0)$ should be relatively large compared to a reference distribution based on permuting the n behavioral instances at random. Obviously, a variety of such transition patterns could be assessed, based on an appropriate redefinition of Q.

5. One basic methodological task in human geography concerns the relationship between a cognitive map for a set of n geographical objects, which an individual is assumed to have represented in memory, and the objective configuration. Here, the matrix P could represent the subjective spatial relationships between the n locales, possibly generated by direct subjective numerical estimates of distance; Q could represent the objective configuration and contain actual interpoint distances between the same locales. A comparison of P and Q is one way of approaching the question of how well a given subjective cognitive map corresponds to a "real" one. For some further comments, see Richardson (1981).

4.4 COMPARISONS BETWEEN RELATED
   QA PROBLEMS (B)

The task of deciding which of two given matrices is closest to a third defines an important comparison problem that probably has no universally acceptable solution. For example, suppose P, Q, and R denote three proximity matrices, where R is the original data and P and Q are reconstructed by different analysis strategies from the information given in R. The two matrices, P and Q, could be obtained using different hierarchical clustering methods, different multidimensional scaling techniques, and so on. Or, possibly, P and Q could denote different structures conjectured a priori such as partitions, seriations, and the like. In this case, our interest would be in evaluating which conjecture is closest to the original data matrix R.

For these and related situations, it may be possible to define a heuristic comparison strategy based on a difference matrix constructed from $\underset{\sim}{P}$ and $\underset{\sim}{Q}$. Assuming that these latter two matrices are commensurable or have been made so by transforming the proximities, say, to mean zero and variance one, this difference could then be evaluated against $\underset{\sim}{R}$ through the comparison statistic $\mathcal{A}(\rho_I)$. Thus, what we have is an immediate matrix extension of the discussion in Section 3.4 on the task of assessing the equality of two-dependent correlation coefficients. We do note, however, that because no degree-of-freedom notion is available when testing for the superiority of a given reconstruction, the comparison method must be treated as a heuristic and interpreted with the appropriate caution.

As a simple illustration that may help clarify the evaluation process in this matrix context, suppose $\underset{\sim}{P}$ and $\underset{\sim}{Q}$ have zero-one entries and represent two different partitions with exactly the same number of classes and objects in each. Thus, $\underset{\sim}{P}$ and $\underset{\sim}{Q}$ are comparable without a transformation because they both contain the same number of ones and zeros. The difference matrix $\underset{\sim}{P} - \underset{\sim}{Q}$ contains (i) a +1, whenever the corresponding object pair is within the same subset for $\underset{\sim}{P}$ and in different subsets for $\underset{\sim}{Q}$; (ii) a -1, when the pair is within the same subset for $\underset{\sim}{Q}$ and in different subsets for $\underset{\sim}{P}$; and (iii) zeros everywhere else. If small proximities are keyed to denote similar objects, then a large and negative value for $\mathcal{A}(\rho_I)$ would indicate that the partition represented by $\underset{\sim}{P}$ is "better than" the partition represented by $\underset{\sim}{Q}$. Large positive values of $\mathcal{A}(\rho_I)$ would suggest the superiority of the partition represented by $\underset{\sim}{Q}$.

By extending this comparison logic one step further, we can even assess the goodness-of-fit for a single reconstruction, say $\underset{\sim}{Q}$, that is based on the original matrix $\underset{\sim}{R}$. Explicitly, if $\underset{\sim}{R}$ and $\underset{\sim}{Q}$ are commensurable or have been made so, and $\underset{\sim}{P} = \underset{\sim}{Q}$, then $\mathcal{A}(\rho_I)$ based on our earlier difference notion in effect compares the original data matrix $\underset{\sim}{R}$ and a residual matrix $\underset{\sim}{R} - \underset{\sim}{Q}$. Thus, the ability of $\underset{\sim}{Q}$ to reconstruct $\underset{\sim}{R}$ is mirrored by the comparison statistic $\mathcal{A}(\rho_I)$ relating $\underset{\sim}{R}$ and $\underset{\sim}{R} - \underset{\sim}{Q}$; if the fit is good, one expects a nonsignificant relationship between the original data matrix and the matrix of residuals. There is one technical difficulty with such an extension, however, that should be mentioned. Because the index $\mathcal{A}(\rho_I)$ can never be negative when $\underset{\sim}{R}$ is compared to $\underset{\sim}{R} - \underset{\sim}{Q}$, we might wish to restrict the permutations we consider to be equally likely to those that lead to positive index values. In a Monte Carlo significance test, for instance, the observed measure would be compared to the set of randomly generated positive indices; in turn, the sample size M used for the significance level would be the number of such randomly generated values (see Section 1.4.2).

Although the scheme we have outlined for comparing $\underset{\sim}{P} - \underset{\sim}{Q}$ and $\underset{\sim}{R}$ may initially appear straightforward, the method is so general that great care must be used when interpreting the results obtained from specific applications. As mentioned earlier, there is no degree-of-freedom notion for the strategy, and consequently, certain comparisons may be biased by the way in which they are constructed. For example, if $\underset{\sim}{R}$ is a raw proximity matrix and $\underset{\sim}{P}$ and $\underset{\sim}{Q}$ represent multidimensional scalings in k + 1 and k dimensions, then we might expect the index $\mathcal{A}(\rho_I)$ to be positive, because intuitively, there is a greater latitude for fitting in k + 1 dimensions rather than in k. It would be of great interest, however, if the statistic could not be declared significantly different from zero.

Obviously, when $\underset{\sim}{P}$ and $\underset{\sim}{Q}$ are generated from data analysis strategies that fit a differential number of "parameters" defined in some broad sense, caution is appropriate in the interpretation of a significant correlation because it may be expected on an a priori basis. No difficulty exists if the structure requiring the fewer "parameters" is the best or if $\mathcal{A}(\rho_I)$ is close to zero when referenced against the permutation distribution. Whenever a positive bias may exist in the comparison under study, a failure to reject could be the most interpretable outcome. Stated somewhat differently, a significant value for $\mathcal{A}(\rho_I)$ can be viewed as a necessary but not sufficient condition when arguing, say, for the superiority of one reconstruction over a second.

It should also be remembered that because the evaluation strategies are conditional on the obtained data, sample size considerations do not enter in any explicit degree-of-freedom calculation. Intuitively, when proximities are based on small sample sizes, then the structure that actually underlies these proximities is not being estimated very well. The comparison measure is attenuated, reducing the sensitivity of our procedure if true differences in structure exist. Conditional randomization inherits some of the same difficulties found in most of the newer data reduction procedures, i.e., there is no generally appropriate way of separating out a failure to identify structure from the lack of sufficient precision in the calculation of the original proximities. The reader is referred to Hubert and Golledge (1981b) and Hubert, Golledge, Kenney, and Costanzo (1982) for a further discussion of these comparison strategies.

Example. To illustrate how the comparison process can be carried out, we return to the data of Section 4.3.4 on the free-sorting of words. Two proximity matrices were given, one for adults and one for children, which we will now denote by $\underset{\sim}{P}_A$ and $\underset{\sim}{P}_C$, respectively. In addition, two different partitions were of interest, defining a "parts of speech" categorization and one based on the ideas from Paivio et al. (1968). The n × n matrices originally based on the unit weights [given as (i) in Section 4.3.4] will be denoted by $\underset{\sim}{Q}_1$ and $\underset{\sim}{Q}_2$, respectively. Furthermore, it will be assumed that all four matrices have been standardized to mean zero and variance 1.

There are four comparisons of major interest: $\underset{\sim}{P}_A$ and $\underset{\sim}{Q}_1 - \underset{\sim}{Q}_2$; $\underset{\sim}{P}_C$ and $\underset{\sim}{Q}_1 - \underset{\sim}{Q}_2$; $\underset{\sim}{P}_A - \underset{\sim}{P}_B$ and $\underset{\sim}{Q}_1$; and $\underset{\sim}{P}_A - \underset{\sim}{P}_B$ and $\underset{\sim}{Q}_2$. The results of these four comparisons are summarized below; the Monte Carlo distributions based on sample sizes of 999 are given in Table 4.4a.

Table 4.4a Monte Carlo Distributions for the Comparisons Discussed in the Text (Sample Size of 999)

	Comparison			
Cumulative frequency	$\underset{\sim}{P}_A$ to $\underset{\sim}{Q}_1$-$\underset{\sim}{Q}_2$	$\underset{\sim}{P}_C$ to $\underset{\sim}{Q}_1$-$\underset{\sim}{Q}_2$	$\underset{\sim}{Q}_1$ to $\underset{\sim}{P}_A$-$\underset{\sim}{P}_C$	$\underset{\sim}{Q}_2$ to $\underset{\sim}{P}_A$-$\underset{\sim}{P}_C$
1	−77.1	−68.5	−48.7	−66.6
5	−59.3	−57.1	−40.4	−43.2
10	−53.6	−51.5	−37.5	−37.7
50	−35.2	−36.2	−25.6	−25.4
100	−27.1	−27.2	−20.9	−20.0
200	−18.2	−17.1	−13.2	−13.0
300	−10.7	− 9.9	− 8.5	− 8.3
400	− 6.0	− 4.4	− 3.3	− 3.8
500	− .4	.0	.8	.2
600	4.7	4.6	5.1	4.2
700	9.8	9.4	8.6	9.3
800	16.1	16.5	14.7	15.5
900	25.4	27.5	22.1	23.2
950	35.3	37.0	29.5	31.6
990	47.7	51.4	43.4	42.0
995	57.0	54.6	46.9	45.5
999	67.8	74.1	67.0	59.2

	Comparison			
	$\underset{\sim}{P}_A$ to $\underset{\sim}{Q}_1$-$\underset{\sim}{Q}_2$	$\underset{\sim}{P}_C$ to $\underset{\sim}{Q}_1$-$\underset{\sim}{Q}_2$	$\underset{\sim}{Q}_1$ to $\underset{\sim}{P}_A$-$\underset{\sim}{P}_C$	$\underset{\sim}{Q}_2$ to $\underset{\sim}{P}_A$-$\underset{\sim}{P}_C$
Observed index	−4.73	−61.76	148.4	91.36
Expectation	0	0	0	0
Variance	469.2	468.1	288.1	283.9
Z	−.029	−2.85	8.74	5.42
γ	−.22	.029	.233	.229
Significance level				
Monte Carlo	.430	.003	.001	.001
Type III	.413	.002	.001	.001

$$r_{\underset{\sim}{P}_A,\underset{\sim}{Q}_1-\underset{\sim}{Q}_2} = -.016 \quad r_{\underset{\sim}{P}_C,\underset{\sim}{Q}_1-\underset{\sim}{Q}_2} = -.208 \quad r_{\underset{\sim}{Q}_1,\underset{\sim}{P}_A-\underset{\sim}{P}_C} = .649 \quad r_{\underset{\sim}{Q}_2,\underset{\sim}{P}_A-\underset{\sim}{P}_C} = .399$$

The comparison of $P_A - P_B$ to Q_1 and to Q_2 suggests that the adult data is "closer" than the children's to both conjectured partitions. Based on the remaining two comparisons, $Q_1 - Q_2$ to P_A and to P_C, it appears that the adult data is about equally represented by either conjecture, but the child matrix is significantly better represented by Q_2. This last observation is true even though Q_2 is closer to the adult data than to the children data. These same conclusions are also reflected by the four correlations, given below, that provide normalized indices for each specific comparison. (We do note, however, that the significance levels reported are all one-tailed, whereas two-tailed values may be more appropriate. One simple alternative is to double the probabilities given; for some additional comments, consult the first footnote in Section 1.2.2.)

4.5 CORRELATION BETWEEN QA (AND LA) INDICES (C)

For many of the comparison strategies we have discussed, some arbitrariness usually exists in the exact choice of at least one of the matrices. For example, in comparing a partition against an empirically derived proximity matrix, there are a number of options for weighting the within-class proximities. Analogously, in developing an index for spatial autocorrelation or for seriation comparisons based on symmetric proximities, squared or absolute differences between variables could be used. Or, when defining measures appropriate for a single cluster of objects, we could depend on within-cluster proximities, those between the cluster and its complement, or some combination of both. In all of these cases, and many others as well, we may wish to obtain the correlation among different QA statistics over all n! permutations, just as we did for LA indices in Section 1.4.3. In this way, we can obtain some indication as to how these choices affect the comparison strategy itself.

To provide some formal notation, suppose $A_1(\rho)$ refers to the comparison of P_1 and Q_1 whereas $A_2(\rho)$ refers to the comparison of P_2 and Q_2. The covariance of $A_1(\rho)$ and $A_2(\rho)$ over all n! possible permutations can be obtained through the same type of indicator strategy that led to the variance terms in Section 4.2.2 (see Section 1.4.1 for an explicit discussion of this method). As might be expected, the expression for $cov(A_1(\rho), A_2(\rho))$ has the same structure as the variance given in Section 4.2.2. Each of the terms in the variance, B_1, B_2, $\ldots$, B_7, is replaced by one of a similar form, using the entries from P_1, P_2, Q_1, and Q_2. The necessary conversions are indicated in the Appendix.

In its general form, the covariance is fairly complicated; however, numerical evaluation for a given problem is fairly straightforward. For some special instances, it also may be possible to obtain fairly simple expressions, after a lot of very tedious algebra, depending on the structure of the matrices. For example, Daniels (1944) has presented several examples based on the skew-symmetric case, e.g., the correlation for untied ranks between Spearman's and Kendall's correlation coefficient is $[2(n + 1)]/\sqrt{2n(2n + 5)}$ which $\longrightarrow 1$ as $n \longrightarrow \infty$. The Appendix also gives a general formula for the covariance between LA and QA statistics, i.e., $\text{cov}(\Gamma(\rho), \mathcal{A}(\rho))$. The above result connecting Spearman's and Kendall's coefficients could also be derived in this way.

We give two simple illustrations of these covariation ideas, the first in the seriation context. Suppose $\underset{\sim}{P}$ refers to a matrix of proximities that represents a perfect spacing of one unit along a continuum, i.e., $\{p_{ij}\} = \{|i - j|\}$. Our concern is with the relationship between the Szczotka and Wilkinson indices; the former is based on $\underset{\sim}{Q_1}$ of the same form as $\underset{\sim}{P}$ and the latter on $\underset{\sim}{Q_2} = \{q_{ij}^{(2)}\}$, where $q_{ij}^{(2)} = 1$ if $|i - j| = 1$ and 0 otherwise. If $\mathcal{A}_1(\rho)$ refers to the comparison of $\underset{\sim}{P}$ and $\underset{\sim}{Q_1}$ and $\mathcal{A}_2(\rho)$ refers to the comparison of $\underset{\sim}{P}$ and $\underset{\sim}{Q_2}$, then

$$\text{correlation } (\mathcal{A}_1(\rho), \mathcal{A}_2(\rho)) = \frac{-(n - 1)(3n - 4)}{\sqrt{(n(n^2 - 1)(n^2 + 38n - 48)(4n - 7))/90}}$$

which, surprisingly, converges to 0 as $n \longrightarrow \infty$.

As a numerical comparison, the short table that follows is informative:

n	Correlation
4	-.89
10	-.59
20	-.40
100	-.12

Because the two indices get more and more unrelated as n gets larger, one should not be surprised if the two indices lead to differing conclusions regarding the adequacy of a conjectured seriation. At least in an intuitive sense, Szczotka's index is based on the evaluation of the more complete pattern that we would expect if a unidimensional scale is being reflected by a conjectured order. For this reason alone, Szczotka's criterion may be preferable to Wilkinson's, particularly in light of the correlational result given above.

A second example provides an interesting comparison of two cluster statistics defined in Section 4.3.3. For convenience and without loss of generality, the symmetric proximity matrix $\underset{\sim}{P}$ is assumed standardized in such a way that $\Sigma_{i,j} \, p_{ij} = 0$, $\Sigma_{i,j} \, p_{ij}^2 = 1$. Our concern is with the sum of within-cluster proximities for a subset D of size K as characterized by $\underset{\sim}{Q_1}$, where

$$q_{ij}^{(1)} = \begin{cases} 1 & \text{if } 1 \leq i \neq j \leq K \\ 0 & \text{otherwise} \end{cases}$$

and the sum of proximities between D and S – D as characterized by Q_2, where

$$q_{ij}^{(2)} = \begin{cases} 1 & \text{if } 1 \leq i \leq K \text{ and } K + 1 \leq j \leq n \\ & \text{or } K + 1 \leq i \leq n \text{ and } 1 \leq j \leq K \\ 0 & \text{otherwise} \end{cases}$$

After some substantial simplification, the correlation between the compactness and isolation measures over all $\binom{n}{K}$ subsets can be represented by

$$\text{correlation } (\mathcal{A}_1(\rho), \mathcal{A}_2(\rho)) =$$

$$\frac{2(K-1)(A(1+n-2K)+(1-n+K))}{\sqrt{2(K-1)[(2(K-2)A+(n-K-1)]\{(n-2)(n-3)A+2(1-2A)[K(n-K)-(n-1)]\}}}$$

where

$$A = \sum_i \left(\sum_j p_{ij} \right)^2$$

If $n \longrightarrow \infty$ and $K/n \longrightarrow p$, then assuming A has size of order n (due to the standardization of $\underset{\sim}{P}$),

$$\text{correlation } (\mathcal{A}_1(\rho), \mathcal{A}_2(\rho)) \longrightarrow \begin{cases} +1 & \text{if } p < 1/2 \\ -1 & \text{if } p > 1/2 \\ 0 & \text{if } p = 1/2 \end{cases}$$

This result is rather counterintuitive but suggests that for large n and $p \neq 1/2$, the compactness and isolation measures are redundant but in opposite ways, depending on whether p is greater than 1/2 or less than 1/2. Thus, in an informal sense it would be more "unusual" to observe a relatively compact and relatively isolated cluster for $p < 1/2$ than for $p > 1/2$. For example, suppose we consider the average within and the average between proximity and obtain a Z value of +1 for the former and –1 for the latter. For large n and $p = 1/4$, the Z value for Johnson's index defined by the difference would be increased to about 2. However, when $p = 3/4$, the Z value for Johnson's index would remain at about 1.

The reason for the disparity in asymptotic correlations becomes some-what more understandable when we note that the sum of within-cluster proximities for D and a similar measure for its complement S - D have an asymptotic correlation of -1. Thus, when $p < 1/2$, the subset D is smaller in size than S - D, which would imply that if D were relatively elevated, then the proximities within S - D should be relatively small. Given the size differential of the two subsets, the proximities within D and S - D, as an aggregate, would also be relatively small. Furthermore, because the between measure is a constant minus the sum of proximities within D and within S - D, the proximities between D and S - D would tend to be relatively large, i.e., a positive correlation would exist between $\mathcal{A}_1(\rho)$ and $\mathcal{A}_2(\rho)$. Similar reasoning when $p > 1/2$ would lead to a conclusion of a negative correlation for $\mathcal{A}_1(\rho)$ and $\mathcal{A}_2(\rho)$.

4.6 PARTIAL ASSOCIATION (B)[†]

The comparison tasks discussed throughout this chapter involve the cor-respondence between two numerical patterns of proximity defined among the objects in S, where the latter patterns are presumably based on two differ-ent informational sources or data sets. As a derivative concern that will be the primary emphasis in this section, our interest changes to the problem of partially attributing the degree of relationship exhibited between these first two patterns to a third. Thus, if the first task is one of assessing some broad notion of correlation or association, the second could be re-phrased through a derivative concept of partial correlation or association.

The general QA framework set up for comparing two proximity matrices can be extended to the related problem of assessing a particular notion of partial association. Again, it is assumed that two proximity measures are available on all pairs in S and represented in the usual way by the matrices $\underset{\sim}{P}$ and $\underset{\sim}{Q}$. As one additional component, there is now a third proximity matrix, $\underset{\sim}{T} = \{t_{ij}\}$, and our task is to relate the pattern of infor-mation present in $\underset{\sim}{T}$ to the pattern of association observed between $\underset{\sim}{P}$ and $\underset{\sim}{Q}$. Assuming throughout this section that the observed permutation ρ_0 is the identity, ρ_I, the general strategy will be to first treat the $n(n-1)$ compo-nents making up $\mathcal{A}(\rho_I)$, which have the form $p_{ij}q_{ij}$ for $1 \leqslant i \neq j \leqslant n$, as indivisible units; a "new" composite measure is then defined between the objects O_i and O_j and $a_{ij} = p_{ij}q_{ij}$. The matrix $\underset{\sim}{A}$ constructed in this manner can itself be compared to $\underset{\sim}{T}$ by another application of the QA model. As a convenience in making the notation somewhat easier to extend, we will denote the usual cross-product measure, $\mathcal{A}(\rho_I)$, between $\underset{\sim}{P}$ and $\underset{\sim}{Q}$, as $\mathcal{A}_{PQ}$, which makes explicit what matrices are being compared.

[†]This section is based on Hubert (1985).

<u>Proximity Interpretations.</u> Before we try to develop this idea of partial association in detail, it may be helpful to digress slightly and develop greater precision regarding the type of proximity matrix we might be dealing with. In particular, it will be relevant, at least conceptually, whether a comparison is carried out using asymmetric dominance matrices or symmetric matrices.

Recalling the discussion of Section 4.2.2, any proximity matrix, say P, possibly contains two distinct types of information about the relationships between the objects in S through the decomposition of P into the sum of a symmetric and a skew-symmetric matrix:

$$P = P^+ + P^-$$

where $p_{ij}^+ = p_{ji}^+$ and $p_{ij}^- = -p_{ji}^-$ for all i and j. Obviously, when P is initially symmetric, then P^- is the null matrix; similarly, if $p_{ij} + p_{ji}$ is equal to a constant for all $i \neq j$, then P^+ is a noninformative matrix of constant off-diagonal entries. In any case, P^+ provides numerical information as to the degree of mutual relationship between the objects in S; P^- specifies the type of asymmetric dominance structure that may exist, with the ± signs providing the directionality of dominance, and with the absolute magnitude of the entries in P^- indicating the degree of dominance.

In a comparison of P and Q, the raw cross-product measure that we have used thus far, $\mathcal{A}_{PQ}$, confounds symmetric and skew-symmetric information because

$$\mathcal{A}_{PQ} = \mathcal{A}_{P^+Q^+} + \mathcal{A}_{P^-Q^-}$$

i.e., the overall raw index $\mathcal{A}_{PQ}$ is a simple sum of two cross-product measures: the first, $\mathcal{A}_{P^+Q^+}$, reflects similarity in the symmetric components of P and Q; the second, $\mathcal{A}_{P^-Q^-}$, reflects similarity in the skew-symmetric components. Obviously, when Q is skew-symmetric (or $q_{ij} + q_{ji}$ is constant) and P is symmetric, then $\mathcal{A}_{PQ}$ is 0, and P and Q are essentially incomparable. Alternatively, when P or Q is symmetric (or skew-symmetric), then the other matrix can be assumed symmetric (or skew-symmetric) without loss of generality because $\mathcal{A}_{P^-Q^-} = 0$ (or $\mathcal{A}_{P^+Q^+} = 0$). In all other cases, $\mathcal{A}_{PQ}$ merely combines the symmetric and skew-symmetric comparison measures together.

Because we may (and would usually) wish to treat these two types of information differently in discussing any overall association between A and B, particularly as it relates to some other source of information, it is assumed from now on that A and B are either both symmetric or both skew-symmetric. In the event that A and B are initially asymmetric with non-trivial symmetric and skew-symmetric components, the latter will be dealt with through two separate comparisons.

Background. When we try to develop an idea of partial association, and, in contrast to a simple comparison of $\underset{\sim}{P}$ and $\underset{\sim}{Q}$ through $\mathcal{A}_{PQ}$, it makes at least some interpretive difference whether $\mathcal{A}_{P^+Q^+}$ or $\mathcal{A}_{P^-Q^-}$ is involved. To be specific, suppose $\underset{\sim}{P}$ and $\underset{\sim}{Q}$ are initially symmetric (or, alternatively, consider P^+ and Q^+) and assume that the entries in both $\underset{\sim}{P}$ and $\underset{\sim}{Q}$ are nonnegative (or have been transformed to be nonnegative) and have a dissimilarity interpretation, i.e., large values of p_{ij} and q_{ij} refer to dissimilar objects O_i and O_j. We wish to argue that the degree of observed association between $\underset{\sim}{P}$ and $\underset{\sim}{Q}$ might be attributed, in a correlational sense, to a second source of proximity (dissimilarity) information as represented by another nonnegative symmetric matrix T. Because $\mathcal{A}_{PQ}$ is additive over the products $p_{ij}q_{ij}$ and both large and small values for the latter contribute to an elevated index in relation to its expectation under randomness, one would hope that the corresponding entry in $\underset{\sim}{T}$, t_{ij}, is also large or small concomitantly. Thus, given the consistency in the keying of t_{ij}, p_{ij}, and q_{ij} as symmetric dissimilarity measures, we would expect a relatively large cross-product index between $\{t_{ij}\}$ and $\{p_{ij}q_{ij}\}$, i.e., large (and small) values of t_{ij} should correspond to large (and small) values of $p_{ij}q_{ij}$; or stated in another way, large (and small) values of t_{ij} should correspond simultaneously to large (and small) values of p_{ij} and q_{ij} whenever some of the association between $\underset{\sim}{P}$ and $\underset{\sim}{Q}$ might be attributed correlationally to $\underset{\sim}{T}$. Conversely, if the pattern of association between $\underset{\sim}{P}$ and $\underset{\sim}{Q}$ is not reflected by $\underset{\sim}{T}$, then relatively large values of the cross-product measure between $\{t_{ij}\}$ and $\{p_{ij}q_{ij}\}$ are not expected. If we would have consistently keyed $\underset{\sim}{P}$, $\underset{\sim}{Q}$, and $\underset{\sim}{T}$ as positive symmetric similarity measures in which large values now reflect similar values, exactly the same discussion would have ensued. Again, if $\underset{\sim}{T}$ reflects the composite of $\underset{\sim}{P}$ and $\underset{\sim}{Q}$, then a relatively large cross-product index might be expected between $\{t_{ij}\}$ and $\{p_{ij}q_{ij}\}$.

Suppose we turn to the case in which $\underset{\sim}{P}$ and $\underset{\sim}{Q}$ are initially skew-symmetric (or, alternatively, consider P^- and Q^-). As a technical convention that will prevent annoying digressions as to how the entries in $\underset{\sim}{P}$ and $\underset{\sim}{Q}$ are keyed, it is assumed from now on that large absolute values of p_{ij} and q_{ij} reflect increased levels of dominance of one object over the other, i.e., the matrices $\{|p_{ij}|\}$ and $\{|q_{ij}|\}$ are symmetric dissimilarity measures, using the interpretation that when two objects are very dissimilar, one will have a large degree of dominance over the other. Secondly, the directionalities implicit in $\underset{\sim}{P}$ and $\underset{\sim}{Q}$ are compatible in the sense that $\mathcal{A}_{PQ}$ is greater than or equal to $E(\mathcal{A}_{PQ})$. Otherwise, the signs in one of the matrices could be reversed, just as in the usual correlational context where we may always deal with positive correlation by reflecting one of the variables.

Given these conventions, our interest is again in relating the degree of (positive) association between $\underset{\sim}{P}$ and $\underset{\sim}{Q}$, as manifested in the cross-product matrix $\{p_{ij}q_{ij}\}$, to a second matrix $\underset{\sim}{T}$. Because $\{p_{ij}q_{ij}\}$ is symmetric even though constructed from skew-symmetric components, $\underset{\sim}{T}$ can be assumed symmetric as well. If some of the (positive) association between $\underset{\sim}{P}$ and $\underset{\sim}{Q}$

can be attributed to $\underset{\sim}{T}$, then one would expect that large values of t_{ij} (reflecting dissimilarity) would correspond to large positive values of $p_{ij}q_{ij}$, i.e., products that have their two components with the same signs and large absolute magnitudes in the degrees of dominance.

Some Formal Details. From the historical perspective developed very nicely by Quade (1974), our interest is in generalizing the idea of "holding some variable constant" as represented through $\underset{\sim}{T}$. [This is in contrast to "adjusting" for some variable through the use of residuals and/or applying the generic product-moment system developed by Somers (1959]. In particular, as a raw measure of how much association in $\{p_{ij}q_{ij}\}$ might be attributable to $\underset{\sim}{T}$, we consider the cross-product

$$\mathcal{A}_{PQ(T)} = \sum_{i,j} t_{ij}[p_{ij}q_{ij}] = \mathcal{A}_{TA} = \sum_{i,j} t_{ij}a_{ij}$$

where as a technical convenience, all the proximities are keyed consistently. In each instance, $\underset{\sim}{T}$ will be treated as a nonnegative, symmetric, dissimilarity matrix. When $\underset{\sim}{P}$ and $\underset{\sim}{Q}$ are both symmetric, they will also be nonnegative dissimilarity mattices; when $\underset{\sim}{P}$ and $\underset{\sim}{Q}$ are skew-symmetric, the association between $\underset{\sim}{P}$ and $\underset{\sim}{Q}$ will be assumed positive [i.e., $\mathcal{A}_{PQ} \geq E(\mathcal{A}_{PQ})$], with the absolute values of p_{ij} and q_{ij} interpreted as dissimilarities for which large values reflect increased levels of dominance. Given the keying of t_{ij} and a_{ij}, we would typically look for a relatively high value of $\mathcal{A}_{PQ(T)}$ if some of the association between $\underset{\sim}{P}$ and $\underset{\sim}{Q}$ could be "attributed to" the information in $\underset{\sim}{T}$. In this latter case, dissimilar objects as reflected by a high value of t_{ij} would be mirrored by a relatively larger value of $a_{ij} = p_{ij}q_{ij}$.

Irrespective of these interpretive concerns, the actual mechanics of evaluating the degree to which the association between $\underset{\sim}{P}$ and $\underset{\sim}{Q}$ can be attributed to the pattern of proximities in $\underset{\sim}{T}$ may be carried out directly using the QA's inference scheme in a comparison of $\underset{\sim}{T}$ and $\underset{\sim}{A}$. In making this latter comparison, we obtain one convenient result which will have some nice implications for the special cases presented below. Under the random-permutation model, the expectation of $\mathcal{A}_{PQ(T)}$ can be rewritten as

$$E(\mathcal{A}_{PQ(T)}(\rho)) = E\left[\sum_{i,j} t_{ij}a_{\rho(i)\rho(j)}\right]$$

$$= \left[\frac{1}{n(n-1)}\sum_{i,j} t_{ij}\right]\sum_{i,j} p_{ij}q_{ij}$$

$$= \left[\frac{1}{n(n-1)}\sum_{i,j} t_{ij}\right]\mathcal{A}_{PQ}$$

Thus, the expectation for the index $\mathcal{A}_{PQ(T)}$ is a constant times the original raw measure $\mathcal{A}_{PQ}$; the multiplicative constant is merely the average proximity in $\underset{\sim}{T}$.

It should be emphasized that this type of comparison is, in effect, a test of the size of $\mathcal{A}_{PQ(T)}$ in relation to the given value of $\mathcal{A}_{PQ}$, and how the observed correspondence between $\underset{\sim}{P}$ and $\underset{\sim}{Q}$ is reflected by the pattern in $\underset{\sim}{T}$. We expect the index, $\mathcal{A}_{PQ(T)}$, under the randomization structure, to be a constant times the original index. Thus, the degree to which $\mathcal{A}_{PQ(T)}$ is greater than its expectation indicates the degree to which some of the original association between $\underset{\sim}{P}$ and $\underset{\sim}{Q}$ might be attributed to (or "explained" by) $\underset{\sim}{T}$. Conversely, the degree to which $\mathcal{A}_{PQ(T)}$ is less than its expectation indicates the magnitude of a suppression effect that $\underset{\sim}{T}$ may have on the observed relation between $\underset{\sim}{P}$ and $\underset{\sim}{Q}$.

We can carry out several further algebraic steps if the values in $\underset{\sim}{T}$ are restricted to lie between 0 and 1, i.e., $0 \leq t_{ij} \leq 1$ for $1 \leq i, j \leq n$. Specifically, a simple decomposition of the raw index $\mathcal{A}_{PQ}$ can be given as

$$\mathcal{A}_{PQ} = \sum_{i,j} t_{ij}(p_{ij}q_{ij}) + \sum_{i,j} \bar{t}_{ij}(p_{ij}q_{ij}) = \mathcal{A}_{PQ(T)} + \mathcal{A}_{PQ(\bar{T})} \tag{4.6a}$$

where

$$\bar{t}_{ij} = \begin{cases} 1 - t_{ij} & \text{for } i \neq j \\ 0 & \text{for } i = j \end{cases}$$

Using equation (4.6a) and extending the terminology used by Davis (1971), the first term $\mathcal{A}_{PQ(T)}$ is the differential index, and $\mathcal{A}_{PQ(\bar{T})}$ is the partial. Thus, the original (zero-order) measure $\mathcal{A}_{PQ}$ is the sum of the partial and differential measures. Given a fixed value for the zero-order measure, the latter are completely intertwined because when one increases, the other must decrease. The differential index suggests how much of the raw index $\mathcal{A}_{PQ}$ might be attributed to $\underset{\sim}{T}$; the partial index measures how much of $\mathcal{A}_{PQ}$ is left when $\underset{\sim}{T}$ is taken into account. (Given the restriction, $0 \leq t_{ij} \leq 1$, we are implicitly assuming that the transformation from t_{ij} to $\bar{t}_{ij}$ merely reflects the directionality of measurement. Thus, when we are faced with an arbitrary nonnegative dissimilarity matrix $\underset{\sim}{T}$, we would typically carry out a rescaling of the entries by dividing each value in $\underset{\sim}{T}$ by the maximum. In those cases where $\underset{\sim}{T}$ is initially a zero-one dichotomous matrix (and no rescaling is necessary) or when the maximum value in $\underset{\sim}{T}$ could be determined theoretically (e.g., using the number of subjects in a free-sort cooccurrence matrix), this standardization is unambiguous and not dependent on the actual entries in $\underset{\sim}{T}$. Unfortunately, in those instances where the maximum in $\underset{\sim}{T}$ must be determined empirically from the entries in $\underset{\sim}{T}$ themselves, the scaling is dependent on the particular data set being used and could make comparisons between different studies problematic. There does not appear to be an obvious solution to this problem whenever the scaling constant must be determined empirically. At best, we must assume that whatever transformation of the entries in $\underset{\sim}{T}$ may have been carried out, t_{ij} and $\bar{t}_{ij}$ represent comparable degrees of relationship, with the first indicating dissimilarity and the second indicating similarity.)

A decomposition similar to equation (4.6a) exists for the expectations:

$$
\mathcal{A}_{PQ} = \left[\frac{1}{n(n-1)} \sum_{i,j} t_{ij} \right] \mathcal{A}_{PQ} + \left[1 - \frac{1}{n(n-1)} \sum_{i,j} t_{ij} \right] \mathcal{A}_{PQ}
$$

$$
= \left[\frac{1}{n(n-1)} \sum_{i,j} t_{ij} \right] \mathcal{A}_{PQ} + \left[\frac{1}{n(n-1)} \sum_{i,j} \bar{t}_{ij} \right] \mathcal{A}_{PQ} \qquad (4.6b)
$$

$$
= E(\mathcal{A}_{PQ(T)}^{(\rho)}) + E(\mathcal{A}_{PQ(\bar{T})}^{(\rho)})
$$

and for a variety of normalized measures that we might consider. For example, consider a normalized index of the form $\mathcal{A}_{PQ}/B_{PQ}$. Then

$$
\frac{\mathcal{A}_{PQ}}{B_{PQ}} = \frac{\mathcal{A}_{PQ(T)}}{B_{PQ}} + \frac{\mathcal{A}_{PQ(\bar{T})}}{B_{PQ}}
$$

$$
= \frac{\mathcal{A}_{PQ(T)}}{B_{PQ(T)}} \frac{B_{PQ(T)}}{B_{PQ}} \qquad (4.6c)
$$

$$
+ \frac{\mathcal{A}_{PQ(\bar{T})}}{B_{PQ(\bar{T})}} \frac{B_{PQ(\bar{T})}}{B_{PQ}}
$$

where $B_{PQ(T)}$ (and $B_{PQ(\bar{T})}$) might have the same structure as B_{PQ} but with a replacement of p_{ij} by $\sqrt{t_{ij}} \, p_{ij}$ and q_{ij} by $\sqrt{t_{ij}} \, q_{ij}$ (or using $\bar{t}_{ij}$). As an illustration, if

$$
B_{PQ} = \sqrt{ \sum_{i,j} p_{ij}^2 \sum_{i,j} q_{ij}^2 }
$$

we could let

$$
B_{PQ(T)} = \sqrt{ \sum_{i,j} t_{ij} p_{ij}^2 \sum_{i,j} t_{ij} q_{ij}^2 } \quad \text{and} \quad B_{PQ(\bar{T})} = \sqrt{ \sum_{i,j} \bar{t}_{ij} p_{ij}^2 \sum_{i,j} \bar{t}_{ij} q_{ij}^2 }
$$

In short, the original zero-order normalized index, $\mathcal{A}_{PQ}/B_{PQ}$, can itself be characterized as a weighted composite of the normalized partial and differential measures.

A Numerical Illustration. As an example of how the mechanics of the partial association comparison can be carried out, we consider a small portion of a data set analyzed in some depth by Cliff and Ord (1981). As extracted from this latter source, Table 4.6a presents information on the deaths from cholera in 34 districts of metropolitan London for the two weeks of September 8 and 15, 1849. Table 4.6b presents a contiguity chart for the districts. Treating the two sets of death rates within Daniels's generalized correlation context using Kendall's ±1 scoring function, we obtain $\mathcal{A}_{PQ}$ = 854. Because of the absence of ties, there is an unambiguous choice for a normalized measure as Kendall's tau. Thus, if B_{PQ} = 34·33 = 1122,

$$\text{tau} = \frac{\mathcal{A}_{PQ}}{B_{PQ}} = \frac{854}{1122} = .76$$

Given the contiguity information of Table 4.6b, the presence of a spatially determined process (e.g., contagion or similarity of water quality in neighboring districts) should be reflected in a lowered value for $\mathcal{A}_{PQ(T)}$, where

$$t_{ij} = \begin{cases} 0 & \text{if districts i and j are contiguous} \\ 1 & \text{otherwise} \end{cases}$$

As can be obtained from the first three moment formulas given in Section 4.2.2 earlier:

$$E(\mathcal{A}_{PQ(\overline{T})}(\rho)) = 117.2$$

$$V(\mathcal{A}_{PQ(\overline{T})}(\rho)) = 109.6$$

$$\gamma(\mathcal{A}_{PQ(\overline{T})}(\rho)) = -.28$$

Thus, given the observed index $\mathcal{A}_{PQ(\overline{T})}$ = 90, a normalized Z value of −2.60 would be obtained, with a lower-tail significance of .009 when referred to a type III distribution based on the given skewness parameter. This is consistent with the Monte Carlo significance level of .011 obtained from the sample cumulative distribution summarized in Table 4.6c.

In any event, $\mathcal{A}_{PQ(\overline{T})}$ is significantly less than what we would expect under the random-relabeling conjecture, suggesting that some of the original association represented through $\mathcal{A}_{PQ}$ can be attributed to or "explained" by spatial contiguity. In terms of the decomposition into a partial and a differential measure using

$$B_{PQ} = \sqrt{\sum_{i,j} p_{ij}^2 \sum_{i,j} q_{ij}^2} = n(n-1) = 34 \cdot 33 = 1122$$

$$B_{PQ(\overline{T})} = \sqrt{\sum_{i,j} \overline{t}_{ij} p_{ij}^2 \sum_{i,j} \overline{t}_{ij} q_{ij}^2} = \sum_{i,j} \overline{t}_{ij} = 154$$

$$B_{PQ(T)} = \sqrt{\sum_{i,j} t_{ij} p_{ij}^2 \sum_{i,j} t_{ij} q_{ij}^2} = \sum_{i,j} t_{ij} = 968$$

we obtain

$$\frac{\mathcal{A}_{PQ}}{B_{PQ}} = \left[\frac{\mathcal{A}_{PQ(\overline{T})}}{B_{PQ(\overline{T})}}\right]\left[\frac{B_{PQ(\overline{T})}}{B_{PQ}}\right] + \left[\frac{\mathcal{A}_{PQ(T)}}{B_{PQ(T)}}\right]\left[\frac{B_{PQ(T)}}{B_{PQ}}\right]$$

$$\frac{854}{1122} = \left[\left(\frac{90}{154}\right)\left(\frac{154}{1122}\right)\right] + \left[\left(\frac{764}{968}\right)\left(\frac{968}{1122}\right)\right]$$

$$.76 = (.58)(.14) + (.79)(.86)$$

Thus, considering the original measure of .76 and the partial of .58, the drop of .18 has been shown to be statistically significant, reflecting the original conjecture of some type of spatial process operating through contiguous districts. Descriptively, there is still a substantial amount of remaining association, given the partial of .58. This is to be expected because contagion and a common water supply exist within districts as well.

Spatial Association. Although we have presented this example in terms of Kendall's scoring function, it should be apparent that any measure of correspondence between two sequences could be handled in a similar manner to define a measure of what may be called "spatial association." For instance, in Cliff and Ord's (1981) discussion of these same data, a similar type of analysis was carried out, using what amounts to a quadrant measure of correlation. In any case, when our concern is spatial association per se, the differential index $\mathcal{A}_{PQ(T)}$ might be a better direct indicator because relatively large values are now keyed positively to reflect the degree to which similar values in the sequence appear in proximal locations. Because of its unique importance in geography, this latter topic of spatial association is developed in more detail in Section 4.3.9. In this context, it is interesting

Table 4.6a Cholera Data for the Thirty-four Metropolitan Districts in London (1849)

District	Population	Cholera deaths			
		September 8		September 15	
		Number	Rate/10,000	Number	Rate/10,000
1. Kensington	61,326	35	5.71	40	6.52
2. Chelsea	40,179	48	11.95	39	9.71
3. St. George, Hanover Square	66,552	18	2.70	21	3.16
4. Westminster	56,712	56	9.87	42	7.41
5. St. Martin-in-the-Fields	25,091	9	3.59	13	5.18
6. St. James, Westminster	37,398	12	3.21	6	1.60
7. Marylebone	138,164	51	3.69	28	2.03
8. Islington	55,690	32	5.75	15	2.69
9. Pancras	129,763	56	4.32	45	3.47
10. Hackney	37,121	16	4.31	7	1.89
11. St. Giles	54,292	27	4.97	45	8.29
12. Strand	43,598	21	4.82	17	3.90
13. Holborn	44,461	20	4.50	15	3.37
14. Clerkenwell	56,708	15	2.65	20	3.53
15. St. Luke	49,829	28	5.62	33	6.62
16. East London	39,655	28	7.06	17	4.29

17.	West London	29,142	42	14.41	43	14.76
18.	London City	55,920	20	3.58	21	3.76
19.	Shoreditch	83,432	109	13.06	91	10.91
20.	Bethnal Green	74,088	96	12.96	91	12.28
21.	Whitechapel	71,765	58	8.08	48	6.69
22.	St. George-in-the-East	41,350	27	6.53	18	4.35
23.	Stepney	90,687	59	6.51	49	5.40
24.	Poplar	31,122	41	13.17	33	10.60
25.	St. Savious	32,975	75	22.74	49	14.86
26.	St. Olave, Southwalk	19,837	44	22.18	33	16.64
27.	Bermendsey	34,947	101	28.90	72	20.60
28.	St. George, Southwalk	46,644	109	23.37	108	23.15
29.	Newington	54,606	157	28.75	137	25.09
30.	Lambeth	12,927	278	215.05	234	181.01
31.	Wordsworth	26,337	45	17.09	28	10.63
32.	Camberwell	37,964	79	20.81	53	13.96
33.	Rotherhithe	13,917	40	28.74	32	22.99
34.	Greenwich	55,212	50	9.06	73	13.22

Table 4.6b Contiguity Table for the Thirty-four Metropolitan
            Districts in London

District	Contiguous districts	District	Contiguous districts
1	2,3,4,7	18	15,16,17,19,21
2	1,3,4	19	8,10,15,16,18,20,21
3	1,2,4,5,6,7	20	10,19,21,23,24
4	1,2,3,5	21	16,18,19,20,22,23
5	3,4,6,11,12	22	21,23
6	3,5,7,12	23	20,21,22,24
7	1,3,6,9,11,12	24	10,20,23
8	9,10,14,15,19	25	26,27,28,30
9	7,8,11,12,13,14	26	25,27
10	8,19,20,24	27	25,26,28,32,33
11	5,7,9,12,13	28	25,27,29,30,32
12	5,6,7,9,11,13,17	29	28,30,32
13	9,11,12,14,17	30	25,28,29,31,32
14	8,9,13,15,16,17	31	30
15	8,14,16,18,19	32	27,28,29,30,33,34
16	14,15,17,18,19,21	33	27,32,34
17	12,13,14,16,18	34	32,33

to note that the population model mentioned earlier in Section 4.2.2 as being appropriate for the spatial autocorrelation problem is also a natural
possibility here as well. The n observations are now bivariate realizations
of a random vector $\underset{\sim}{Y}^t = [Y_1, Y_2]$, and the null conjecture is one of independence of these n realizations. The latter hypothesis would induce an
equally likely distribution over the n! possible values of $A_{PQ(T)}(\rho)$.

A Special Case for Dichotomous $\underset{\sim}{T}$ Representing a Partition of S. As
one important special case of $A_{PQ(T)}$ that may help make some of the
ideas just introduced a little clearer, suppose $\underset{\sim}{T}$ is a symmetric zero-one
dichotomous matrix constructed from a set of $\tilde{n}$ observations on the objects

in S. These observations are denoted by $\underset{\sim}{t} = \{t_1, \ldots, t_n\}$ and may contain numerical labels that have only nominal significance. Explicitly, let

$$t_{ij} = \begin{cases} 0 & \text{if } t_i = t_j \\ 1 & \text{if } t_i \neq t_j \end{cases}$$

then $\underset{\sim}{T}$ represents a partition of S into K classes where the sizes of the classes depend on the number of values in $\underset{\sim}{t}$ that are tied at particular values. This specific structure leads to the interpretation of $\mathcal{A}_{PQ(\overline{T})}$ as the sum of all products of the form $p_{ij}q_{ij}$, where both O_i and O_j are tied in $\underset{\sim}{t}$, i.e., $t_i = t_j$.

If some of the association between P and Q can be attributed to $\underset{\sim}{T}$, and thus, to the values in $\underset{\sim}{t} = \{t_1, \ldots, t_n\}$, we would expect this effect to show

Table 4.6c   Monte Carlo Distribution for the Index $\mathcal{A}_{PQ(\overline{T})}$
Based on a Sample Size of 999

Index	Cumulative frequency
82	1
90	10
98	50
102	100
106	200
114	300
114	400
118	500
118	600
122	700
126	800
130	900
134	950
138	990
146	999

up in the size of $\mathcal{A}_{PQ(\overline{T})}$. Intuitively, if the original measure $\mathcal{A}_{PQ}$ is additively decomposed into two parts, $\mathcal{A}_{PQ(T)}$ and $\mathcal{A}_{PQ(\overline{T})}$ as in equation (4.6a), the index $\mathcal{A}_{PQ(\overline{T})}$ tells us what association between the matrices $\underset{\sim}{P}$ and $\underset{\sim}{Q}$ still exists within the tied values of the sequence $\underset{\sim}{t}$, i.e., what association is left over once the information in $\underset{\sim}{t}$ is taken into account. The differential measure $\mathcal{A}_{PQ(T)}$ is based on information for those object pairs that have differing values in $\underset{\sim}{t}$ and is an indication of what association between $\underset{\sim}{P}$ and $\underset{\sim}{Q}$ might be attributed to the data in $\underset{\sim}{t}$.

Partial γ. As one particular example of this special case, suppose we consider the Kendall specification of $\underset{\sim}{P}$ and $\underset{\sim}{Q}$ through ± 1 scoring functions, using the two sequences $\underset{\sim}{x}$ and $\underset{\sim}{y}$, respectively. Letting $B_{PQ} = \Sigma_{i,j} |p_{ij}q_{ij}|$ and applying the decomposition in equation (4.6c), we have

$$\gamma = \frac{\mathcal{A}_{PQ}}{B_{PQ}} = \left[\frac{\mathcal{A}_{PQ(T)}}{B_{PQ(T)}} \cdot \frac{B_{PQ(T)}}{B_{PQ}} \right] + \left[\frac{\mathcal{A}_{PQ(\overline{T})}}{B_{PQ(\overline{T})}} \cdot \frac{B_{PQ(\overline{T})}}{B_{PQ}} \right]$$

where

$$B_{PQ(T)} = \sum_{i,j} t_{ij} |p_{ij}q_{ij}| \quad \text{and} \quad B_{PQ(\overline{T})} = \sum_{i,j} \overline{t}_{ij} |p_{ij}q_{ij}|$$

The term $\mathcal{A}_{PQ(\overline{T})}/B_{PQ(\overline{T})}$ is Davis's partial γ coefficient; $\mathcal{A}_{PQ(T)}/B_{PQ(T)}$ is the differential γ coefficient. Thus, the "zero-order" measure γ is a weighted combination of two others. Supposedly, partial γ indicates what association is "left over" between the $\underset{\sim}{x}$ and $\underset{\sim}{y}$ sequences once $\underset{\sim}{t}$ has been "controlled for" or "taken into account"; differential γ is the association between $\underset{\sim}{x}$ and $\underset{\sim}{y}$ that can be "attributed" to $\underset{\sim}{t}$. The intuitive meanings that might be attached to various numerical combinations of the partial and differential γ measures are discussed in detail by Davis (1971).

In the partial γ context, there are a number of generalizations that could be pursued. For example, the values in the sequence $\underset{\sim}{t}$ could actually be constructed from several other variables by letting all possible numerical combinations of the latter define the observations in $\underset{\sim}{t}$. In this way, we can extend the notion of partial γ (or for that matter, we can extend any partial measure) to more than one control variable. As a second extension, the partial index $\mathcal{A}_{PQ(\overline{T})}$ or its normalized version could itself be decomposed into K separate components corresponding to the subsets of objects in S tied at particular values in $\underset{\sim}{t}$. Each of these components would be characterized by a matrix $\underset{\sim}{\overline{T}}(d) = \{ \overline{t}_{ij}(d) \}$, where d is one of the K possible values in $\underset{\sim}{t}$, and

$$\overline{t}_{ij}(d) = \begin{cases} 0 & \text{if } t_i = t_j = d \\ 1 & \text{otherwise} \end{cases}$$

If the K values of d are denoted by $d_1, \ldots, d_K$, the decomposition of $\mathcal{A}_{PQ(\overline{T})}$ and $\mathcal{A}_{PQ(\overline{T})}/B_{PQ(\overline{T})}$ can be rewritten explicitly as

$$\mathcal{A}_{PQ(\overline{T})} = \sum_{k=1}^{K} \mathcal{A}_{PQ(\overline{T}(d_k))}$$

and

$$\frac{\mathcal{A}_{PQ(\overline{T})}}{B_{PQ(\overline{T})}} = \sum_{k=1}^{K} \left[\frac{\mathcal{A}_{PQ(\overline{T}(d_k))}}{B_{PQ(\overline{T}(d_k))}} \right] \left[\frac{B_{PQ(\overline{T}(d_k))}}{B_{PQ(\overline{T})}} \right] \tag{4.6d}$$

Given d, the raw index $\mathcal{A}_{PQ(\overline{T}(d))}$ could be evaluated for significance by use of the same random relabeling strategy as before. Here, $\{p_{ij}q_{ij}\}$ would be compared to a zero-one dichotomous matrix $\underset{\sim}{T}(d)$ that now represents a single subset; $\mathcal{A}_{PQ(\overline{T}(d))}$ is evaluated against the original index $\mathcal{A}_{PQ}$, using the expectation

$$\left[\frac{1}{n(n-1)} \sum_{i,j} \overline{t}_{ij}(d) \right] \mathcal{A}_{PQ}$$

(In effect, $\mathcal{A}_{PQ(\overline{T}(d))}$ is compared, as in Section 4.3.3, to all similar indices that could be constructed by randomly selecting subsets of the same size from S.) As an alternative question discussed later, $\mathcal{A}_{PQ(\overline{T}(d))}$ could be evaluated in an absolute sense by considering only those objects in S that have values in $\underset{\sim}{t}$ tied at d and restricting the randomization model to the corresponding submatrices of $\underset{\sim}{P}$ and $\underset{\sim}{Q}$. In any case, on the basis of Davis's terminology, the use of these latter indices falls under the topic of specification, i.e., the degree to which the terms making up a partial measure may themselves vary, depending on the specific values observed in $\underset{\sim}{t}$.

Although we have emphasized partial γ here, it should be apparent that similar discussions could be developed in exact analogy, using the Pearson and Kendall score functions. In addition, there are a variety of possibly different normalization strategies that could be used, producing a wide variety of different alternatives, depending on what descriptive indices the researcher wishes to develop.

As a further type of possible extension, it should be noted that when $\underset{\sim}{T}$ represents a partition of S, an analog of one-way analysis of variance is obtained, as discussed within Section 4.3.4. Now, however, the proximity

between objects O_i and O_j is not defined, say, as a simple squared difference, $(x_i - x_j)^2$, based on a single sequence, but rather, in terms of $p_{ij}q_{ij}$,
where p_{ij} and q_{ij} may themselves be constructed from two distinct sequences.
Thus, the indices $\mathcal{A}_{PQ(\overline{T})}$ and $\mathcal{A}_{PQ(T)}$ are within- and between-class
measures, respectively, which decompose the total measure $\mathcal{A}_{PQ}$. A decision as to how the classes of the partition contribute to the measure
$\mathcal{A}_{PQ(T)}$ [or $\mathcal{A}_{PQ(\overline{T})}$] depends on the weighting structure chosen for $\underset{\sim}{T}$. We
have selected the obvious zero-one alternative in which the classes con-
tribute as a direct function of the number of object pairs they contain. But,
as mentioned in the earlier discussion of one-way analysis of variance, a
number of other possibilities exist which are defined by replacing the value
of 1 by c_k, where c_k may depend on the number of objects in the partition.
Consequently, in addition to the descriptive normalizations that we might
consider in decomposing an index as $\mathcal{A}_{PQ}/B_{PQ}$, there is an additional
choice as to how various types of weighting schemes are incorporated
directly into the definitions of the raw indices $\mathcal{A}_{PQ(T)}$ and $\mathcal{A}_{PQ(\overline{T})}$.

Some Comments on Other Possible Applications. The concept of partial
association extends to all uses of the QA strategy. For example, it is
straightforward to develop the notion of partial spatial autocorrelation within the framework we have developed using, say, Geary's coefficient. The
analog of the decomposition in equation (4.6c) (assuming the bounds of 0
and 1 on the entries in $\underset{\sim}{T}$) would be

$$\frac{\mathcal{A}_{PQ}}{E(\mathcal{A}_{PQ}(\rho))} = \left[\frac{\mathcal{A}_{PQ(T)}}{E(\mathcal{A}_{PQ(T)}(\rho))}\right]\left[\frac{E(\mathcal{A}_{PQ(T)}(\rho))}{E(\mathcal{A}_{PQ}(\rho))}\right]$$

$$+ \left[\frac{\mathcal{A}_{PQ(\overline{T})}}{E(\mathcal{A}_{PQ(\overline{T})}(\rho))}\right]\left[\frac{E(\mathcal{A}_{PQ(\overline{T})}(\rho))}{E(\mathcal{A}_{PQ}(\rho))}\right]$$

where $\mathcal{A}_{PQ(\overline{T})}/E(\mathcal{A}_{PQ(\overline{T})}(\rho))$ is a partial Geary spatial autocorrelation
measure, and $\mathcal{A}_{PQ(T)}/E(\mathcal{A}_{PQ(T)}(\rho))$ is a corresponding differential index.

In an analogous manner, our concern may be in comparing two empirically obtained proximity (e.g., correlation) matrices on the same objects
and relating the observed degree of correspondence to some a priori partition of the object set S [as represented by a (0-1) dichotomous matrix, $\underset{\sim}{T}$].
Or, suppose $\underset{\sim}{P}$ represents an empirically generated proximity structure
that is to be compared to $\underset{\sim}{Q}$ (e.g., a partition). The lack (or presence) of
the structure represented by $\underset{\sim}{Q}$ in $\underset{\sim}{P}$, as measured by $\mathcal{A}_{PQ}$, can be referred
to other data on the objects in S as manifested in $\underset{\sim}{T}$. From a more specific
substantive perspective, for instance, we could compare the degree of
correspondence between the interpoint distances for a subjectively estimated
cognitive map (P) and an objective map (Q) to the familiarity of the geographic

locations (e.g., absolute differences in familiarity as represented by $\underset{\sim}{T}$).
Or, we might try to control for the presence of a particular (weighted)
communication network $(\underset{\sim}{T})$ in assessing the correspondence between two
sociometric matrices. In all cases, the basic principle of relating some
matrix $\underset{\sim}{T}$ to the components comprising the raw cross-product measure $\mathcal{A}_{PQ}$
remain the same, although there may be variations as to how the measures
are normalized and interpreted.

A Restricted Randomization Strategy. When the QA comparison method is
used to evaluate T against $\{p_{ij}q_{ij}\}$, the index $\mathcal{A}_{PQ(T)}$ is being assessed in
relation to $\mathcal{A}_{PQ}$. Thus, the concern is with the degree to which the pattern
of association between P and Q could or could not be attributed to $\underset{\sim}{T}$. Now,
however, consider the usual partial correlation context for skew-symmetric
matrices $\underset{\sim}{P}$ and $\underset{\sim}{Q}$ and a (zero-one) dichotomous matrix $\underset{\sim}{T}$ representing a
partition constructed from the equality of values in a sequence $\underset{\sim}{t} = \{t_1, \ldots, t_n\}$.
An alternative question can be phrased as to whether any partial association
exists per se, i.e., whether $\mathcal{A}_{PQ(\overline{T})}$ is different from 0.

As one solution to this problem, Quade (1974) has developed the idea
of partial association within this same context of comparing two sequences
through ± 1 scoring functions. Depending on how the measure is to be normalized by the number of "relevant" pairs, Quade has given a general
expression for the asymptotic (nonnull) variance for the normalized measure
through the theory of U statistics. This latter variance can then be used to
construct an approximate confidence interval for the corresponding population quantity, and, as a special case, a test of the normalized partial association against zero. Unfortunately, given the arbitrariness of the matrices
$\underset{\sim}{P}, \underset{\sim}{Q}$, and $\underset{\sim}{T}$, the U-statistic approach does not appear capable of a completely general application to measures of the form $\mathcal{A}_{PQ(\overline{T})}$. This is in
contrast to the philosophy behind the QA strategy, at least under the usual
null hypothesis of randomness.

One approach to the evaluation of the absolute level of $\mathcal{A}_{PQ(\overline{T})}$ in the
special case that $\underset{\sim}{T}$ represents a partition demands a restricted randomization model. Specifically, if $\mathcal{A}_{PQ(\overline{T})}$ is decomposed as in equation (4.6d):

$$\mathcal{A}_{PQ(\overline{T})} = \sum_{k=1}^{K} \mathcal{A}_{PQ(\overline{T}(d_k))}$$

we can "block" on the K subsets by treating each of the components separately and applying the randomization model to the individual subsets. From
the perspective of suggesting an appropriate population model, we would
now have K separate populations, and our interest is in aggregating the
separate evidence for the given null conjecture, i.e., a factorization criterion holds for each of the joint cumulative distributions that characterize
the K-independent data-generating mechanisms. This notion of blocking in

the randomization framework is discussed by Klauber (1975) and in Chapter 7 under various random/fixed restrictions on the K components of the overall index. We now develop this topic in more detail for our partial association problem.

Evaluating $\mathcal{A}_{PQ(\overline{T})}$. In developing a test of $\mathcal{A}_{PQ(\overline{T})}$ in an absolute sense, it may help to introduce a more precise notational system. Specifically, the decomposition

$$\mathcal{A}_{PQ(\overline{T})} = \sum_{k=1}^{K} \mathcal{A}_{PQ(\overline{T}(d_k))}$$

will be rewritten as

$$\mathcal{A}_{PQ(\overline{T})} = \sum_{k=1}^{K} c_k \mathcal{A}_{P_k Q_k}$$

In analogy with an analysis-of-variance discussion, c_k is now allowed to be an arbitrary weight, i.e.,

$$\overline{t}_{ij}(d_k) = \begin{cases} c_k & \text{if } t_i = t_j = d_k \text{ and } i \neq j \\ 0 & \text{otherwise} \end{cases}$$

$\mathcal{A}_{P_k Q_k}$ refers to Mantel's cross-product statistic restricted to the $n_k \times n_k$ submatrices of $\underset{\sim}{P}$ and $\underset{\sim}{Q}$ that correspond to n_k objects in S tied at the value d_k in $\underset{\sim}{t}$.

From an inference point of view, the random-relabeling conjecture is now restricted within the K classes defined by tied values in t. In particular, suppose ρ_k is a permutation on the n_k objects tied at d_k in $\underset{\sim}{t}$, $1 \leqslant k \leqslant K$. The observed index $\mathcal{A}_{PQ(\overline{T})}$ is now compared to a reference distribution constructed from the random variable

$$\mathcal{A}_{PQ(\overline{T})}(\rho_1, \ldots, \rho_K) = \sum_{k=1}^{K} c_k \mathcal{A}_{P_k Q_k}(\rho_k)$$

where $\rho_1, \ldots, \rho_K$ are picked independently; the permutation ρ_K is chosen uniformly from the $n_k!$ possible permutations of the n_k objects tied at d_k in t. Thus, there are $n_1! \ldots n_K!$ (possibly nondistinct) values of $\mathcal{A}_{\underset{\sim}{PQ(\overline{T})}}(\rho_1, \ldots, \rho_K)$ that are equally likely under the restricted random-relabeling conjecture.

Because of the independence of the K permutations, $\rho_1, \ldots, \rho_K$, the moment formulas are relatively simple:

$$E(\mathcal{A}_{PQ(\overline{T})}(\rho_1, \ldots, \rho_K)) = \sum_{k=1}^{K} c_k E(\mathcal{A}_{P_k Q_k}(\rho_k))$$

$$V(\mathcal{A}_{PQ(\overline{T})}(\rho_1, \ldots, \rho_K)) = \sum_{k=1}^{K} c_k^2 V(\mathcal{A}_{P_k Q_k}(\rho_k))$$

The terms $E(\mathcal{A}_{P_k Q_k}(\rho_k))$ and $V(\mathcal{A}_{P_k Q_k}(\rho_k))$ can be obtained from the QA mean and variance expressions. The skewness parameter for $\mathcal{A}_{PQ(\overline{T})}(\rho_1, \ldots, \rho_K)$ would be obtained by summing the numerators for each of the individual skewness parameters for $\mathcal{A}_{P_k Q_k}(\rho_k)$ and dividing by the $3/2$ power of $V(\mathcal{A}_{PQ(\overline{T})}(\rho_1, \ldots, \rho_K))$. Thus, type III approximations as well as a Monte Carlo strategy may be used to construct a significance level.

In general, we compare $\mathcal{A}_{PQ(\overline{T})}$ to its expectation under the restricted random-relabeling conjecture. If the original matrices $\underset{\sim}{P}$ and $\underset{\sim}{Q}$ are skew-symmetric, as in Daniels's generalized correlation context, this later expectation is zero. Thus, a significance test of $\mathcal{A}_{PQ(\overline{T})}$ is, in effect, a simultaneous test of its K constituent components against zero. If we wish, each of the single indices, $\mathcal{A}_{P_k Q_k}$, could be evaluated separately by comparing the appropriate submatrices $\underset{\sim}{P_k}$ and $\underset{\sim}{Q_k}$.

REFERENCES

Abe, O. A central limit theorem for the number of edges in the random intersection of two graphs, Ann. Math. Stat. 49 (1969), 144–151.
Ager, J. W., and S. B. Brent. An index of agreement between a hypothesized partial order and an empirical rank order, J. Am. Stat. Assoc. 73 (1978), 827–830.
Anglin, J. M. The Growth of Word Meaning, MIT Press, Cambridge, 1970.
Arabie, P., and S. A. Boorman. Multidimensional scaling of measures between partitions, J. Math. Psychol. 10 (1973), 148–203.
Ascher, S. Moments of the Mantel-Valand statistic, Paper presented at the annual meeting of the American Statistical Association, Houston, Texas, August 1980.
Ascher, S., and J. Bailar. Moments of the Mantel-Valand procedure, J. Stat. Comput. Simul. 14 (1982), 101–111.
Baker, F. B., and L. J. Hubert. Inference procedures for ordering theory, J. Educ. Stat. 2 (1977), 217–233.

Baker, F. B., and L. J. Hubert. A nonparametric technique for the analysis of social interaction data, Sociol. Methods Res. 9 (1981), 339-361.

Barton, D. E., and F. N. David. The random intersection of two graphs, in F. N. David (Ed.), Research Papers in Statistics, Wiley, New York, 1966.

Berry, K. J., K. L. Kvamme, and P. W. Mielke. A permutation technique for the spatial analysis of the distribution of artifacts into classes, Am. Antiquity 45 (1980), 55-59.

Besag, J., and P. J. Diggle. Simple Monte Carlo tests for spatial pattern, Appl. Stat. 26 (1977), 327-333.

Bloemena, A. R. Sampling From a Graph, Mathematisch Centrum, Amsterdam, 1964.

Born, R., and C. Nevison. A probabilistic analysis of roll call cohesion measures, Polit. Methodol. 2 (1975), 517-528.

Bradley, J. V. Distribution-Free Statistical Tests, Prentice-Hall, New York, 1968.

Brennan, R. L., and R. J. Light. Measuring agreement when two observers classify people into categories not defined in advance, Br. J. Math. Stat. Psychol. 27 (1974), 154-163.

Brockwell, P. J., P. W. Mielke, and J. Robinson. On non-normal invariance principles for multi-response permutation procedures, Aust. J. Stat. 24 (1982), 33-41.

Campbell, D. T., W. H. Kruskal, and W. P. Wallace. Seating aggregation as an index of attitude, Sociometry 29 (1966), 1-15.

Carroll, J. D., and P. Arabie. Multidimensional scaling, Annu. Rev. Psychol. 31 (1980), 607-649.

Carroll, J. D., A. K. Romney, C. Farmer, and W. F. Delvac. The distribution of a path length statistic with illustrative applications, Unpublished manuscript, Bell Labs, 1976.

Chang, J. J., and J. D. Carroll. How to use PROFIT, a computer program for property fitting by optimizing nonlinear or linear correlation, Unpublished manuscript, Bell Labs, 1968.

Cliff, A. D., and J. K. Ord. Spatial Autocorrelation, Pion, London, 1973.

Cliff, A. D., and J. K. Ord. Spatial Processes: Models and Applications, Pion, London, 1981.

Cliff, N. Snake races and other ordering methods, Paper presented at the annual meeting of the Psychometric Society, Iowa City, Iowa, May 1980.

Coombs, C. H. A Theory of Data, Wiley, New York, 1964.

Costanzo, C. M., L. J. Hubert, and R. G. Golledge. A higher moment for spatial statistics, Geogr. Anal. 15 (1983), 347-351.

Court, A. Map comparisons, Econ. Geogr. 46 (1970), 436-438.

Cox, D. R., and A. Stuart. Some quick sign tests for trend in location and dispersion, Biometrika 42 (1955), 80-95.

Daniels, H. E. The relation between measures of correlations in the universe of sample permutations, Biometrika 33 (1944), 129-135.

David, F. N., and D. E. Barton. Combinatorial Chance, Hafner, New York, 1962.

David, H. A. The Method of Paired Comparisons, Griffin, London, 1963.

Davis, J. A. Elementary Survey Analysis, Prentice-Hall, Englewood Cliffs, N.J., 1971.

Edmonds, J. Maximum matching and a polyhedron with 0,1-vertices, J. Res. Nat. Bur. Standards, B 69B (1965), 125-130.

Epp, R. J., J. W. Tukey, and G. S. Watson. Testing unit vectors for correlation, J. Geophys. Res. 76 (1971), 8480-8483.

Fillenbaum, S., and A. Rapoport. Structures in the Subjective Lexicon, Academic Press, New York, 1971.

Frank, O. Statistical Inference in Graphs, FOA Repro, Stockholm, 1971.

Freeman, L. C. Segregation in social networks, Sociol. Methods Res. 6 (1978), 411-429.

Friedman, J. H., and L. C. Rafsky. Multivariate generalizations of the Wald-Wolfowitz and Smirnov two-sample tests, Ann. Stat. 7 (1979), 697-717.

Garner, W. R. Uncertainty and Structure as Psychological Concepts, Wiley, New York, 1962.

Gibbons, J. D. Nonparametric Statistical Inference, McGraw-Hill, New York, 1971.

Gilbert, E. J. The matching problem, Psychometrika 21 (1956), 253-266.

Glick, B. J. Tests for space-time clustering used in cancer research, Geogr. Anal. 11 (1979), 202-208.

Glushko, R. J. Pattern goodness and redundancy revisited: Multidimensional scaling and hierarchical clustering analyses, Percept. Psychophys. 17 (1975), 158-162.

Goodman, L. A., and Y. Grunfeld. Some nonparametric tests for co-movements between time series, J. Am. Stat. Assoc. 56 (1961), 11-26.

Goodman, L. A., and W. H. Kruskal. Measures of association for cross classifications, IV: Simplification of asymptotic variances, J. Am. Stat. Assoc. 67 (1972), 415-421.

Graves, G. W., and A. B. Whinston. An algorithm for the quadratic assignment problem, Management Sci. 16 (1970a), 692-707.

Graves, G. W., and A. B. Whinston. Derivation of the mean and variance for the quadratic assignment problem, Unpublished manuscript, 1970b.

Guttman, L. A new approach to factor analysis: The radex, in P. F. Lazarsfeld (Ed.), Mathematical Thinking in the Social Sciences, The Free Press, Glencoe, Ill., 1954.

Hanan, M., and J. M. Kurtzberg. A review of the placement and quadratic assignment problems, SIAM Rev. 14 (1972), 324-342.

Hare, A. P., and R. F. Bales. Seating positions and small group interaction, Sociometry 26 (1963), 480-486.

Hartigan, J. A. Clustering Algorithms, Wiley, New York, 1975.

Hawkes, R. K. The multivariate analysis of ordinal measures, Am. J. Sociol. 76 (1971), 908-926.

Hildebrand, D. K., J. D. Laing, and H. Rosenthal. Prediction Analysis of Cross-Classifications, Wiley, New York, 1977.

Hodson, F. R., D. G. Kendall, and P. Tǎutu (Eds.). Mathematics in the Archaeological and Historical Sciences, Edinburgh University Press, Edinburgh, 1971.

Holland, P. W., and S. Leinhardt (Eds.). Perspectives on Social Network Research, Academic Press, New York, 1979.

Hubert, L. J. Some applications of graph theory and related nonmetric techniques to problems of approximate seriation: The case of symmetric proximity measures, Br. J. Math. Stat. Psychol. 27 (1974), 133-153.

Hubert, L. J. Seriation using asymmetric proximity measures, Br. J. Math. Stat. Psychol. 29 (1976), 32-52.

Hubert, L. Kappa revisited, Psychol. Bull. 84 (1977a), 289-297.

Hubert, L. J. Nominal scale response agreement as a generalized correlation, Br. J. Math. Stat. Psychol. 30 (1977b), 98-103.

Hubert, L. J. Nonparametric tests for patterns in geographical variation: Possible generalizations, Geograph. Anal. 10 (1978), 86-88.

Hubert, L. J. Alternative inference models based on matching for a weighted index of nominal scale response agreement, Qual. Quant. 14 (1980), 711-725.

Hubert, L. J. Data theory and problems of analysis, in R. G. Golledge and J. N. Rayner (Eds.), Proximity and Preference, University of Minnesota Press, Minneapolis, 1982, pp. 111-130.

Hubert, L. J. Inference procedures for the evaluation and comparison of proximity matrices, in J. Felsenstein (Ed.), Numerical Taxonomy, Springer-Verlag, New York, 1983, pp. 209-228.

Hubert, L. J. Combinatorial data analysis: Association and partial association, Psychometrika 50 (1985), 449-467.

Hubert, L. J., and F. B. Baker. Analyzing distinctive features, J. Educ. Stat. 2 (1977a), 79-98.

Hubert, L. J., and F. B. Baker. The comparison and fitting of given classification schemes, J. Math. Psychol. 16 (1977b), 233-253.

Hubert, L. J., and F. B. Baker. Evaluating the conformity of sociometric measurements, Psychometrika 43 (1978a), 31-41.

Hubert, L. J., and F. B. Baker. Analyzing the multitrait-multimethod matrix, Multivar. Behav. Res. 13 (1978b), 163-179.

Hubert, L. J., and F. B. Baker. Evaluating the symmetry of a proximity matrix, Qual. Quant. 13 (1979), 77-84.

Hubert, L. J., and R. G. Golledge. Inference models for roll-call cohesion measures, Qual. Quant. 15 (1981a), 425-432.

Hubert, L. J., and R. G. Golledge. A heuristic method for the comparison of related structures, J. Math. Psychol. 23 (1981b), 214-226.

Hubert, L. J., and J. R. Levin. Evaluating object set partitions: Free-sort analysis and some generalizations, J. Verbal Learning Verbal Behav. 15 (1976a), 459-470.

Hubert, L. J., and J. R. Levin. A general statistical framework for assessing categorical clustering in free recall, Psychol. Bull. 83 (1976b), 1072-1080.

Hubert, L. J., and J. R. Levin. Inference models for categorical clustering, Psychol. Bull. 84 (1977), 878-887.

Hubert, L. J., and J. V. Schultz. Quadratic assignment as a general data analysis strategy, Br. J. Math. Stat. Psychol. 29 (1976a), 190-241.

Hubert, L. J., and J. V. Schultz. A note on seriation and quadratic assignment, Classific. Soc. Bull. 3 (1976b), 16-24.

Hubert, L. J., and M. J. Subkoviak. Confirmatory inference and geometric models, Psychol. Bull. 86 (1979), 361-370.

Hubert, L. J., R. G. Golledge, and C. M. Costanzo. Generalized procedures for evaluating spatial autocorrelation, Geograph. Anal. 13 (1981), 224-233.

Hubert, L. J., R. G. Golledge, and C. M. Costanzo. Analysis of variance procedures based on a proximity measure between subjects, Psychol. Bull. 91 (1982), 424-430.

Hubert, L. J., R. G. Golledge, T. Kenney, and C. M. Costanzo. A methodology for evaluating classification schemes used in reporting criminal justice data, Evaluation Rev. 6 (1982), 505-520.

Hubert, L. J., R. G. Golledge, C. M. Costanzo, N. Gale, and W. C. Halperin. Nonparametric tests for directional data, in G. Bahrenberg, M. Fischer, and P. Nijkamp (Eds.), Recent Developments in Spatial Analysis: Methodology, Measurement, Models, Gower, Aldershot, U.K., 1984, pp. 171-190.

Hubert, L. J., R. G. Golledge, C. M. Costanzo, and N. Gale. Measuring association between spatially defined variables: An alternative procedure, Geograph. Anal. 17 (1985a), 36-46.

Hubert, L. J., R. G. Golledge, C. M. Costanzo, and N. Gale. Tests of randomness: Unidimensional and multidimensional, Envir. Planning, A 17 (1985), 373-385.

Johnson, S. C. Hierarchical clustering schemes, Psychometrika 32 (1967), 241-254.

Johnson, S. C. A simple cluster statistic, Unpublished manuscript, Bell Labs, 1968a.

Johnson, S. C. Metric clustering, Unpublished manuscript, Bell Labs, 1968b.

Kaiser, H. F. A second generation little jiffy, Psychometrika 35 (1970), 401-405.

Katz, L., and J. H. Powell. A proposed index of the conformity of one sociometric measurement to another, Psychometrika 18 (1953), 249-256.

Katz, L., and T. R. Wilson. The variance of the number of mutual choices in sociometry, Psychometrika 21 (1956), 299-304.

Kawabata, H. M., J. P. Mulherin, and J. A. Sonquist. Quadratic assignment procedures: Uses in social network analysis, Paper presented at the Pacific Sociological Association Meetings, Portland, Oregon, March 20, 1981.

Kendall, D. G. Seriation from abundance matrices, in F. R. Hodson, D. G. Kendall, and P. Tǎutu (Eds.), Mathematics in the Archaeological and Historical Sciences, Edinburgh University Press, 1971, pp. 215-252.

Kendall, M. G. Rank Correlation Methods, 4th ed., Hafner, New York, 1970.

Klauber, M. R. Space-time clustering for more than two samples, Biometrics 31 (1975), 719-726.

Knight, W. A run-like statistic for ecological transects, Biometrics 30 (1974), 553-555.

Krippendorff, K. Bivariate agreement coefficients for reliability of data, in E. F. Borgatta (Ed.), Sociological Methodology 1970, Jossey-Bass, San Francisco, 1970.

Kruskal, J. B. On the shortest spanning subtree of a graph and the traveling salesman problem, Proc. Am. Math. Soc. 7 (1956), 48-50.

Kruskal, J. B., and M. Wish. Multidimensional Scaling, Sage, Beverly Hills, 1978.

Lehmann, E. L. Nonparametrics: Statistical Methods Based on Ranks, Holden-Day, San Francisco, 1975.

Lerman, I. C. Combinatorial analysis in the statistical treatment of behavioral data, Qual. Quant. 14 (1980), 431-469.

Levene, H. On a matching problem arising in genetics, Ann. Math. Stat. 20 (1949), 91-94.

Light, R. J. Measures of agreement for qualitative data: Some generalizations and alternatives, Psychol. Bull. 76 (1971), 365-377.

Light, R. J., and B. H. Margolin. An analysis of variance for categorical data, J. Am. Stat. Assoc. 66 (1971), 534-544.

MacRae, D. Issues and Parties in Legislative Voting, Harper and Row, New York, 1970.

Mandler, G. Organization and memory, in K. W. Spence and J. T. Spence (Eds.), The Psychology of Learning and Motivation, Vol. 1, Academic Press, New York, 1967, pp. 327-372.

Mantel, N. The detection of disease clustering and a generalized regression approach, Cancer Res. 27 (1967), 209-220.

Mantel, N., and J. C. Bailar. A class of permutational and multinomial tests arising in epidemiological research, Biometrics 26 (1970), 687-700.

Mantel, N., and R. S. Valand. A technique of nonparametric multivariate analysis, Biometrics 26 (1970), 547-558.

Margolin, B. H., and R. J. Light. An analysis of variance for categorical data, II: Small sample comparisons with chi-square and other computations, J. Am. Stat. Assoc. 69 (1974), 755-764.

Mielke, P. W. Clarification and appropriate inference for Mantel and Valand's nonparametric multivariate analysis technique, Biometrics 34 (1978), 277-282.

Mielke, P. W. On asymptotic non-normality of null distributions of MRPP statistics, Commun. Stat.: Theory Methods A8 (1979), 1541-1550; errata: A10 (1981), 1795 and A11 (1982), 847.

Mielke, P. W., K. J. Berry, and G. W. Brier. Application of multi-response permutation procedures for examining seasonal changes in monthly sea-level pressure patterns, Monthly Weather Rev. 109 (1981), 120-126.

Mielke, P. W., K. J. Berry, P. J. Brockwell, and J. S. Williams. A class of nonparametric tests based on multi-response permutation procedures, Biometrika 68 (1981), 720-724.

Mielke, P. W., K. J. Berry, and E. S. Johnson. Multi-response permutation procedures for a priori classifications, Commun. Stat. A5 (1976), 1409-1424.

Miller, G. A., and P. E. Nicely. Analysis of perceptual confusions among some English consonants, J. Acoust. Soc. Am. 27 (1955), 338-352.

Mirkin, B. G. Group Choice, Wiley, New York, 1979.

Moore, G. H., and W. A. Wallis. Time series significance tests based on signs of differences, J. Am. Stat. Assoc. 38 (1943), 153-164.

Norcliffe, G. B. Inferential Statistics for Geographers, Wiley, New York, 1977.

O'Reilly, F. J., and P. W. Mielke. Asymptotic normality of MRPP statistics from invariance principles of U-statistics, Commun. Stat.: Theory Methods A9 (1980), 629-637.

Paivio, A., J. C. Yuille, and S. Madigan. Concreteness, imagery, and meaningfulness for 925 nouns, J. Exp. Psychol. Monograph Suppl. 76 (1, Part 2) (1968).

Pellegrino, J., and L. J. Hubert. The analysis of organization and structure in free recall, in C. R. Puff (Ed.), Handbook of Methods in Memory and Cognition, Academic Press, New York, 1982, pp. 129-172.

Pielou, E. C. Interpretation of paleoecological similarity matrices, Paleobiology 5 (1979), 435-443.

Pollard-Gott, L., M. McCloskey, and A. K. Todres. Subjective story structure, Discourse Processes 2 (1979), 251-281.

Puri, M. L., and P. K. Sen. Nonparametric Methods in Multivariate Analysis, Wiley, New York, 1971.

Quade, D. Nonparametric partial correlation, in H. M. Blalock, Jr. (Ed.), Measurement in the Social Sciences: Theories and Strategies, Aldine, Chicago, 1974, pp. 369-398.

Ramsay, J. O. Confidence regions for multidimensional scaling analysis, Psychometrika 43 (1978), 145-160.

Rand, W. M. Objective criteria for the evaluation of clustering methods, J. Am. Stat. Assoc. 66 (1971), 846-850.

Reynolds, H. T. The Analysis of Cross-Classifications, The Free Press, New York, 1977.

Richardson, G. D. Comparing two cognitive mapping methodologies 13 (1981), 325-331.

Roenker, D. L., C. P. Thompson, and S. C. Brown. A comparison of measures for the estimation of clustering in free recall, Psychol. Bull. 76 (1971), 45-48.

Rothkopf, E. A measure of stimulus similarity and errors in some paired-associate learning tasks, J. Exp. Psychol. 53 (1957), 94-101.

Royalty, H. H., E. Astrachan, and R. R. Sokal. Tests for patterns in geographical variation, Geogr. Anal. 7 (1975), 369-395.

Rushton, G. The scaling of locational preferences, in K. R. Cox and R. G. Golledge (Eds.), Behavioral Problems in Geography: A Symposium, Department of Geography, Northwestern University, Studies in Geography, No. 17, 1969, pp. 197-227.

Sattath, S., and A. Tversky. Additive similarity trees, Psychometrika 42 (1977), 319-345.

Schiffman, S., M. L. Reynolds, and F. W. Young. Introduction to Multidimensional Scaling, Academic Press, New York, 1981.

Schultz, J. V., and L. J. Hubert. A nonparametric test for the correspondence between two proximity matrices, J. Educ. Stat. 1 (1976), 59-67.

Scott, W. A. Reliability of content analysis: The case of nominal scale coding, Publ. Opinion Q. 19 (1955), 321-325.

Shapiro, C. P., and L. J. Hubert. Asymptotic normality of permutation statistics derived from weighted sums of bivariate functions, Ann. Stat. 7 (1979), 788-794.

Shepard, R. N. Analysis of proximities as a technique for the study of information processing in man, Human Factors 5 (1963), 19-34.

Shepard, R. N. The circumplex and related topological manifolds in the study of perception, in S. Shye (Ed.), Theory Construction and Data Analysis in the Behavioral Sciences, Jossey-Bass, San Francisco, 1978, pp. 29-80.

Shirahata, S. Intraclass rank tests for independence, Biometrika 68 (1981), 451-456.

Shuell, T. J. Clustering and organization in free recall, Psychol. Bull. 72 (1969), 353-374.

Siemiatycki, J. Mantel's space-time clustering statistic: Computing higher moments and a comparison of various data transforms, J. Stat. Comput. Simul. 7 (1978), 13-31.

Sneath, P. H. A., and R. R. Sokal. Numerical Taxonomy, Freeman, San Francisco, 1973.

Sokal, R. R. Testing statistical significance of geographical variation patterns, Syst. Zool. 28 (1979), 227-232.

Somers, R. H. The rank analogue of product-moment partial corelation and regression, with application to manifold, ordered contingency tables, Biometrika 46 (1959), 241-246.

Somers, R. H. A new asymmetric measure for ordinal variables, Am. Sociol. Rev. 27 (1962), 799-811.

Steinzor, B. The spatial factor in face-to-face discussion groups, J. Abnorm. Soc. Psychol. 45 (1950), 552-555.

Stuart, A. The power of two difference-sign tests, J. Am. Stat. Assoc. 47 (1952), 416-424.

Szczotka, F. On a method of ordering and clustering of objects, Zastosowania Mathemetyki 13 (1972), 23-33.

Tobler, W. R. Spatial interaction patterns, J. Envir. Stud. 6 (1976), 271-301.

Tobler, W. R. Estimation of attractivities from interactions, Envir. Planning [A] 11 (1979), 121-127.

Togerson, W. S. Theory and Methods of Scaling, Wiley, New York, 1958.

Tukey, J. W. The future of data analysis, Ann. Math. Stat. 33 (1962), 1-67.

Wald, A., and A. Wolfowitz. On exact tests for randomness in the non-parametric case based on serial correlation, Ann. Math. Stat. 14 (1943), 387-388.

Wampold, B. C., and G. Margolin. Nonparametric strategies to test the independence of behavioral states in sequential data, Psychol. Bull. 92 (1982), 755-765.

Watson, G. S., and R. J. Beran. Testing a sequence of unit vectors for serial correlation, J. Geophys. Res. 72 (1967), 5655-5659.

Wilkinson, E. M. Archaeological seriation and the traveling salesman problem, in F. R. Hodson, D. G. Kendall, and P. Tǎutu (Eds.), Mathematics in the Archaeological and Historical Sciences, Edinburgh University Press, Edinburgh, 1971, pp. 276-283.

Winsborough, H. H., E. L. Quarantelli, and D. Yutzky. The similarity of connected observations, Am. Sociol. Rev. 28 (1963), 977-983.

APPENDIX

Part A

The terms $B_1, \ldots, B_7$ are needed for $V(\mathcal{A}(\rho))$, based on arbitrary matrices $\underset{\sim}{P}$ and $\underset{\sim}{Q}$ (with zero diagonals)—see Section 4.2.2.

$$B_1 = \left(\sum_{i,j} p_{ij}\right)^2 \left(\sum_{i,j} q_{ij}\right)^2$$

$$B_2 = \left(\sum_{i,j} p_{ij}^2 \; \sum_{i,j} q_{ij}^2\right)$$

$$B_3 = \left[\sum_i \left(\sum_j p_{ij} p_{ji}\right)\right]\left[\sum_i \left(\sum_j q_{ij} q_{ji}\right)\right]$$

$$B_4 = \left[\sum_i \left(\sum_j p_{ij}\right)^2 - \sum_{i,j} p_{ij}^2\right]\left[\sum_i \left(\sum_j q_{ij}\right)^2 - \sum_{i,j} q_{ij}^2\right]$$

$$B_5 = \left\{\sum_i \left[\left(\sum_j p_{ij}\right)\left(\sum_j p_{ji}\right)\right] - \sum_{i,j} p_{ij} p_{ji}\right\}$$

$$\times \left\{\sum_i \left[\left(\sum_j q_{ij}\right)\left(\sum_j q_{ji}\right)\right] - \sum_{i,j} q_{ij} q_{ji}\right\}$$

$$B_6 = \left[\sum_j \left(\sum_i p_{ij}\right)^2 - \sum_{i,j} p_{ij}^2\right]\left[\sum_j \left(\sum_i q_{ij}\right)^2 - \sum_{i,j} q_{ij}^2\right]$$

$$B_7 = \left[\left(\sum_{i,j} p_{ij}\right)^2 - \sum_i \left(\sum_j p_{ij}\right)^2 - 2 \sum_i \left(\sum_j p_{ij}\right)\left(\sum_j p_{ji}\right)\right.$$
$$\left. - \sum_i \left(\sum_j p_{ij}\right)^2 + \sum_i \left(\sum_j p_{ij} p_{ji}\right) + \sum_{i,j} p_{ij}^2\right]$$
$$\times \left[\left(\sum_{i,j} q_{ij}\right)^2 - \sum_i \left(\sum_j q_{ij}\right)^2 - 2 \sum_i \left(\sum_j q_{ij}\right)\left(\sum_j q_{ji}\right)\right.$$
$$\left. - \sum_i \left(\sum_j q_{ij}\right)^2 + \sum_i \left(\sum_j q_{ij} q_{ji}\right) + \sum_{i,j} q_{ij}^2\right]$$

Part B

The raw third moment of $\mathcal{A}(\rho)$ is given by the following (complicated) formula. The index γ is obtained from the expression given in Section 1.4.1. See Mielke (1979, and errata).

$$
E(\mathcal{A}(\rho)^3) = \frac{1}{n(n-1)}\left(\sum p_{i_1 i_2}^3 \sum q_{i_1 i_2}^3 + 3\sum p_{i_1 i_2}^2 p_{i_2 i_1} \sum q_{i_1 i_2}^2 q_{i_2 i_1}\right)
$$

$$
+ \frac{1}{n(n-1)(n-2)}\left(\sum p_{i_1 i_2}^2 p_{i_1 i_3} \sum q_{i_1 i_2}^2 q_{i_1 i_3}\right.
$$

$$
+ 2\sum p_{i_1 i_2} p_{i_2 i_1} p_{i_1 i_3} \sum q_{i_1 i_2} q_{i_2 i_1} q_{i_1 i_3}
$$

$$
+ \sum p_{i_2 i_1}^2 p_{i_1 i_3} \sum q_{i_2 i_1}^2 q_{i_1 i_3}
$$

$$
+ \sum p_{i_1 i_2}^2 p_{i_3 i_1} \sum q_{i_1 i_2}^2 q_{i_3 i_1}
$$

$$
+ 2\sum p_{i_1 i_2} p_{i_2 i_1} p_{i_3 i_1} \sum q_{i_1 i_2} q_{i_2 i_1} q_{i_3 i_1}
$$

$$
\left.+ \sum p_{i_2 i_1}^2 p_{i_3 i_1} \sum q_{i_2 i_1}^2 q_{i_3 i_1}\right)
$$

$$
+ \frac{3}{n(n-1)(n-2)(n-3)}\left(\sum p_{i_1 i_2}^2 p_{i_3 i_4} \sum q_{i_1 i_2}^2 q_{i_3 i_4}\right.
$$

$$
\left.+ \sum p_{i_1 i_2} p_{i_2 i_1} p_{i_3 i_4} \sum q_{i_1 i_2} q_{i_2 i_1} q_{i_3 i_4}\right)
$$

$$
+ \frac{1}{n(n-1)(n-2)}\left(6\sum p_{i_1 i_2} p_{i_1 i_3} p_{i_2 i_3} \sum q_{i_1 i_2} q_{i_1 i_3} q_{i_2 i_3}\right.
$$

$$
\left.+ 2\sum p_{i_1 i_2} p_{i_2 i_3} p_{i_3 i_1} \sum q_{i_1 i_2} q_{i_2 i_3} q_{i_3 i_1}\right)
$$

$$
+ \frac{6}{n(n-1)(n-2)(n-3)}\left(\sum p_{i_1 i_2} p_{i_1 i_3} p_{i_2 i_4} \sum q_{i_1 i_2} q_{i_1 i_3} q_{i_2 i_4}\right.
$$

$$
+ \sum p_{i_1 i_2} p_{i_1 i_3} p_{i_4 i_2} \sum q_{i_1 i_2} q_{i_1 i_3} q_{i_4 i_2}
$$

$$+ \sum p_{i_1 i_2} p_{i_3 i_1} p_{i_2 i_4} \sum q_{i_1 i_2} q_{i_3 i_1} q_{i_2 i_4}$$

$$+ \sum p_{i_1 i_2} p_{i_3 i_1} p_{i_4 i_2} \sum q_{i_1 i_2} q_{i_3 i_1} q_{i_4 i_2} \Bigg)$$

$$+ \frac{1}{n(n-1)(n-2)(n-3)} \Bigg(\sum p_{i_1 i_2} p_{i_1 i_3} p_{i_1 i_4} \sum q_{i_1 i_2} q_{i_1 i_3} q_{i_1 i_4}$$

$$+ 3\sum p_{i_1 i_2} p_{i_1 i_3} p_{i_4 i_1} \sum q_{i_1 i_2} q_{i_1 i_3} q_{i_4 i_1}$$

$$+ 3\sum p_{i_2 i_1} p_{i_3 i_1} p_{i_1 i_4} \sum q_{i_2 i_1} q_{i_3 i_1} q_{i_1 i_4}$$

$$+ \sum p_{i_2 i_1} p_{i_3 i_1} p_{i_4 i_1} \sum q_{i_2 i_1} q_{i_3 i_1} q_{i_4 i_1} \Bigg)$$

$$+ \frac{1}{n(n-1)(n-2)(n-3)(n-4)} \Bigg(3\sum p_{i_1 i_2} p_{i_1 i_3} p_{i_4 i_5} \sum q_{i_1 i_2} q_{i_1 i_3} q_{i_4 i_5}$$

$$+ 6\sum p_{i_1 i_2} p_{i_2 i_3} p_{i_4 i_5} \sum q_{i_1 i_2} q_{i_2 i_3} q_{i_4 i_5}$$

$$+ 3\sum p_{i_1 i_3} p_{i_2 i_3} p_{i_4 i_5} \sum q_{i_1 i_3} p_{i_2 i_3} q_{i_4 i_5} \Bigg)$$

$$+ \frac{1}{n(n-1)(n-2)(n-3)(n-4)(n-5)} \Bigg(\sum p_{i_1 i_2} p_{i_3 i_4} p_{i_5 i_6} \sum q_{i_1 i_2} q_{i_3 i_4} q_{i_5 i}$$

Each of the sums used above are based on distinct indices, as indicated by the subscripts, and can be represented in terms of other quantities. We give these for the expressions based on $\underset{\sim}{P}$; similar expressions could be obtained for $\underset{\sim}{Q}$.

Let (more or less following Mielke's notation):

$$a_{Ki} = \sum_j p_{ij}^K, \quad b_{Ki} = \sum_j p_{ji}^K, \quad (K = 1, 2, 3)$$

$$d_i = \sum_j p_{ij} p_{ji}, \quad e_i = \sum_j p_{ij}^2 p_{ji}$$

$$A_1 = \sum_i a_{1i}, \quad A_2 = \sum_i a_{2i}, \quad A_3 = \sum_i d_i$$

$$A_4 = \sum_i a_{3i}, \quad A_5 = \sum_i e_i, \quad B_1 = \sum_i a_{1i}^2$$

$$B_2 = \sum_i a_{1i} b_{1i}, \quad B_3 = \sum_i b_{1i}^2$$

$$D_1 = \sum_i a_{2i} a_{1i}, \quad D_2 = \sum_i d_i a_{1i}$$

$$D_3 = \sum_i b_{2i} a_{1i}, \quad D_4 = \sum_i a_{2i} b_{1i}$$

$$D_5 = \sum_i d_i b_{1i}, \quad D_6 = \sum_i b_{2i} b_{1i}$$

$$E_1 = \sum_{i,j,k} p_{ij} p_{ik} p_{jk}, \quad E_2 = \sum_{i,j,k} p_{ij} p_{jk} p_{ki}$$

$$F_1 = \sum_i a_{1i}^3, \quad F_2 = \sum_i a_{1i}^2 b_{1i}$$

$$F_3 = \sum_i a_{1i} b_{1i}^2, \quad F_4 = \sum_i b_{1i}^3$$

$$G_1 = \sum_{i,j} p_{ij} a_{1i} a_{1j}, \quad G_2 = \sum_{i,j} p_{ij} a_{1i} b_{1j}$$

$$G_3 = \sum_{i,j} p_{ij} b_{1i} a_{1j}, \quad G_4 = \sum_{i,j} p_{ij} b_{1i} b_{1j}$$

Then, the sums can be represented as:

$$\sum p_{i_1 i_2} = A_1$$

$$\sum p_{i_1 i_2}^2 = A_2$$

$$\sum p_{i_1 i_2} p_{i_2 i_1} = A_3$$

$$\sum p_{i_1 i_2} p_{i_1 i_3} = B_1 - A_2$$

$$\sum p_{i_1 i_2} p_{i_2 i_3} = B_2 - A_3$$

$$\sum p_{i_1 i_3} p_{i_2 i_3} = B_3 - A_2$$

$$\sum p_{i_1 i_2} p_{i_3 i_4} = A_1^2 - (B_1 + 2B_2 + B_3) + (A_2 + A_3)$$

$$\sum p_{i_1 i_2}^3 = A_4$$

$$\sum p_{i_1 i_2}^2 p_{i_2 i_1} = A_5$$

$$\sum p_{i_1 i_2}^2 p_{i_1 i_3} = D_1 - A_4$$

$$\sum p_{i_1 i_2} p_{i_2 i_1} p_{i_1 i_3} = D_2 - A_5$$

$$\sum p_{i_2 i_1}^2 p_{i_1 i_3} = D_3 - A_5$$

$$\sum p_{i_1 i_2}^2 p_{i_3 i_1} = D_4 - A_5$$

$$\sum p_{i_1 i_2} p_{i_2 i_1} p_{i_3 i_1} = D_5 - A_5$$

$$\sum p_{i_2 i_1}^2 p_{i_3 i_1} = D_6 - A_4$$

$$\sum p_{i_1 i_2}^2 p_{i_3 i_4} = A_1 A_2 - (D_1 + D_3 + D_4 + D_6) + (A_4 + A_5)$$

$$\sum p_{i_1 i_2} p_{i_2 i_1} p_{i_3 i_4} = A_1 A_3 - 2(D_2 + D_5) + 2A_5$$

$$\sum p_{i_1 i_2} p_{i_1 i_3} p_{i_2 i_3} = E_1$$

$$\sum p_{i_1 i_2} p_{i_2 i_3} p_{i_3 i_1} = E_2$$

$$\sum p_{i_1 i_2} p_{i_1 i_3} p_{i_2 i_4} = G_1 - E_1 - (D_2 + D_3) + A_5$$

$$\sum p_{i_1 i_2} p_{i_1 i_3} p_{i_4 i_2} = G_2 - E_1 - (D_1 + D_6) + A_4$$

$$\sum p_{i_1 i_2} p_{i_3 i_1} p_{i_2 i_4} = G_3 - E_2 - (D_2 + D_5) + A_5$$

$$\sum p_{i_1 i_2} p_{i_3 i_1} p_{i_4 i_2} = G_4 - E_1 - (D_4 + D_5) + A_5$$

$$\sum p_{i_1 i_2} p_{i_1 i_3} p_{i_1 i_4} = F_1 - 3D_1 + 2A_4$$

$$\sum p_{i_1 i_2} p_{i_1 i_3} p_{i_4 i_1} = F_2 - (2D_2 + D_4) + 2A_5$$

$$\sum p_{i_2 i_1} p_{i_3 i_1} p_{i_1 i_4} = F_3 - (D_3 + 2D_5) + 2A_5$$

$$\sum p_{i_2 i_1} p_{i_3 i_1} p_{i_4 i_1} = F_4 - 3D_6 + 2A_4$$

$$\sum p_{i_1 i_2} p_{i_1 i_3} p_{i_4 i_5} = A_1(B_1 - A_2) - 2(G_1 + G_2) - (F_1 + F_2) + 2E_1$$
$$+ (3D_1 + 2D_2 + 2D_3 + D_4 + 2D_6) - 2(A_4 + A_5)$$

$$\sum p_{i_1 i_2} p_{i_2 i_3} p_{i_4 i_5} = A_1(B_2 - A_3) - (G_1 + 2G_3 + G_4) - (F_2 + F_3)$$
$$+ (E_1 + E_2) + (4D_2 + D_3 + D_4 + 4D_5) - 4A_5$$

$$\sum p_{i_1 i_3} p_{i_2 i_3} p_{i_4 i_5} = A_1(B_3 - A_2) - 2(G_2 + G_4) - (F_3 + F_4) + 2E_1$$
$$+ (2D_1 + D_3 + 2D_4 + 2D_5 + 3D_6) - 2(A_4 + A_5)$$

$$\sum p_{i_1 i_2} p_{i_3 i_4} p_{i_5 i_6} = A_1^3 - 3A_1(B_1 + 2B_2 + B_3 - A_2 - A_3)$$
$$+ 6(G_1 + G_2 + G_3 + G_4) + 2(F_1 + 3F_2 + 3F_3 + F_4)$$
$$- 2(3E_1 + E_2) - 6(D_1 + 2D_2 + D_3 + D_4 + 2D_5 + D_6)$$
$$+ 4(A_4 + 3A_5)$$

Part C

The general expression for $\text{cov}(\mathcal{A}_1^*(\rho), \mathcal{A}_2(\rho))$ has the same structure as the variance in Section 4.2.2. The components comprising $B_1, \ldots, B_7$ need replacement, however, and we give these below for terms involving $\underset{\sim}{P}$ (a similar replacement would be necessary for terms involving $\underset{\sim}{Q}$):

$$\left(\sum_{i,j} p_{ij}\right)^2 \rightarrow \sum_{i,j} p_{ij}^{(1)} \sum_{i,j} p_{ij}^{(2)}$$

$$\sum_{i,j} p_{ij}^2 \rightarrow \sum_{i,j} p_{ij}^{(1)} p_{ij}^{(2)}$$

$$\sum_i \left(\sum_j p_{ij} p_{ji}\right) \rightarrow \sum_i \left[\sum_j p_{ij}^{(1)} p_{ji}^{(2)}\right]$$

$$\sum_i \left(\sum_j p_{ij}\right)^2 \rightarrow \sum_i \left[\sum_j p_{ij}^{(1)}\right]\left[\sum_j p_{ij}^{(2)}\right]$$

$$\sum_i \left[\left(\sum_j p_{ij}\right)\left(\sum_j p_{ji}\right)\right] \rightarrow \sum_i \left[\sum_j p_{ij}^{(1)}\right]\left[\sum_j p_{ji}^{(2)}\right] \quad \text{and}$$

$$\sum_i \left[\sum_j p_{ij}^{(2)}\right]\left[\sum_j p_{ji}^{(1)}\right]$$

(The "$2B_5$" term is split into two separate expressions.)

$$\sum_j \left(\sum_i p_{ij}\right)^2 \rightarrow \sum_j \left[\sum_i p_{ij}^{(1)}\right]\left[\sum_i p_{ij}^{(2)}\right]$$

Part D

If an LA index, $\Gamma(\rho)$, is based on $\underset{\sim}{C}$ and a QA index, $\mathcal{A}(\rho)$, is based on $\underset{\sim}{P}$ and $\underset{\sim}{Q}$, then $\text{cov}(\Gamma(\rho), \mathcal{A}(\rho)) =$

$$-\left[\frac{1}{n^2(n-1)}\right]\left(\sum_{i,j} c_{ij}\right)\left(\sum_{i,j} p_{ij}\right)\left(\sum_{i,j} q_{ij}\right)$$

$$+ \left[\frac{1}{n(n-1)}\right] \sum_{i,j} c_{ij} \left[\left(\sum_k p_{ik}\right)\left(\sum_k q_{jk}\right) + \left(\sum_k p_{ki}\right)\left(\sum_k q_{kj}\right)\right]$$

$$+ \left[\frac{1}{n(n-1)(n-2)}\right] \sum_{i,j} c_{ij} \left[\left(\sum_{i,j} p_{ij} - \sum_k p_{ik} - \sum_k p_{ki}\right)\right.$$

$$\times \left.\left(\sum_{i,j} q_{ij} - \sum_k q_{jk} - \sum_k q_{kj}\right)\right]$$

5
Extensions of the QA Model

5.1 INTRODUCTION (A)[†]

In the discussion of LA in Chapter 1, a matrix of assignment scores, denoted by $C = \{c_{ij}\}$, was initially used to obtain the raw index $\Gamma(\rho)$. A variety
of special cases were then introduced, including one important alternative
based on a multiplicative structure for C, defined as $c_{ij} = x_i y_j$, and constructed from two given numerical sequences $\{x_1, \ldots, x_n\}$ and $\{y_1, \ldots, y_n\}$.
As apparent from Chapter 4, we chose to develop the QA measure, $\mathcal{A}(\rho)$,
in a different manner because a multiplicative form was assumed at the outset. This special structure obviously has a wide variety of applications;
nevertheless, various generalizations of this basic QA index could also be
pursued, which might be used to approach different data analysis tasks, or
in some instances, suggest a better method than available with the simpler
form of QA. These extensions are discussed in the beginning sections of
this chapter, first by removing the multiplicative restriction inherent in
$\mathcal{A}(\rho)$ and then by broadening the notion of matrix correspondence to measures
that are naturally rephrased as multiplicative forms of cubic or quartic
assignment. Again, various applications and examples of how these higher-
order assignment ideas could be used will be presented in the course of our
discussion. We will also make a few comments about indices that are not
based on sums but rather on the maximum or minimum of the cross-products
defining a (possibly higher-order) assignment statistic.

5.2 NONMULTIPLICATIVE QA INDICES (B)

In the optimization context we have emphasized, the QA index is based on a
cross-product measure between two $n \times n$ matrices, P and Q, and the task

---

[†] The reader is referred to the Preface for an explanation of the use of (A),
(B), and (C) in text headings.

is to locate a permutation ρ to maximize or minimize $\mathcal{A}(\rho) = \Sigma_{i,j}\, p_{\rho(i)\rho(j)}q_{ij}$. Although this formulation is by far the most common, it is of some interest to note that the statement of the original QA optimization task in the operations research literature is phrased in a nonmultiplicative form (e.g., see Koopmans and Beckman, 1957). As might be expected, this extension can also be used to obtain a corresponding inference method paralleling the simpler cross-product measure.

To carry out such a generalization, we define an index, $\mathcal{B}(\rho)$, based on a permutation ρ and a four-place function r_{ghij}, $1 \leqslant g, h, i, j \leqslant n$:

$$\mathcal{B}(\rho) = \sum_{i,j} r_{\rho(i)\rho(j)ij}$$

Once the index $\mathcal{B}(\rho)$ is defined, the inference procedure remains exactly as before. Some identified permutation, ρ_0, defines the observed index, $\mathcal{B}(\rho_0)$, which is then compared to a reference distribution constructed from an evaluation of $\mathcal{B}(\rho)$ over all $n!$ equally likely permutations. Clearly, when $r_{ghij} = p_{gh}q_{ij}$, $\mathcal{B}(\rho)$ reduces to the simpler multiplicative form used throughout Chapter 4.

Given such an extension of the QA index to a nonmultiplicative form, we are limited only by our ability to select an appropriate measure of correspondence between $\underset{\sim}{P}$ and $\underset{\sim}{Q}$ that could be rephrased as a four-place function. There are, however, several more or less natural scoring functions that could be placed in this framework starting from two $n \times n$ matrices, $\underset{\sim}{P}$ and $\underset{\sim}{Q}$. For instance, absolute differences could be used:

$$r_{ghij} = |p_{gh} - q_{ij}|$$

or strict equalities:

$$r_{ghij} = \begin{cases} 1 & \text{if } p_{gh} = q_{ij}, \ g \neq h \text{ and } i \neq j \\ 0 & \text{otherwise} \end{cases}$$

This latter measure would be particularly appropriate when the entries in $\underset{\sim}{P}$ and $\underset{\sim}{Q}$ are on a nominal scale and reflect only qualitative relationships among the objects. In fact, we can move beyond a strictly numerical interpretation for the proximities p_{gh} and q_{ij}. As an example, suppose our concern is with shared properties and p_{gh} denotes a set of attributes held in common by objects O_g and O_h under one circumstance and q_{ij} refers to a second set obtained under another for objects O_i and O_j. One natural way to obtain r_{ghij} would be through the number of attributes in the intersection of p_{gh} and q_{ij}, i.e., $|p_{gh} \cap q_{ij}|$. If deemed appropriate, more complicated

weighting functions or scoring systems could be applied to construct the
four-place functions.[†]

Depending on how the four-place function r_{ghij} is defined, some minor
reduction in the complexity of $\mathcal{B}(\rho)$ may be possible. Specifically, $\mathcal{B}(\rho)$ can
be decomposed into two parts

$$\mathcal{B}(\rho) = \sum_{i,j} r_{\rho(i)\rho(j)ij} = \sum_{i,j} r^o_{\rho(i)\rho(j)ij} + \sum_g s_{g\rho(g)}$$

where

$$r^o_{ghij} = \begin{cases} r_{ghij} & \text{when } g \neq h \text{ and } i \neq j \\ 0 & \text{otherwise} \end{cases}$$

and

$$s_{gi} = r_{ggii}$$

Thus, $\mathcal{B}(\rho)$ can be split into a four-place QA index in which the scoring
function r^o_{ghij} is zero when $g = h$ or $i = j$, and a separate LA index (in most
typical applications, s_{gi} would be constant over all g and i). If $\mathcal{B}^o(\rho)$ and
$\Gamma(\rho)$ denote $\Sigma_{i,j} r^o_{\rho(i)\rho(j)ij}$ and $\Sigma_g s_{g\rho(g)}$, respectively, then

$$\mathcal{B}(\rho) = \mathcal{B}^o(\rho) + \Gamma(\rho)$$

$$E(\mathcal{B}(\rho)) = E(\mathcal{B}^o(\rho)) + E(\Gamma(\rho)) = \left[\frac{1}{n(n-1)}\right] \sum_{g,h,i,j} r^o_{ghij} + \left(\frac{1}{n} \sum_{g,i} r_{ggii}\right)$$

and

$$V(\mathcal{B}(\rho)) = V(\mathcal{B}^o(\rho)) + V(\Gamma(\rho)) + 2\mathrm{cov}(\mathcal{B}^o(\rho), \Gamma(\rho))$$

The formulas for the variance and covariance involving $\mathcal{B}^o(\rho)$ are available
in Hubert (1979) and will not be given here. Unfortunately, the third moment for $\mathcal{B}(\rho)$ is extremely complicated and has not been obtained in any
computationally usable form, suggesting that, at least for the present,
Monte Carlo significance testing would be the inference strategy to follow.

[†]We might recall that the use of squared proximity differences would
be statistically equivalent to the multiplicative QA model, e.g., see
Section 4.2.2.

5.3 HIGHER-ORDER ASSIGNMENT INDICES (A)

As apparent from Chapter 4, the index $\mathcal{A}(\rho_0)$ has a wide variety of possible applications, but it also has one feature that makes its use more or less inconsistent with several current emphases in psychological data analysis, particularly in the areas of multidimensional scaling and cluster analysis. The statistic $\mathcal{A}(\rho_0)$ is really an unnormalized Pearson product-moment correlation coefficient between corresponding entries in two $n \times n$ matrices $\underset{\sim}{P}$ and $\underset{\sim}{Q}$; consequently, the comparison strategy itself must rely on more than the rank orderings of the object pairs as defined by the two proximity functions.[†]

Phrased differently, $\mathcal{A}(\rho_0)$ is not invariant under all monotone transformations of the proximity values, and different significance statements could result if different monotone transformations are considered. Viewed more generally, the index $\mathcal{A}(\rho)$ is based on a one-to-one map of the entries in $\underset{\sim}{P}$ to the entries in $\underset{\sim}{Q}$; thus, the comparison process does not involve internal comparisons within either of the two matrices. This last property is the basic reason for the lack of a possible monotone invariance for a simple index of the cross-product form.

The aim of this section is to suggest analogs for the matrix comparison statistic $\mathcal{A}(\rho)$ that can allow certain internal comparisons within $\underset{\sim}{P}$ and/or $\underset{\sim}{Q}$ to contribute to a final index of correspondence. The discussion will initially emphasize general-purpose measures of correspondence between $\underset{\sim}{P}$ and $\underset{\sim}{Q}$ that depend only on proximity rank order. These measures could be used for most of the applications of the QA index mentioned in Chapter 4. Later sections will provide a number of other extensions beyond this basic comparison task and to higher-order assignment statistics that are tailor-made for very specific comparison tasks.

<u>Cubic Assignment Measures.</u> As one possible generalization of $\mathcal{A}(\rho_0)$, suppose we are given the set $S = \{O_1, \ldots, O_n\}$ and the two proximity matrices $\underset{\sim}{P}$ and $\underset{\sim}{Q}$, two new functions are defined on $S \times S \times S$ (i.e., over all object triples), and denoted by $\overline{p}_{ijk}$ and $\overline{q}_{ijk}$, respectively. A raw index of correspondence between the original two matrices is now redefined in terms of the three-place functions $\overline{p}_{ijk}$ and $\overline{q}_{ijk}$:

$$\mathcal{A}_3(\rho) = \sum_{i,j,k} \overline{p}_{\rho(i)\rho(j)\rho(k)}\,\overline{q}_{ijk}$$

[†] This assumes that no canonical transformation of the proximities is consistently carried out prior to the calculation of $\mathcal{A}(\rho_0)$, e.g., to ranks or normal scores. Transformations of this type can always achieve a trivial form of monotone invariance.

where, for convenience, the identity permutation ρ_I is assumed to characterize the observed index that we wish to evaluate for relative size. In
short, the index $\mathcal{A}_3(\rho_I)$ defines a measure of correspondence between $\underset{\sim}{P}$ and
$\underset{\sim}{Q}$ in a cubic assignment framework using cross-products of three-place
functions rather than of two-place as in QA. In general, if $\bar{p}_{ijk}$ and $\bar{q}_{ijk}$
rely solely on the ordering of the proximity values in $\underset{\sim}{P}$ and $\underset{\sim}{Q}$, the index
$\mathcal{A}_3(\rho)$ itself will depend solely on the ordinal properties of the original
proximities. In fact, the three-place indices we emphasize will depend
only on the ordinal properties of the proximities <u>within</u> the rows of P and Q.

As one example that will be attributed to Hartigan (1975, p. 13), suppose we define

$$
\bar{p}_{ijk} = \begin{cases} 1 & \text{if } p_{ij} < p_{ik} \text{ for distinct } i,j,k \\ 0 & \text{otherwise} \end{cases} \tag{5.3a}
$$

Among all $n(n-1)(n-2)$ triads that could be formed from distinct objects
using the definitions given above, the raw index $\mathcal{A}_3(\rho_I)$ counts the number
of inconsistencies in the ordering of proximities between $\underset{\sim}{P}$ and $\underset{\sim}{Q}$. Because
the triads are compared only <u>within</u> the rows of $\underset{\sim}{P}$ and $\underset{\sim}{Q}$, the proximities
have to be commensurable within each row but not across different rows.
(An index to compare triads only within columns is immediate if we use the
transposes of $\underset{\sim}{P}$ and $\underset{\sim}{Q}$ to define the three-place functions. To make this
distinction clear, we could refer to the comparison of row triads or column
triads whatever the appropriate case may be. In general, however, it will
be assumed that row triads are of interest unless stated otherwise, and the
qualifier of "row" will usually be omitted.) In short, the type of definition
given in (5.3a) provides a way of comparing conditional proximity matrices
in the sense of Coombs (1979) or Shepard (1972); only the entries within the
same row need to be strictly comparable.

If one of the inequalities were reversed in (5.3a), $\mathcal{A}_3(\rho_I)$ would then
count consistent rather than inconsistent row triads between $\underset{\sim}{P}$ and $\underset{\sim}{Q}$. In
turn, this suggests a nice parallel to the discussion of generalized correla-
tion coefficients in Section 4.3.2 if we combine the number of consistent
and inconsistent triads into a single measure. Following the discussion in
this latter section, suppose we define

$$
\text{sign}(x) = \begin{cases} +1 & \text{if } x > 0 \\ 0 & \text{if } x = 0 \\ -1 & \text{if } x < 0 \end{cases}
$$

Then, if

$$\overline{p}_{ijk} = \begin{cases} \text{sign}(p_{ik} - p_{ij}) & \text{for distinct } i,j,k \\ 0 & \text{otherwise} \end{cases}$$

$$\overline{q}_{ijk} = \begin{cases} \text{sign}(q_{ik} - q_{ij}) & \text{for distinct } i,j,k \\ 0 & \text{otherwise} \end{cases} \qquad (5.3b)$$

the index

$$\mathcal{A}_3(\rho_I) = \sum_{i,j,k} \overline{p}_{ijk}\overline{q}_{ijk}$$

is twice the number of consistent row triads (P_c) minus the number of inconsistent row triads (P_I). As an alternative representation that will be convenient later on, $\mathcal{A}_3(\rho_I) = 2(P_c - P_I) = p_a - p_b$, where $p_a = 2P_c$ and $p_b = 2P_I$. Thus, p_a (or p_b) refers to the number of consistent (or inconsistent) ordered triads (O_i, O_j, O_k) for distinct i,j,k. Both (O_i, O_j, O_k) and (O_i, O_k, O_j) contribute to the counts represented by p_a or p_b, whereas only one of these two triads can contribute to P_c or P_I.

As we will see, the problems of normalizing $\mathcal{A}_3(\rho_I)$ are almost identical to that encountered in the context of Kendall's coefficient of correlation (see Section 4.3.2). In fact, the same type of descriptive measures that have operational meaning in terms of conditional probabilities of picking consistent (concordant) and inconsistent (discordant) pairs will be proposed here as well. The major difference is that now the interpretations will be based on picking consistent and inconsistent triads. Because of this convenient probabilistic interpretation, a triad measure based on (5.3b) will be our index of choice. A comparable four-place index mentioned below would be another natural alternative, but the latter requires more computational effort when a Monte Carlo significance testing strategy is followed. In fact, some work by Dietz (1983) suggests that the move to four-place indices may not be advantageous even if it were no more difficult computationally. Based on some preliminary Monte Carlo comparisons of power, the crucial components even for a four-place measure appear to be based on object triples and the associated proximity comparisons that are mediated by a common object. For more detail, the reader is referred to Dietz (1983).

It may help to clarify what the general triad measure does in one simple case. Suppose $\underset{\sim}{P}$ is an arbitrary proximity matrix, but $\underset{\sim}{Q}$ contains dichotomous zero-one entries representing a partition as in Section 4.3.4. In constructing $\overline{q}_{ijk}$ from the sign function, nonzero values will be obtained only when $q_{ij} = 0$ and $q_{ik} = 1$ or $q_{ij} = 1$ and $q_{ik} = 0$. Thus, in the triple of objects (O_i, O_j, O_k), the object O_i must belong to one of the classes of the partition and either O_j or O_k (but not both) must be outside of this class. For example, when $q_{ik} = 1$ and $q_{ij} = 0$, then O_i and O_k belong to the same class and O_j does not; moreover, $\text{sign}(p_{ik} - p_{ij}) = 1$ if the proximity from O_i to the outside object O_j is less than the proximity from O_i to the inside object O_k. If large proximities denote similar objects and the partition represents the proximity data well, we would expect this condition to obtain. Conversely, a value of -1 for $\text{sign}(p_{ik} - p_{ij})$ would represent an inconsistency for the given partition in terms of an adequate representation for the data in $\underset{\sim}{P}$. To summarize, when large proximities denote similar objects, a consistent (inconsistent) triad is one in which a within-class proximity p_{ik} is strictly greater (less) than a between-class proximity p_{ij}, using the common object O_i as a key. The raw index $\mathcal{A}_3(\rho_I)$ merely combines these two types of triads into a single overall measure of the adequacy of the given partition, which would be represented through large and positive values for $\mathcal{A}_3(\rho_I)$. If proximity were keyed oppositely so that small values denoted similar objects, then negative values of $\mathcal{A}_3(\rho_I)$ would be desired. In any case, an obvious extension of this same strategy could be followed for assessing a single cluster of objects as in Section 4.3.3. Here, $\underset{\sim}{Q}$ would be the zero-one matrix representing membership in the given subset D.

From a somewhat broader perspective, we can be very flexible in our characterization of the three-place functions because the latter do not have to arise from specific measures defined on S × S nor is their use confined to the data analysis problem of comparing two matrices $\underset{\sim}{P}$ and $\underset{\sim}{Q}$. For instance, $\overline{p}_{ijk}$ could denote the outcome of a triad comparison obtained experimentally from a subject who specifies the degree to which O_i is more different from O_j than from O_k. As we will see in the sections to follow, applications of this latter type present no additional difficulties, and in fact, may provide some of the more interesting uses of $\mathcal{A}_3(\rho)$. With or without an invariance restriction and irrespective of the origin of the functions $\overline{p}_{ijk}$ and $\overline{q}_{ijk}$, the distribution of $\mathcal{A}_3(\rho)$ under randomness can be generated by assuming a uniform distribution over all permutations on the first n integers. Again, Monte Carlo significance testing would provide an obvious operational strategy for evaluating the relative size of $\mathcal{A}_3(\rho_I)$.

Quartic Assignment Measures. On the basis of the general form of $\mathcal{A}_3(\rho)$, it is relatively straightforward to develop extensions to k-place functions defined, say, from $\underset{\sim}{P}$ and $\underset{\sim}{Q}$. For our purposes, however, only the three-place alternative mentioned above and the next obvious four-place extension to a quartic assignment index will be considered in any detail. Because the index based on three-place functions was denoted by $\mathcal{A}_3(\rho)$, indices based on four-place functions will be denoted by $\mathcal{A}_4(\rho)$, and so on. [For notational consistency, $\mathcal{A}(\rho)$ could now be denoted by $\mathcal{A}_2(\rho)$.]

In terms of four-place functions, there is one obvious analog to the triads index that has been suggested by Lerman (1969, 1970). If we define

$$
\overline{p}_{ijkl} = \begin{cases} 1 & \text{if } p_{ij} < p_{kl} \text{ for } i \neq j; \, k \neq l \\ 0 & \text{otherwise} \end{cases}
$$

$$
\overline{q}_{ijkl} = \begin{cases} 1 & \text{if } q_{ij} > q_{kl} \text{ for } i \neq j; \, k \neq l \\ 0 & \text{otherwise} \end{cases}
$$

then $\mathcal{A}_4(\rho_I)$ counts the number of inconsistent quadruples between $\underset{\sim}{P}$ and $\underset{\sim}{Q}$. As in the case for the analogous three-place function, a change in one of the inequalities would count consistent quadruples. Moreover, both the inconsistent and consistent quadruples could be combined to produce a parallel to the coefficients related to Kendall's tau but now with comparisons across the rows of $\underset{\sim}{P}$ and of $\underset{\sim}{Q}$:

$$
\overline{p}_{ijkl} = \begin{cases} \text{sign}(p_{kl} - p_{ij}) & \text{for } i \neq j; \, k \neq l \\ 0 & \text{otherwise} \end{cases}
$$

$$
\overline{q}_{ijkl} = \begin{cases} \text{sign}(q_{kl} - q_{ij}) & \text{for } i \neq j; \, k \neq l \\ 0 & \text{otherwise} \end{cases}
$$

Viewed in a slightly different way, the index $\mathcal{A}_4(\rho_I)$ provides the numerator of a Kendall's tau statistic between all $n(n - 1)$ off-diagonal entries in $\underset{\sim}{P}$ and $\underset{\sim}{Q}$, i.e., twice the number of consistent quadruples minus the number of inconsistent quadruples (or in analogy to what was done in the three-place context, $\mathcal{A}_4(\rho_I)$ is the number of consistent <u>ordered</u> quadruples minus the number of such inconsistencies).

In the type of partition comparison task mentioned for row triads in which q_{ij} is a dichotomous zero-one function, somewhat the same interpretation applies. The exception is that now all within-class proximities are compared to all between-class proximities and no common object needs to be present to define a consistency or an inconsistency. Again, the raw index merely aggregates these two counts into a single overall measure. For a further discussion of the Kendall index in this same context, the reader is referred to Jackson (1969).

<u>Moments for $\mathcal{A}_k(\rho)$</u>. On the basis of the indicator method of Section 1.4.1, it is (theoretically) possible to obtain the moments for $\mathcal{A}_k(\rho)$. Because these will tend to be excessively complicated, however, we will be content with providing the expectations for (i) $\mathcal{A}_3(\rho)$, using arbitrary three-place functions; (ii) $\mathcal{A}_4(\rho)$, when the four-place functions are zero for $i = j$ and/or $k = l$; and finally, (iii) $\mathcal{A}_k(\rho)$, when the k-place functions are zero if any of the k subscripts are tied. In general, significance testing will be carried out by Monte Carlo sampling.

i. For $\mathcal{A}_3(\rho)$ based on arbitrary three-place functions,

$$E(\mathcal{A}_3(\rho)) = \left(\frac{1}{n}\right)\sum_i p_{iii}q_{iii} + \left[\frac{1}{n(n-1)}\right]$$

$$\times\left[\left(\sum_{i,j}\bar{p}_{iji} - \sum_i \bar{p}_{iii}\right)\left(\sum_{i,j}\bar{q}_{iji} - \sum_i \bar{q}_{iii}\right)\right.$$

$$+ \left(\sum_{i,k}\bar{p}_{iik} - \sum_i \bar{p}_{iii}\right)\left(\sum_{i,k}\bar{q}_{iik} - \sum_i \bar{q}_{iii}\right)$$

$$+ \left.\left(\sum_{i,j}\bar{p}_{ijj} - \sum_i \bar{p}_{iii}\right)\left(\sum_{i,j}\bar{q}_{ijj} - \sum_i \bar{q}_{iii}\right)\right]$$

$$+ \left[\frac{1}{n(n-1)(n-2)}\right]\left[\sum_{i,j,k}\bar{p}_{ijk} - \sum_{i,k}\bar{p}_{iik} + 2\sum_i \bar{p}_{iii}\right.$$

$$- \sum_{i,j}\bar{p}_{iji} - \sum_{i,j}\bar{p}_{ijj}\right]\left[\sum_{i,j,k}\bar{q}_{ijk}\right.$$

$$- \sum_{i,j}\bar{q}_{iik} + 2\sum_i \bar{q}_{iii} - \sum_{i,j}\bar{q}_{iji} - \sum_{i,j}\bar{q}_{ijj}\right]$$

ii. In the four-place instance in which p_{ijkl} and q_{ijkl} are zero when $i = j$ and/or $k = l$,

$$E(\mathcal{A}_4(\rho)) = \frac{1}{n(n-1)(n-3)}\left[\left(\sum_{i,j,l}\bar{p}_{ijil}\right)\left(\sum_{i,j,l}\bar{q}_{ijil}\right)\right.$$

$$+ \left(\sum_{i,j,k}\bar{p}_{ijki}\right)\left(\sum_{i,j,k}\bar{q}_{ijki}\right)$$

$$+ \left(\sum_{i,j,l}\bar{p}_{ijjl}\right)\left(\sum_{i,j,l}\bar{q}_{ijjl}\right)$$

$$+ \left.\left(\sum_{i,j,k}\bar{p}_{ijkj}\right)\left(\sum_{i,j,k}\bar{q}_{ijkj}\right)\right]$$

$$+ \frac{1}{n(n-1)(n-2)(n-3)}\left[\left(\sum_{i,j,k,l}\bar{p}_{ijkl}\right.\right.$$

$$- \sum_{i,j,l} \bar{p}_{ijil} - \sum_{i,j,k} \bar{p}_{ijki} - \sum_{i,j,l} \bar{p}_{ijjl}$$

$$- \sum_{i,j,k} \bar{p}_{ijkj} \Bigg) \Bigg(\sum_{i,j,k,l} \bar{q}_{ijkl} - \sum_{i,j,l} \bar{q}_{ijil}$$

$$- \sum_{i,j,k} \bar{q}_{ijki} - \sum_{i,j,l} \bar{q}_{ijjl} - \sum_{i,j,k} \bar{q}_{ijkj} \Bigg) \Bigg]$$

iii. When the k-place functions are 0 unless the indices are all distinct, the formula for $E(\mathcal{A}_k(\rho))$ is reasonably simple:

$$E(\mathcal{A}_k(\rho)) = \left[\frac{1}{n(n-1)\cdots(n-k+1)} \right] \sum_{i_1,\ldots,i_k} \bar{p}_{i_1\cdots i_k} \sum_{i_1,\ldots,i_k} \bar{q}_{i_1\cdots i_k}$$

In the special cases in which the two measures are defined by the difference between the consistent and inconsistent triads or quadruples, the expectation is 0.

Spearman's Rank Order Correlation. There is one particularly interesting application of a three-place index and the moment formula for an arbitrary three-place function to Spearman's rank order correlation discussed in Chapter 1. Suppose $\{x_1,\ldots,x_n\}$ and $\{y_1,\ldots,y_n\}$ consist of the untied ranks from 1 to n and let $a_{ij} = \text{sign}(x_j - x_i)$ and $b_{ij} = \text{sign}(y_j - y_i)$, which are the same scoring functions used to define Kendall's tau in Section 4.3.2. For the present, however, our interests are in obtaining an alternative representation for Spearman's rank order correlation as

$$\rho_s = \frac{3}{(n^3 - n)} \sum_{i,j,k} a_{ij} b_{ik}$$

Now, suppose we treat this index as a population value and sample K objects at random from the total pool of size n. Defining

$$\bar{p}_{ijk} = \begin{cases} a_{ij}b_{ik} & \text{for all i,j, and k for which } i \neq j, \, i \neq k \\ 0 & \text{otherwise} \end{cases}$$

and

$$\bar{q}_{ijk} = \begin{cases} 1 & \text{for } 1 \leq i,j,k \leq K \text{ and } i \neq j, \, i \neq k \\ 0 & \text{otherwise} \end{cases}$$

then

$$r_s = \frac{3}{(K^3 - K)} \sum_{i,j,k} \bar{p}_{\rho(i)\rho(j)\rho(k)} \bar{q}_{ijk}$$

is the Spearman rank order correlation for the sample of K objects $\{O_{\rho(1)}, \ldots, O_{\rho(K)}\}$.

The index $\mathcal{A}_3(\rho) = \Sigma_{i,j,k} \bar{p}_{\rho(i)\rho(j)\rho(k)} \bar{q}_{ijk}$ has expectation

$$\left[\frac{1}{n(n-1)}\right] \left(\sum_{i,j} \bar{p}_{ijj}\right)\left(\sum_{i,j} \bar{q}_{ijj}\right) + \left[\frac{1}{n(n-1)(n-2)}\right]$$

$$\times \left[\left(\sum_{i,j,k} \bar{p}_{ijk} - \sum_{i,j} \bar{p}_{ijj}\right)\left(\sum_{i,j,k} \bar{q}_{ijk} - \sum_{i,j} \bar{q}_{ijj}\right)\right]$$

$$= \left(\frac{K^3 - K}{3}\right)\left\{\left[\frac{3(n-K)}{(K+1)(n-2)}\right]\tau + \left[\frac{(K-2)(n+1)}{(K+1)(n-2)}\right]\rho_s\right\}$$

where Kendall's tau is obtained as

$$\tau = \left[\frac{1}{n(n-1)}\right]\sum_{i,j} a_{ij} b_{ij}$$

Thus, the expectation of the sample Spearman rank order correlation is

$$E(r_s) = \left[\frac{3(n-K)}{(K+1)(n-2)}\right]\tau + \left[\frac{(K-2)(n+1)}{(K+1)(n-2)}\right]\rho_s$$

This result (also derived in Kendall, 1970, p. 93) points out the biasedness of the sample rank order correlation when used to estimate a population correlation. We recall from Section 4.3.11 that a sample Kendall tau coefficient is unbiased under these same circumstances. Frank (1978) has discussed and extended some of these sampling ideas in a graph-theoretic framework to the problem of estimating triad counts of a particular kind.

Normalization. Given a raw measure of the type $\mathcal{A}_k(\rho_I)$, various normalizations could be considered, having the general form $\mathcal{A}_k^*(\rho_0)$ and $\mathcal{A}_k^{**}(\rho_0)$, from Section 1.2.4. For example, if we count inconsistent row triads as in equation (5.3a) and assume that a small number of such inconsistencies are desirable, one obvious measure would be $\mathcal{A}_3^{**}(\rho_0)$. Because the minimum index is zero, at least as a bound, $\mathcal{A}_3^{**}(\rho_0)$ would be defined as

$$1 - \frac{\mathcal{A}_3(\rho_0)}{E(\mathcal{A}_3(\rho))}$$

Obviously, a similar measure could be constructed for inconsistent quadruples as well.

The two general measures we have mentioned based on ± 1 scoring functions have a particularly obvious relationship to the various normalizations we encountered in the context of Kendall's coefficient of correlation. For instance, the obvious analog for tau_a would take the form

$$\frac{\sum_{i,j,k} \bar{p}_{ijk} \bar{q}_{ijk}}{n(n-1)(n-2)}$$

and for tau_b

$$\frac{\sum_{i,j,k} \bar{p}_{ijk} \bar{q}_{ijk}}{\sqrt{\sum_{i,j,k} \bar{p}_{ijk}^2 \sum_{i,j,k} \bar{q}_{ijk}^2}}$$

i.e.,

$$n(n-1)(n-2) \geq \sqrt{\sum_{i,j,k} \bar{p}_{ijk}^2 \sum_{i,j,k} \bar{q}_{ijk}^2}$$

$$= \sqrt{\sum_{i,j,k} |\bar{p}_{ijk}| \sum_{i,j,k} |\bar{q}_{ijk}|} \geq \max_{\rho} \mathcal{A}_3(\rho)$$

In fact, all of the variations in Section 4.3.2 on conditionalizations regarding tied and untied pairs could be adopted directly and given corresponding descriptive probabilistic interpretations. Now, triples based on distinct objects are selected at random, and we conditionalize on triples that are or are not tied in terms of their associated proximities in $\underset{\sim}{P}$ and/or $\underset{\sim}{Q}$. Operationally, the denominators used to standardize the raw cross-product index would be given by expressions that have the same general structure given in Section 4.3.2 for the Somers and Goodman-Kruskal statistics, except that triples would replace the pairs in the defining quantities. Specifically, one of Somers's analogs could be interpreted as the difference between the probability of a consistent triad among all triads whose defining proximities are not tied in $\underset{\sim}{P}$ (i.e., p_{ij} and p_{ik}) minus the probability of an inconsistent triad among all triads whose defining proximities are not tied in $\underset{\sim}{P}$ (a denominator of $\Sigma_{i,j,k} |\bar{p}_{ijk}|$ would be used). If we conditionalize on proximities untied in $\underset{\sim}{Q}$, the denominator would be $\Sigma_{i,j,k} |\bar{q}_{ijk}|$. A Goodman-Kruskal analog would depend on a divisor of $\Sigma_{i,j,k} |\bar{p}_{ijk}\bar{q}_{ijk}|$; the conditionalization would be on triads not tied in $\underset{\sim}{P}$ or in $\underset{\sim}{Q}$ (the type of comment made in Section 4.3.2 on how this latter quantity may not necessarily be an upper bound on $\max_{\rho} \mathcal{A}_3(\rho)$ would apply). An extension to the ± 1 scoring function for quadruples would follow exactly this same pattern all the way through.

Throughout this chapter we routinely rely on analogs for the Somers and Goodman-Kruskal measures and will refer to them by these names. If both $\underset{\sim}{P}$ and $\underset{\sim}{Q}$ are empirically based, the Goodman-Kruskal index is a natural one to use. Alternatively, if $\underset{\sim}{Q}$ represents a structure constructed by the researcher, we typically depend on the Somers alternative of conditionalizing on the absence of ties in $\underset{\sim}{Q}$. In this latter case, the normalized index will be relatively lower whenever the structure matrix makes a particular distinction (represented by untied entries) that corresponds to tied entries in $\underset{\sim}{P}$ (for a further development of these parallels, the reader should consult Hubert, Golledge, Costanzo, and Gale, 1985).

An Example. As a numerical illustration of how a three-place comparison could be carried out, we consider the two 16×16 Miller-Nicely confusion matrices discussed in Section 4.3.8, where $\underset{\sim}{P}$ and $\underset{\sim}{Q}$ represent Matrix I and Matrix II, respectively. Each matrix was obtained under a different signal/ noise condition and contains the conditional probabilities of responding with a particular consonant phoneme, given the presentation of another. Based on the comparison of row triads through the sign functions of equation (5.3b), a Monte Carlo significance of .001 could be reported for the observed index $\mathcal{A}_3(\rho_I) = 710$, when a sample size of 999 and the distribution summarized in Table 5.3a are used.

To normalize this raw index, suppose as notation we let P_C and P_I be the number of consistent and inconsistent row triads, respectively, then $710 = 2(P_C - P_I)$, where $P_C = 933$ and $P_I = 578$. As possible denominators discussed earlier, we could choose

1. $n(n-1)(n-2) = 3360$

2. $\sqrt{\sum_{i,j,k} \bar{p}_{ijk}^2 \sum_{i,j,k} \bar{q}_{ijk}^2} = \sqrt{\sum_{i,j,k} |\bar{p}_{ijk}| \sum_{i,j,k} |\bar{q}_{ijk}|} = 3184.97$

3. $\sum_{i,j,k} |\bar{p}_{ijk}| = 3198$

Table 5.3a Monte Carlo Distributions for the
Triad Comparison Measure Based
on the Miller-Nicely Data—Sample
Size of 999

Cumulative frequency	Index
1	-466
5	-400
10	-358
50	-244
100	-198
200	-134
300	- 84
400	- 50
500	- 14
600	24
700	66
800	118
900	198
950	264
990	380
995	490
999	688

4. $\displaystyle\sum_{i,j,k} |\bar{q}_{ijk}| = 3172$

5. $\displaystyle\sum_{i,j,k} |\bar{p}_{ijk}\bar{q}_{ijk}| = 3022$

For instance, the Goodman-Kruskal analog based on (5) would give a value of

$$.235 = \frac{710}{3022} = \frac{\displaystyle\sum_{i,j,k} \bar{p}_{ijk}\bar{q}_{ijk}}{\displaystyle\sum_{i,j,k} |\bar{p}_{ijk}\bar{q}_{ijk}|} = \frac{2(P_c - P_I)}{2(P_c + P_I)}$$

$$= \frac{P_c}{P_c + P_I} - \frac{P_I}{P_c + P_I} = .617 - .383$$

Thus, the conditional probability of picking at random a consistent (inconsistent) row triad untied in both matrices is .617(.383). Although the raw index $\mathcal{A}_3(\rho_I)$ is significant at a small level, the size of this latter index (.235) reflects a substantial number of inconsistencies; there is a fairly large disparity in the order information provided by the two matrices.

Some Descriptive Comparisons. In the QA context, it was suggested that an inspection of the matrix $E = \{(p_{ij} - \bar{p})(q_{ij} - \bar{q})\}$ might help in a descriptive understanding of where the correspondence (or lack of it) between P and Q comes from. In a similar fashion, a matrix that is an analog of E can be constructed in the k-place context as well. For instance, if we define F to be the n × n matrix

$$\underset{\sim}{F} = \left\{ \sum_k \bar{p}_{ijk}\bar{q}_{ijk} \right\}$$

then the sum of entries in F produces $\mathcal{A}_3(\rho_I)$. Each entry in F indexes the contribution for that particular row and column position across P and Q.

As a numerical example based on comparing the two Miller-Nicely matrices, we would obtain Table 5.3b. The entries can range from –14 to +14, with negative values suggesting a discrepancy in pattern for that given position in P and Q; positive values would indicate a consistency. It is apparent from Table 5.3b that certain conditional probabilities are very consistent in relative size (within rows) between P and Q, and certain others are not. Because large negative values indicate strong reversals in the relative size of the conditional probabilities, they may be the most important to note and try to explain. Judging by the row sums given in Table 5.3b, five of the phonemes are particularly troublesome and contribute the least to the correspondence index: $(d, z, \math3, m, n)$. As a group, their aggregate contribution to the overall index of 710 is only 28, or about 4% of the total. To provide a contrast, the five phonemes $(p, t, k, b, \eth)$ produce 57% of the overall measure; these are the most consistent in terms of the pattern of conditional probabilities across the two matrices.

Table 5.3b Matrix F for the Comparison of the Two Miller–Nicely Matrices

	p	t	k	f	θ	s	ʃ	b	d	g	v	ð	z	ʒ	m	n	Row sums
p	×	12	7	10	10	9	4	3	-1	9	5	6	5	4	7	6	96
t	10	×	14	11	9	10	12	5	4	10	-6	4	2	2	3	2	92
k	10	10	×	8	7	6	7	-1	5	6	5	3	-5	6	7	6	80
f	11	12	11	×	-7	-2	-2	2	4	4	1	1	4	12	3	-2	52
θ	4	5	5	-6	×	10	1	3	3	3	2	1	3	11	3	2	50
s	6	8	5	-8	5	×	4	5	-2	4	4	-1	1	4	-4	11	42
ʃ	1	7	8	4	-4	10	×	6	1	0	0	-4	2	5	2	4	42
b	8	10	-10	2	0	7	8	×	6	5	13	-6	4	7	4	4	62
d	-2	-10	-5	6	2	1	0	2	×	6	5	4	5	-3	2	3	16
g	10	-3	-10	12	2	2	4	6	13	×	0	-5	-3	4	4	4	40
v	11	-6	-7	5	3	2	4	14	2	2	×	2	7	4	6	1	50
ð	-2	4	5	-2	9	6	6	14	5	8	2	×	7	8	4	4	78
z	-1	-5	-3	8	-2	4	0	5	0	-9	4	-3	×	-8	-7	-1	-18
ʒ	3	-3	-3	4	6	-2	-2	2	9	-7	0	-2	2	×	-1	-2	4
m	-6	1	-5	-1	-1	-1	7	2	1	-2	-2	-8	6	1	×	14	6
n	4	-8	-9	2	6	-3	6	1	1	0	-4	2	4	2	14	×	18

5.3.1 Seriation for Symmetric Proximities (B)

As mentioned in Section 4.3.1, one of the standard data analysis problems encountered in the social and behavioral sciences deals with the sequencing of objects from a set S along a one-dimensional continuum. Although this area has been discussed in detail in Chapter 4 using the index $\mathcal{A}(\rho)$, some of the more popular statistics proposed in the literature require measures of the form $\mathcal{A}_3(\rho)$ and $\mathcal{A}_4(\rho)$. For this reason, the topic of seriation provides an excellent area of application for higher-order assignment indices. Several of these alternatives are listed below, along with a brief interpretation and references for further discussion. For convenience, all these statistics measure the degree to which the matrix $\underset{\sim}{P}$ has an anti-Robinson form, i.e., if $\underset{\sim}{P}$ is symmetric and small values of p_{ij} denote similar objects, then $\underset{\sim}{P}$ is anti-Robinson if the entries within a particular row never decrease when moving away from the main diagonal in either direction. Thus, we assume implicitly that the conjectured ordering is given by the identity permutation ρ_I.

For the present, it will be assumed that the ordering we wish to evaluate is complete in the sense that there are no tied locations along the underlying continuum. Toward the end of this section, a generalization will be presented to deal explicitly with the evaluation task when some of the objects may be identically placed by the conjecture. This latter problem is one of assessing a hypothesized partial ordering of the n objects.

One obvious approach to testing the correspondence of $\underset{\sim}{P}$ to an anti-Robinson pattern would define $\underset{\sim}{Q}$ as, say, $\{|i-j|\}$ and carry out one of the general matrix comparison methods discussed in the previous section. Unfortunately, this strategy is a little too comprehensive for the seriation problem. The anti-Robinson gradient condition does not explicitly involve comparing proximities within a row on the opposite sides of the main diagonal, but a general comparison index would rely heavily on such comparisons. For this reason, we will propose a variety of more specific measures that explicitly assess the degree to which an anti-Robinson pattern is present in $\underset{\sim}{P}$ without making across-diagonal evaluations. Moreover, the indices we suggest will depend only on the conjectured ordering of the objects along the continuum (assumed to be defined by ρ_I) and not on any particular spacing. This does not preclude, however, statistics that may depend on more than the ordinal information in the proximity matrix $\underset{\sim}{P}$.

For uniformity, all of the initial examples given below are treated in such a way that low values of the index are desirable, even though the cited reference may not phrase the problem exactly in this manner. Also, the arbitrary self-proximities defining the main diagonal of $\underset{\sim}{P}$ are not used in any of the internal matrix comparisons although the reference cited may incorporate these in defining the raw index.

1. Hole and Shaw (1967):

$$\bar{p}_{ijk} = \begin{cases} p_{ij} - p_{ik} & \text{if } p_{ij} > p_{ik}; \ i,j,k \text{ distinct} \\ 0 & \text{otherwise} \end{cases}$$

$$\bar{q}_{ijk} = \begin{cases} 1 & \text{if } k = j + 1, \ i < j \text{ or } k = j - 1, \ j < i \\ 0 & \text{otherwise} \end{cases}$$

Using these two functions, $\mathcal{A}_3(\rho_I)$ measures how "close" $\underset{\sim}{P}$ is to an anti-Robinson form, and does so in such a way that we depend on more than the ordinal information available in the proximity measures. Within each row, a sum is taken over the differences between off-diagonal adjacent proximities whenever these adjacent proximities differ from the anti-Robinson pattern. Thus, the function $\bar{p}_{ijk}$ serves to define all positive proximity differences using the object O_i as a "key"; i.e., in $\underset{\sim}{P}$, we are working within the row corresponding to O_i. The function $\bar{q}_{ijk}$, on the other hand, selects those positive differences to include in the index $\mathcal{A}_3(\rho_I)$ that represent violations of the anti-Robinson form for two off-diagonal <u>adjacent</u> proximities within the row corresponding to O_i.

As a simplification of (1), a related measure is defined below in (2) that merely counts the number of such discrepancies, i.e., the number of violations of the anti-Robinson form for off-diagonal adjacent proximities within the rows of $\underset{\sim}{P}$. Because the severity of a particular violation is not considered (as opposed to (1) when it is measured by the magnitude of a positive difference), the dichotomous transformation defining $\bar{p}_{ijk}$ produces an index depending only on the ranking of the proximity values within the rows of $\underset{\sim}{P}$.

2. Kuzara, Mead, and Dixon (1966):

$$\bar{p}_{ijk} = \begin{cases} 1 & \text{if } p_{ij} > p_{ik}, \ i,j,k \text{ distinct} \\ 0 & \text{otherwise} \end{cases}$$

$$\bar{q}_{ijk} = \begin{cases} 1 & k = j + 1, \ i < j; \text{ or } k = j - 1, \ j < i \\ 0 & \text{otherwise} \end{cases}$$

In assessing the adequacy of an anti-Robinson form, both of the indices in (1) and (2) represent particular improvements over the use of the QA index $\mathcal{A}(\rho)$ discussed in Section 4.3.1. When the QA measure is considered for this application, we typically define q_{ij} as $|i - j|$ and view larger values of $\mathcal{A}(\rho_I)$ as denoting greater degrees of anti-Robinson patterning. To a certain extent, however, the definition of q_{ij} is arbitrary, and the evaluation of patterning could vary, depending on changes in the definition of q_{ij}, e.g., by picking a different monotone function of $|i-j|$ as a "target." The indices given in (1) and (2) do not require this type of initial specification, and, for this reason alone, these measures may be viewed as less arbitrary. The further choice between (1) and (2) depends on the desire for ordinal invariance with respect to the entries in P.

Because the last two measures in (1) and (2) are defined only from neighboring off-diagonal entries within a row, there are two natural extensions that include more than the simple adjacency information. The indices in (3) and (4) are direct analogs of those in (1) and (2), respectively. The sole difference exists in the definition of q_{ijk}, which now picks out those positive differences to include in the index $\mathcal{A}_3(\rho_I)$ that represent any off-diagonal violation of the anti-Robinson gradient within the row corresponding to O_i, including nonadjacent violations. Thus, the function $\bar{q}_{ijk}$ in (1) and (2) selects only "local" violations of the anti-Robinson condition (and includes no redundant comparisons), whereas the corresponding function in (3) and (4) can select additional off-diagonal violations of a more global nature.

3. No reference:

$$\bar{p}_{ijk} = \begin{cases} p_{ij} - p_{ik} & \text{if } p_{ij} > p_{ik}, \ i,j,k \text{ distinct} \\ 0 & \text{otherwise} \end{cases}$$

$$\bar{q}_{ijk} = \begin{cases} 1 & \text{if } i < j < k \text{ or } k < j < i \\ 0 & \text{otherwise} \end{cases}$$

4. Sibson (1971):

$$\bar{p}_{ijk} = \begin{cases} 1 & \text{if } p_{ij} > p_{ik}, \ i,j,k \text{ distinct} \\ 0 & \text{otherwise} \end{cases}$$

$$\bar{q}_{ijk} = \begin{cases} 1 & \text{if } i < j < k \text{ or } k < j < i \\ 0 & \text{otherwise} \end{cases}$$

The measures just presented are all motivated by a particular necessary condition on three correctly ordered objects in a perfect anti-Robinson matrix, e.g., if $i < j < k$ or $k < j < i$, then p_{ij} should be no greater than p_{ik}. In a similar manner, a perfect anti-Robinson pattern implies a necessary condition on four correctly placed objects, e.g., if $k < i < j < l$, then p_{ij} should be no greater than p_{kl}. As an illustration of an index based on this latter four-object condition, the measure in (5) requires a four-place index $\mathcal{A}_4(\rho_I)$ and includes a comparison of proximities of a particular type across rows. Again, only the rank ordering of the proximities in $\underset{\sim}{P}$ is required in the first version considered:

5. Craytor and Johnson (1968); Sibson (1971); Johnson (1972):

$$\bar{p}_{ijkl} = \begin{cases} 1 & \text{if } p_{ij} > p_{kl}, \ i \neq j, \ k \neq l \\ 0 & \text{otherwise} \end{cases}$$

$$\bar{q}_{ijkl} = \begin{cases} 1 & \text{if } k \leqslant i < j \leqslant l \\ 0 & \text{otherwise} \end{cases}$$

It should be noted that the index $\mathcal{A}_4(\rho_I)$ based on (5) [and (6) below] would also include triad comparisons within each row on either side on the main diagonal. Pielou (1979) has proposed this same inconsistent quadruples index to measure the "gradedness" of a matrix—see the numerical example in Section 4.3.1.

Generalizing further, the four-place function in (5) could be modified in the same way that (3) modifies (4), i.e.,

6. No reference:

$$\bar{p}_{ijkl} = \begin{cases} p_{ij} - p_{kl} & \text{if } p_{ij} > p_{kl}, \ i \neq j, \ k \neq l \\ 0 & \text{otherwise} \end{cases}$$

$$\bar{q}_{ijkl} = \begin{cases} 1 & \text{if } k \leqslant i < j \leqslant l \\ 0 & \text{otherwise} \end{cases}$$

The last two statistics include the same discrepancies as in (3) and (4), but additional comparisons are evaluated as well. In fact, Sibson (1971) has argued that four-place indices, such as (5) and (6), are overly redundant and require an inordinate amount of computation. This same redundancy argument, however, can be used against his suggested index given in (4) because it is redundant when compared to (2). Although some redundancy may be desirable for a reasonably sensitive and stable statistic, it is unclear whether a four-place alternative should be considered overly redundant or too computationally burdensome (but see Dietz, 1983).

The indices in (2), (4), and (5) all depend on a zero-one scoring function to define $\bar{p}_{ijk}$ or $\bar{p}_{ijkl}$, and in turn, count triads or quadruples that have a particular type of inconsistency with an anti-Robinson form. Obviously, a simple reversal of the inequality characterizing $\bar{p}_{ijk}$ or $\bar{p}_{ijkl}$ would count consistent triads or quadruples of the same type; furthermore, the use of the sign function as in Section 5.3 would lead to an index defined by the difference between the consistent and inconsistent triads or quadruples. Several convenient normalizations could also be defined. For example, in (2), suppose $\overline{p}_{ijk} = \text{sign}(p_{ik} - p_{ij})$ for distinct i, j and k, and 0 otherwise. Defining q_{ijk} in the same way as before, the normalized index

$$\frac{\sum_{i,j,k} \overline{p}_{ijk}\overline{q}_{ijk}}{(n-1)(n-2)}$$

would correspond to a Somers statistic in which the maximum number of triad comparisons, $(n-1)(n-2) = \sum_{i,j,k} \overline{q}_{ijk}$, is used as a denominator. Alternatively,

$$\frac{\sum_{i,j,k} \overline{p}_{ijk}\overline{q}_{ijk}}{\sum_{i,j,k} |\overline{p}_{ijk}\overline{q}_{ijk}|}$$

could be viewed as an analog of a Goodman-Kruskal gamma coefficient discussed in Section 4.3.2. An interpretation of this latter index can be given as the difference between two conditional probabilities. Suppose a triple of distinct objects (O_i, O_j, O_k) is chosen at random, subject to the constraints that $p_{ij} \neq p_{ik}$ and $k = j + 1$, $i < j$ or $k = j - 1$, $j < i$. Then the two relevant conditional probabilities are of a consistency and an inconsistency with respect to an adjacency condition for the anti-Robinson form. Similar interpretations could be developed starting with the measures given by (4) and (5). We do note, however, that $\sum_{i,j,k} |\overline{p}_{ijk}\overline{q}_{ijk}|$ is not necessarily an upper bound on $\max_\rho \mathcal{A}_3(\rho)$ (cf. Section 4.3.2), in contrast to the quantity, $(n-1)(n-2)$, used for the analog of a Somers measure.

In addition to the six variations given above, other three- and four-place measures could be considered in a variety of modifications, e.g., we could rely on the square of the proximity differences rather than on simple differences, or, because the measures suggested in (1) to (6) weight all discrepancies equally, it is possible to redefine $\overline{q}_{ijk}$ and $\overline{q}_{ijkl}$, at least in (3) to (6), to weight the discrepancies according to their severity. For instance, in (3) and (4), we could redefine the $\overline{q}_{ijk}$ and $\overline{q}_{ijkl}$ measures as

$$\overline{q}_{ijk} = \begin{cases} ||i-j| - |i-k|| & \text{if } i < j < k \text{ or } k < j < i \\ 0 & \text{otherwise} \end{cases}$$

and in (5) and (6) as

$$\overline{q}_{ijkl} = \begin{cases} ||i-j| - |k - 1|| & \text{if } k \leqslant i < j \leqslant l \\ 0 & \text{otherwise} \end{cases}$$

Obviously, there is an enormous number of possible indices that could be used to assess the correspondence between P and an anti-Robinson pattern, and it is hoped that at some later time formal arguments could be developed as to which are the most preferable and under what conditions. For the average researcher, however, the two simple indices given in (1) and (2) will probably suffice, and the choice between these will be based on the desirability of monotone invariance. On the other hand, if some degree of redundancy is considered appropriate or if weights are desired on the discrepancies, the various alternatives given above provide a rather extensive range of possibilities, including the natural analog of the rank correlation measures based on ± 1 scoring functions discussed in Section 4.3.1. We will emphasize these latter measures in the illustrations that follow.

An Example. As illustrations of how a higher-order assignment measure can be used in a real data-analysis context, we reconsider the two data matrices, P_a and P_b, given by Pielou (1979) and discussed earlier in Section 4.3.1. Based on the sign scoring function for $\overline{p}_{ijk}$:

$$\overline{p}_{ijk} = \begin{cases} \text{sign}(p_{ik} - p_{ij}) & \text{for } i, j, \text{ and } k \text{ distinct} \\ 0 & \text{otherwise} \end{cases}$$

and

$$\overline{q}_{ijk} = \begin{cases} 1 & \text{if } i < j < k \text{ or } k < j < i \\ 0 & \text{otherwise} \end{cases}$$

the "keying" of the proximities would suggest that large negative values of $\mathcal{A}_3(\rho_I)$ are of interest; i.e., a Robinson form rather than anti-Robinson is expected. In any event, using these three-place functions, we obtain the following results (the Monte Carlo distributions are given in Table 5.3.1a):

	P_a	P_b
$\mathcal{A}_3(\rho_I)$	-170	-39
Monte Carlo significance	.003	.174

Table 5.3.1a Monte Carlo Distributions for
the Pielou Data

Cumulative frequency	Index: $\underset{\sim}{P}_b$	Index: $\underset{\sim}{P}_a$
1	-132	-199
2	-117	-175
5	-113	-128
10	-102	-119
50	- 71	- 80
100	- 53	- 54
173	- 39	- 34
200	- 35	- 28
300	- 21	- 16
400	- 11	- 6
500	- 1	2
600	9	12
700	19	21
800	32	31
900	49	46
950	59	58
990	79	79
995	84	83
999	112	105

The two obvious normalizations would correspond to a Somers index, which uses a denominator of $\Sigma_{i,j,k} \bar{q}_{ijk} = [n(n-1)(n-2)]/3$ (440 for $\underset{\sim}{P}_a$ and 330 for $\underset{\sim}{P}_b$), or to a Goodman-Kruskal index, which uses a denominator of $\Sigma_{i,j,k} |\bar{p}_{ijk}\bar{q}_{ijk}| = $ (334 for $\underset{\sim}{P}_a$ and 249 for $\underset{\sim}{P}_b$). For example, relying on the first Somers alternative because $\bar{q}_{ijk}$ is constructed by the researcher, let $p_1 (p_2)$ be the probability of a consistency (inconsistency) in picking a triple of distinct objects O_i, O_j, and O_k at random subject to the constraint that $i < j < k$ or $k < j < i$. For $\underset{\sim}{P}_a$, we would obtain $p_1 - p_2 = .186 - .573 = (82/440) - (252/440) = -170/440 = -.386$; for $\underset{\sim}{P}_b$, $p_1 - p_2 = .318 - .436 =$

$(105/330) - (144/330) = -39/330 = -.118$. This latter value, as we have noted above, cannot be considered significantly different from zero at any of the usual levels.

<u>Partial Seriation</u>. The discussion thus far in this section has assumed that the particular order of the n objects under test is complete in the sense that no locations are tied along the underlying continuum. At least for the indices that depend on a zero-one k-place function for the second component defining $\mathcal{A}_k(\rho_I)$, e.g., $\overline{q}_{ijk}$ or $\overline{q}_{ijkl}$, simple extensions can be proposed to include this more general problem of evaluating a partial ordering (or seriation). All that is required is to restrict, say, $\overline{q}_{ijk}$ to be 1 when the original constraints hold, plus the condition that the locations j and k are not the same.

An example dealing with the figural goodness conjecture and Glushko's data from Section 4.3.1 may help clarify the mechanics. Here, there are only distinct locations along the continuum. To obtain a descriptive index, suppose we define $\overline{p}_{ijk}$ through the sign function as before:

$$\overline{p}_{ijk} = \begin{cases} \text{sign}(p_{ik} - p_{ij}) & \text{for distinct i, j, and k} \\ 0 & \text{otherwise} \end{cases}$$

but redefine $\overline{q}_{ijk}$ as

$$\overline{q}_{ijk} = \begin{cases} 1 & \text{if } i < j < k \text{ and } k < j < i, \text{ and the positions} \\ & \text{corresponding to j and k are not tied} \\ 0 & \text{otherwise} \end{cases}$$

There are 701 triad comparisons that can now be made through $\overline{q}_{ijk}$, i.e., $\Sigma_{i,j,k}\, \overline{q}_{ijk} = 701$ (if the ordering were complete, 1360 triad comparisons would be defined from $\overline{q}_{ijk}$). Thus, a Somers index could be constructed with this latter value as a denominator. Specifically, given the observed index $\mathcal{A}_3(\rho_I) = 561$, this normalized measure would take the form

$$.800 = \frac{561}{701} = p_1 - p_2 = \frac{606}{701} - \frac{45}{701} = .864 - .064$$

where $p_1(p_2)$ is the probability of a consistency (inconsistency) in the random selection of a distinct triple (O_i, O_j, O_k) subject to the conditions that define a value of 1 for $\overline{q}_{ijk}$. This latter value has a significance level of .001, based on the Monte Carlo distribution of Table 5.3.1b. (A Goodman-Kruskal measure would be based on $\Sigma_{i,j,k}\, |\overline{p}_{ijk}\overline{q}_{ijk}| = 651$.)

<u>Some Descriptive Considerations</u>. In Section 5.3, we indicated how an overall measure for the comparison of two matrices could be decomposed into

Table 5.3.1b Monte Carlo Distributions for
            the Glushko Data

Cumulative frequency	Index
1	-153
5	-138
10	-128
50	- 92
100	- 74
200	- 52
300	- 35
400	- 21
500	- 6
600	10
700	28
800	51
900	89
950	122
990	211
995	249
999	328

the contributions attributable to certain rows and columns through the definition of a matrix $\underset{\sim}{F}$. In a similar fashion, it is possible to isolate descriptively what is producing the aggregate measure for evaluating a given seriation. In effect, we assign a measure of discrepancy for each entry in the data matrix $\underset{\sim}{P}$. For example, in the type of index used in the Glushko example, suppose we construct an $n \times n$ matrix $\underset{\sim}{G} = \{g_{ij}\}$, where g_{ij} is defined from all possible comparisons that p_{ij} is involved with. Explicitly, g_{ij} is the number of consistencies minus the number of inconsistencies among all triads that contribute to $A_3(\rho_I)$ and involve p_{ij}. Because $A_3(\rho_I)$ is defined in such a way that each triad is counted only once, the sum of all entries in $\underset{\sim}{G}$ is actually twice $A_3(\rho_I)$. Moreover, because $\underset{\sim}{P}$ is symmetric and $p_{ij} = p_{ji}$,

Table 5.3.1c Matrix $\tilde{G}^o$ for the Glushko Data

Configuration	1	2	3	4	5	6	7	8	9	10	11	12	13	14	15	16	17
1	×	14	7	9	10	11	8	13	11	10	17	14	15	20	19	22	23
2		×	7	8	9	10	11	11	10	12	14	16	17	18	19	21	22
3			×	9	9	9	6	9	9	4	6	9	8	10	13	14	13
4				×	9	9	8	9	9	6	7	6	9	9	9	13	12
5					×	9	9	9	-4	8	4	6	8	9	9	12	10
6						×	4	6	5	5	-2	-2	7	8	10	10	11
7							×	9	7	7	1	5	7	7	9	9	9
8								×	1	-4	0	3	5	6	5	3	5
9									×	2	1	0	5	4	7	6	8
10										×	2	1	1	3	0	0	5
11											×	4	7	4	8	3	-4
12												×	9	10	9	8	10
13													×	10	9	8	9
14														×	8	10	10
15															×	9	10
16																×	10
17																	×

the matrix we give in Table 5.3.1c, $\underset{\sim}{G}^{O}$, is upper-triangular and obtained by summing across the main diagonal of $\underset{\sim}{G}$, i.e., $\underset{\sim}{G}^{O} = \{g_{ij}^{O}\} = \{g_{ij} + g_{ji}\}$. Now, the sum of the upper triangular values in $\underset{\sim}{G}^{O}$ is twice $\mathcal{A}_3(\rho_I)$; the low (high) entries in $\underset{\sim}{G}^{O}$ indicate proximities in $\underset{\sim}{P}$ that depress (elevate) the index $\mathcal{A}_3(\rho_I)$. As apparent from Table 5.3.1c, the least contribution is from objects 10 and 11. We pointed out earlier in Section 4.3.1 that these two objects produce some clear inconsistencies to the pattern we expect. (To a certain degree, objects such as 9 and 12 are also suspect, given the aggregate contributions for these two objects.)

Multidimensional Extensions. The measures we have just discussed in the seriation context, and thus, for a single dimension, can be extended to multidimensional representations by use of the index generalization developed by Holman (1978) in an optimization framework. As one example, suppose $\underset{\sim}{P}$ is a symmetric proximity matrix and we define

$$\bar{p}_{ijk} = \begin{cases} \text{sign}(p_{ik} - p_{ij}) & \text{for } i,j,k \text{ distinct} \\ 0 & \text{otherwise} \end{cases}$$

Each of the n objects $O_1, \ldots, O_n$ is assumed to have a coordinate representation in T-dimensional space, i.e., object O_i can be represented as $[x_i^{(1)}, \ldots, x_i^{(T)}]$. If we define

$$\bar{q}_{ijk} = \begin{cases} 1 & \text{if } x_i^{(t)} < x_j^{(t)} < x_k^{(t)} \text{ or } x_k^{(t)} < x_j^{(t)} < x_i^{(t)} \\ & \text{for all } t, 1 \leqslant t \leqslant T \\ 0 & \text{otherwise} \end{cases}$$

then $\mathcal{A}_3(\rho_I)$ defines the number of consistencies minus the number of inconsistencies with respect to a perfect multidimensional representation for the n objects. Specifically, a consistency (inconsistency) for an ordered triple (O_i, O_j, O_k) occurs whenever O_j is strictly between O_i and O_k on all dimensions and p_{ik} is greater (less) than p_{ij}.

5.3.2 Seriation for Asymmetric Proximities (B)

Much of the discussion of seriation for symmetric proximities in the previous section extends immediately to asymmetric proximities. As we recall from Chapter 4, our attention can be restricted to the skew-symmetric portion of a proximity matrix $\underset{\sim}{P}$, denoted by $\underset{\sim}{P}^{-} = \{p_{ij}^{-}\}$ and to joint consideration of sign and magnitude information. Sign information alone can best be approached by the simpler QA measures mentioned in Section 4.3.1; absolute value information can be approached by the methods for symmetric proximities of Section 5.3.1.

Assuming that large proximities in $\underset{\sim}{P}$ denote greater degrees of dominance, an appropriately seriated matrix $\underset{\sim}{P}^{-}$ would have (i) an anti-Robinson form above the main diagonal and a Robinson form below; and (ii) large entries above the main diagonal and small entries below. Thus, indices are desired that would assess the degree to which these two conditions are present.

As a first example that assesses condition (ii), suppose $Q = \{q_{ij}\}$, where $q_{ij} = 1$ if $i < j$ and 0 otherwise. In the QA framework, $\mathcal{A}(\rho_I)$ would be the sum of entries above the main diagonal of P^{-}, and is one measure suggested in Section 4.3.1. For the general triad comparison measure of (5.3b), $\mathcal{A}_3(\rho_I)$ would be twice the number of instances in which an entry to the right of the main diagonal is greater than an entry to the left and within the same row minus the number of instances in which it is less.

In evaluating the gradient conditions, (i), of an anti-Robinson (Robinson) form above (below) the main diagonal, all of the measures for symmetric proximities of the previous section can be modified merely by distinguishing the entries above and below the main diagonal. For example, suppose we define

$$\bar{p}_{ijk} = \begin{cases} \text{sign}(p_{ik} - p_{ij}) & \text{for distinct } i,\ j,\ \text{and } k \\ 0 & \text{otherwise} \end{cases}$$

and

$$\bar{q}_{ijk} = \begin{cases} +1 & \text{if } i < j < k \\ -1 & \text{if } k < j < i \\ 0 & \text{otherwise} \end{cases} \tag{5.3.2a}$$

Then, $\mathcal{A}_3(\rho_I)$ is the number of distinct row triads that satisfy either the anti-Robinson form above the main diagonal or the Robinson form below minus the number of such inconsistencies.

5.3.3 Generalized Cluster Statistics (B)

Most data reduction procedures currently used in the behavioral sciences can be characterized very broadly as formal methods for choosing a particular structure from some given class that will describe an available data set the "best." For example, starting with a measure of proximity between objects from some set S, hierarchical clustering is concerned with locating ultrametric, additive tree analysis searches for a function on S × S satisfying a certain four-point condition (Sattath and Tversky, 1977; Cunningham,

1978; Section 4.3.9), seriation is keyed to finding a matrix having the Robinson or anti-Robinson pattern, and so on. These properties that distinguish the various data reduction techniques are algebraic and are typically presented as defining characteristics of a final representation. Only in a very heuristic sense, however, can these properties also be used to describe data subject to error, because error-free conditions will usually fail, even for a data set only minimally different from some ideal characterization.

Given the complexity and type of structure we commonly wish to study, and even if we could define a measure of how far a particular data set is from satisfying an error-free condition, it is difficult, to say the least, to formulate a reasonable reference distribution based on acceptable statistical principles that could then be used in evaluating the size of the measure. Thus, for instance, no general techniques have been proposed that could test the complex conjecture that a given proximity measure is really an ultrametric up to an acceptable level of error. It may be obvious that statistical schemes for evaluating complete data sets and the degree to which they conform to a given algebraic property are desirable, but at the same time, they would probably be very difficult to develop in any generally convincing manner. In short, even though truly comprehensive methods would obviously be of value, we may be forced to design evaluation strategies subject to very stringent constraints on the type of proximities and structures that are considered.

As an alternative to attacking the difficult task of assessing global structure, it is possible to evaluate variation in structure <u>within</u> an object set if all the available data could be used to define an operational population. The technique we present can be viewed as a generalized form of a cluster statistic of the type discussed in Section 4.3.3, and which could be used with most data reduction schemes that have a combinatorial base or when conditions for an error-free representation are algebraic. As a cryptic summary, our major emphasis will be on a confirmatory inference strategy for evaluating the degree to which a subset of S satisfies some conjectured property. This latter property is defined in terms of the proximities between objects within the given subset. In particular, the reference could be to some simple characteristic such as symmetry or to a more complex algebraic condition that is necessary for a perfect representation in terms of some given structure (cf. Shepard, 1978).

In Section 4.3.3 on simple cluster statistics, we were interested in some notion of compactness and/or isolation for a subset D of S that was operationalized through a particular QA measure. The distribution of this index over all $\binom{n}{K}$ possible subsets served as a reference distribution to evaluate the size of the observed value of the measure for the subset of size K; thus, we hoped to reject the hypothesis that D was actually chosen by chance from the $\binom{n}{K}$ possible alternatives. Using higher-order assignment statistics, this same scheme of index evaluation can be extended to a variety of other data analysis areas that depend on specific underlying algebraic conditions by merely varying the meaning of compactness and/or isolation and how it is measured. In all cases, the evaluation of the salience of a subset with respect to a property assumes that the subset was identified independently of the given proximities, e.g., from theory or possibly from a second data set. The key to the inference strategy itself lies in the fact that proximity data are assumed to be available between all elements in S, even though the initial conjecture related to only a smaller subset. Using this broader background information, we can then carry out a statistical evaluation by comparing the degree to which the property holds in the given subset to what could be expected from other subsets of the same size that might have been formed from the larger parent set.

As one example, suppose a given subset is conjectured to be well represented by a hierarchical clustering, and thus, the (symmetric) proximities within this subset should more or less have an ultrametric form (see Section 4.3.9). For a measure of the degree of which the structure within D corresponds to that of an ultrametric, we could use the sum, over all triples of distinct objects, of the absolute values of the differences between the maximum proximities. If the representation as an ultrametric is perfect, this sum should be zero. Formally, let

$$\bar{q}_{ijk} = \begin{cases} 1 & \text{for distinct } i, j, \text{ and } k \text{ and } 1 \leq i, j, k \leq K \\ 0 & \text{otherwise} \end{cases}$$

and

$$\bar{p}_{ijk} = \begin{cases} \text{absolute value of the difference between the} \\ \text{maximum proximities } p_{ij}, p_{ik}, \text{ and } p_{jk} \text{ for} \\ \text{distinct } i, j, \text{ and } k \\ 0 \quad \text{otherwise} \end{cases}$$

Assuming that the first K rows and columns of P define the subset D, the sum of the required absolute differences can be represented as

$$\mathcal{A}_3(\rho_I) = \sum_{i,j,k} \bar{p}_{ijk}\bar{q}_{ijk} = \sum_{1 \leq i,j,k \leq K} \bar{p}_{ijk} \tag{5.3.3a}$$

As always, the relative size of $\mathcal{A}_3(\rho_I)$ can be assessed in relation to a reference distribution constructed from an evaluation of $\mathcal{A}_3(\rho)$ over all n! possible permutations, or equivalently, over all $\binom{n}{K}$ subsets of size K.

In the case that the original proximity measure is zero-one, $\mathcal{A}_3(\rho_I)$ is merely the number of intransitivities defined by the subset D when the proximity measure is interpreted as a binary relation on D × D. Thus, for this special case, the measure indicates how far the binary relation on D × D is from being an equivalence relation.

Seriation. There are some obvious applications of the subset salience notion to the seriation context based on the type of index given in equation (5.3.3a). For example, suppose the proximities in $\underset{\sim}{P}$ are symmetric, small values denote similar objects, and our conjecture is that the ordered subset D (defined for convenience by the first K rows and columns of D) has a good representation along a unidimensional continuum. If $\bar{p}_{ijk}$ is defined as

$$\bar{p}_{ijk} = \begin{cases} \text{sign}(p_{ik} - p_{ij}) & \text{for distinct i, j, and k; and} \\ & i < j < k \text{ or } k < j < i \\ 0 & \text{otherwise} \end{cases} \qquad (5.3.3b)$$

then for the ordered subset, $D = (O_1, \ldots, O_K)$, $\mathcal{A}_3(\rho_I)$ in equation (5.3.3a) is the number of triads consistent with the anti-Robinson form minus the number of inconsistent triads. In this instance, the distribution of $\mathcal{A}_3(\rho)$ over all n! equally likely permutations reduces to an evaluation of the index over all $\binom{n}{K}$ K! equally likely ordered subsets of size K. Thus, in this confirmatory context, the null conjecture has two parts: The K objects defining D were selected at random, and, within the subset, the particular ordering was chosen at random.

In an asymmetric context in which a large proximity value, p_{ij}, denotes, say, a greater degree of preference of O_i over O_j or a greater degree of "flow" from O_i to O_j, suppose in analogy to Section 5.3.2 we define $\bar{p}_{ijk}$ as

$$\bar{p}_{ijk} = \begin{cases} \text{sign}(p_{ik} - p_{ij}) & \text{for distinct i, j, and k; and } j < i < k \\ 0 & \text{otherwise} \end{cases}$$

Here, for the ordered subset we use the number of triads in which an entry to the right of the main diagonal is greater than an entry within the same row to the left of the main diagonal minus the number of times it is less. Given the general hope of finding the smaller proximities to the left of the main diagonal, a large value of $\mathcal{A}_3(\rho_I)$ would be hoped for.

Example. As an illustration of how a generalized cluster statistic could be used in a seriation context, we return to the Rothkopf Morse Code data of Section 4.3.3. Our interests will be in a symmetric proximity matrix, and consequently, the data in Table 4.4.3a are first converted by summing the corresponding entries across the main diagonal to produce $\underset{\sim}{P}$. As a reasonable conjecture for parts of the 36 × 36 matrix, we may hypothesize that a submatrix of $\underset{\sim}{P}$ for any subset of five symbols, each defined by a different number of dots and dashes, should display a Robinson form when the rows and columns are ordered according to the number of dots and dashes. Thus, when the definition of p_{ijk} in equation (5.3.3b) is used, large negative values of the index would be expected if this conjecture were correct.

The Monte Carlo distribution of the index, given in Table 5.3.3a, is based on a sample size of 999. For each of the two representative 5 × 5 submatrices given below, the observed index is sufficiently small:

<table>
<tr><td colspan="6">(i)</td><td colspan="6">(ii)</td></tr>
<tr><td></td><td>E</td><td>I</td><td>S</td><td>H</td><td>5</td><td></td><td>T</td><td>N</td><td>W</td><td>X</td><td>1</td></tr>
<tr><td>·E</td><td>×</td><td>27</td><td>18</td><td>13</td><td>7</td><td>–T</td><td>×</td><td>15</td><td>8</td><td>5</td><td>7</td></tr>
<tr><td>··I</td><td>27</td><td>×</td><td>51</td><td>22</td><td>13</td><td>–·N</td><td>15</td><td>×</td><td>18</td><td>12</td><td>12</td></tr>
<tr><td>···S</td><td>18</td><td>51</td><td>×</td><td>96</td><td>55</td><td>·––W</td><td>8</td><td>18</td><td>×</td><td>43</td><td>24</td></tr>
<tr><td>····H</td><td>13</td><td>22</td><td>96</td><td>×</td><td>137</td><td>–·· ·X</td><td>5</td><td>12</td><td>43</td><td>×</td><td>29</td></tr>
<tr><td>·····5</td><td>7</td><td>13</td><td>55</td><td>137</td><td>×</td><td>·––––1</td><td>7</td><td>12</td><td>24</td><td>29</td><td>×</td></tr>
</table>

Specifically, for submatrices (i) and (ii), respectively, we obtain indices of –18 and –17; the associated Monte Carlo significance levels are .003 and .004. A Somers-type normalized index would be based on the number of triad comparisons made: $K(K - 1)(K - 2)/3 = 20$; this denominator would lead to an index of $-.90 = .05 - .95 = 1/20 - 19/20$ for submatrix (i) and $-.85 = .05 - .90 = 1/20 - 18/20$ for submatrix (ii).

Extensions and Other Applications. The possibility of different index specification is almost endless, and if we wish, the function $\bar{p}_{ijk}$ in equation (5.3.3a) could be tailored to very specific questions of importance in a particular substantive area. Even more generally, three-place functions could be replaced by four-place functions and used over all possible pairs of proximities, and not only over those pairs that contain a common entity. In this way, the four-point condition used in a perfect representation as an additive tree could be placed within our framework as well as could the four-place functions proposed by Johnson (1973) in his approach to nonmetric muldimensional scaling. In fact, $\mathcal{A}_k(\rho_I)$ could be based on a goodness-of-fit statistic calculated after an optimization process is carried out on the subset D and on the associated within-subset proximities, e.g., some type of scaling or cluster analysis, or alternatively, could be based on some graph-theoretic property of the subset D. For example, suppose the asymmetric proximity matrix P is zero-one, and $p_{ij} + p_{ji} = 1$ and we define

$$\bar{p}_{ijk} = \begin{cases} 1 & \text{if } O_i, O_j \text{ and } O_k \text{ form a circular triad for } i < j < k \\ 0 & \text{otherwise} \end{cases}$$

Here, we use Kendall's (1970) definition and count a circular triad as being present for O_i, O_j, and O_k if $p_{ij} + p_{jk} + p_{ki} = 3$ or $p_{ji} + p_{ik} + p_{kj} = 3$. Thus, for the unordered subset D, the index in equation (5.3.3a) is the number of circular triads in D, and indicates the degree to which the objects in D can be ordered consistently along a continuum. A perfect ordering is present for the objects in D if the number of circular triads is zero.

Table 5.3.3a Monte Carlo Distributions for the
Seriation Statistic Based on 5×5
Submatrices for the Rothkopf Data

Cumulative frequency	Index
1	-20
2	-18
3	-17
4	-16
5	-16
10	-16
50	-12
100	- 9
200	- 6
300	- 4
400	- 2
500	0
600	2
700	4
800	6
900	8
950	9
990	12
999	18

<u>Evaluating Other Conjectures of Combinatorial Structure for a Subset D.</u>
In addition to the within-subset sum defined through equation (5.3.3a), a
variety of other conjectures of combinatorial structure for a subset D can
be represented through the judicious choice of three-place functions. For
instance, suppose $\underset{\sim}{P}$ represents a set of "flows" between the objects in S,
where larger values of p_{ij} represent greater degrees of interaction <u>from</u> O_i
to O_j. If the three-place function $\bar{p}_{ijk}$ is defined as

$$\bar{p}_{ijk} = \begin{cases} \operatorname{sign}(p_{ik} - p_{ij}) & \text{for distinct } i,j,k \\ 0 & \text{otherwise} \end{cases} \qquad (5.3.3c)$$

then, depending on how $\bar{q}_{ijk}$ is characterized, several different measures of
subset salience can be defined that depend on more than comparisons strictly
within a subset:

$$\bar{q}_{ijk} = \begin{cases} +1 & \text{if } O_i, O_j \in D \text{ and } O_k \notin D \\ -1 & \text{if } O_i, O_k \in D \text{ and } O_j \notin D \\ 0 & \text{otherwise} \end{cases} \qquad (5.3.3d)$$

Here, large (small) values of $\mathcal{A}_3(\rho_I)$ indicate large (small) degrees of flow
<u>out</u> of D to S - D, as compared to flow within D. In particular, $\mathcal{A}_3(\rho_I) =$
$p_a - p_b$, where $p_a(p_b)$ refers to the number of distinct triads (O_i, O_j, O_k) for
which $O_i \in D$ and either O_j or $O_k \in S - D$ but not both, and the flow from O_i
to the object outside D is larger (or smaller) than the flow to the object
inside D. If we use

$$\bar{q}_{ijk} = \begin{cases} +1 & \text{if } O_i, O_j \in D \text{ and } O_k \notin D \\ -1 & \text{if } O_i, O_k \in D \text{ and } O_j \notin D \\ 0 & \text{otherwise} \end{cases}$$

our interpretation would change to flow <u>into</u> D from S - D as compared to
flow within S - D. Generalizing in a slightly different way, we can redefine
$\bar{p}_{ijk}$ as

$$\bar{p}_{ijk} = \begin{cases} \operatorname{sign}(p_{ki} - p_{ji}) & \text{for distinct } i,j,k \\ 0 & \text{otherwise} \end{cases}$$

Then, using the same three-place functions as above, the concern would be
to flow into D as compared to flow within D, and to flow out of D compared
to flow within S - D, respectively. In short, depending on our interests, we
can concentrate on flow out of or into D and in comparison to flow either
within D or S - D.

If the original proximity matrix $\underset{\sim}{P}$ were symmetric with small proximities denoting the more similar objects, the most obvious example of subset salience would be based on $\bar{p}_{ijk}$ in equation (5.3.3c) and $\bar{q}_{ijk}$ in equation (5.3.3d). The index $\mathcal{A}_3(\rho_I)$ would be large or small, depending on whether the proximities within D were relatively large or small compared to those between D and S - D.

5.3.4 Some Additional Applications (B)

There are a number of specific applications of QA we have mentioned thus far in the book that can be approached with k-place functions tailor-made for the given problem. We mention a few others here in summary form and refer the reader to Hubert (1978, 1980) for a further discussion of higher-order assignment indices in general.

Comparisons Between Related QA Problems. In Section 4.4, our interests centered on the task of comparing an n × n matrix $\underset{\sim}{P}$ to a second n × n matrix, $\underset{\sim}{Q} - \underset{\sim}{R}$. The choice of a cross-product measure to carry out this comparison is somewhat arbitrary, and other alternatives based on k-place functions may be more appropriate for certain applications, e.g., those based on nonmetric analyses. As one example, suppose we are given the two matrices $\underset{\sim}{P}$, $\underset{\sim}{Q}$, and $\underset{\sim}{R}$ and define two three-place functions:

$$\bar{p}_{ijk} = \begin{cases} +1 & \text{for } i,j,k \text{ distinct and } p_{ij} > p_{ik} \\ -1 & \text{for } i,j,k \text{ distinct and } p_{ij} < p_{ik} \\ 0 & \text{otherwise} \end{cases}$$

$$\bar{q}_{ijk} = \begin{cases} +1 & \text{for } i,j,k \text{ distinct and } q_{ij} > q_{ik}, \ r_{ij} < r_{ik} \\ -1 & \text{for } i,j,k \text{ distinct and } q_{ij} < q_{ik}, \ r_{ij} > r_{ik} \\ 0 & \text{otherwise} \end{cases}$$

Then, the raw cross-product statistic $\mathcal{A}_3(\rho_I)$ is also a measure of similar patterning between $\underset{\sim}{P}$ and a difference between $\underset{\sim}{Q}$ and $\underset{\sim}{R}$. For instance, within the ith row of $\underset{\sim}{P}$, we score a +1 if p_{ij} and p_{ik} bear the same order relation as q_{ij} and q_{ik} and the opposite order relation to r_{ij} and r_{ik}; conversely, we score a -1 if p_{ij} and p_{ik} bear the same order relation as r_{ij} and r_{ik} and the opposite order relation to q_{ij} and q_{ik}. These scores are then summed within each row, and finally, across rows. Given this scoring system for $\mathcal{A}_3(\rho_I)$, positive values reflect a greater similarity of $\underset{\sim}{P}$ to $\underset{\sim}{Q}$ than of $\underset{\sim}{P}$ to $\underset{\sim}{R}$, and negative values would reflect the opposite. We note that no standardization of the entries in $\underset{\sim}{Q}$ and $\underset{\sim}{R}$ is now necessary, and furthermore, the comparisons are carried out only within the rows of $\underset{\sim}{P}$, $\underset{\sim}{Q}$, and $\underset{\sim}{R}$. This latter property suggests that $\mathcal{A}_3(\rho_I)$ would be appropriate for comparing conditional matrices in which possibly different "metrics" are used in obtaining the proximities within the different rows in each of the three matrices. Furthermore, given the ±1 scoring function, obvious analogs exist for Kendall's tau_a and tau_b and for the rank correlation measures of Somers and Goodman-Kruskal, using the appropriate denominators mentioned in Section 5.3.

As an illustration, suppose $\underset{\sim}{P}$ is an empirically generated symmetric proximity matrix with the usual interpretation that the smaller proximities denote the more similar objects. Moreover, let Q and R denote two different partitions of S that we believe may represent the data in $\underset{\sim}{P}$ but possibly to varying degrees, i.e., q_{ij} (and r_{ij}) takes on the values 0 or 1, depending on whether the objects O_i and O_j are within the same class or not, respectively. The index $\mathcal{A}_3(\rho I)$ can be represented in the usual way as a difference. Here, a consistency refers to an ordered triple, (O_i, O_j, O_k), for which (i) $p_{ik} < p_{ij}$ and O_i and O_j belong to the same class, and O_i and O_k to different classes in the partition represented by Q, and the opposite for $\underset{\sim}{R}$; or (ii) $p_{ik} > p_{ij}$ and O_i and O_k belong to the same class and O_i and O_j to different classes for Q, and the opposite for R. An inconsistent triple merely reverses the inequality between p_{ik} and p_{ij}. Thus, large positive values of $\mathcal{A}_3(\rho I)$ reflect the superiority of the partition represented by Q; large negative values of $\mathcal{A}_3(\rho I)$ reflect the superiority of the partition represented by $\underset{\sim}{R}$.

Free Recall. In the free-recall paradigm discussed in Section 4.3.4, the k-place functions provide a way of evaluating, for units larger than pairs, the correspondence between, say, the input and output structure of a protocol or the correspondence between two different trials of a free-recall experiment. The latter topic is concerned with what is referred to as subjective organization, i.e., how a subject organizes memory (and presumably recall) when no explicit, experimenter-defined structure is intended. We measure subjective organization from the consistency in structure shown by a subject over a number of free-recall trials of the same stimulus list.

As an example, suppose a set of n items are recalled, and the following two three-place functions are defined:

$$\overline{p}_{ijk} = \begin{cases} 1 & \text{for distinct i, j, k and the objects are presented} \\ & \text{adjacently (or appear adjacently in the previous} \\ & \text{trial) and in the order } O_i \longrightarrow O_j \longrightarrow O_k \\ \\ 0 & \text{otherwise} \end{cases}$$

$$\overline{q}_{ijk} = \begin{cases} 1 & \text{for distinct i, j, k and when the objects were recalled} \\ & \text{adjacently and in the order } O_i \longrightarrow O_j \longrightarrow O_k \\ \\ 0 & \text{otherwise} \end{cases}$$

Then, $A_3(\rho_I)$ is the number of unidirectional (intertrial) repetitions of size 3. If a score of one is given for $\overline{p}_{ijk}$ and $\overline{q}_{ijk}$ when the order is $O_k \longrightarrow O_j \longrightarrow O_i$ as well, $A_3(\rho_I)$ is twice the number of bidirectional repetitions; finally, if any ordering of three adjacently ordered objects is counted as a 1, then $A_3(\rho_I)$ is 3! times the number of unordered repetitions. Similar extensions for units of size k could be developed in the same fashion. The reader is referred to Hubert and Levin (1980) for a more thorough discussion of how these k-place measures could be used in the free-recall context.

Coombsian Unfolding. Given an asymmetric paired comparison matrix $\underset{\sim}{P}$ as in Section 4.3.1, the seriation model we have considered up to this point assumes that (i) the n objects are ordered in some way along the continuum, and (ii) each proximity in $\underset{\sim}{P}$, say p_{ij}, reflects directly the distances between the placement for objects O_i and O_j. Thus, an index that assesses the adequacy of a conjectured ordering relies on one of two properties: an anti-Robinson (Robinson) pattern above (below) the main diagonal and the presence of the large entries above the main diagonal. As an alternative spatial model, we could conjecture that a Coombsian model is appropriate in which subjects are also placed at ideal points along the single dimension in addition to the objects, and subjects respond as a function of their distance from the objects. In this latter case, a different gradient condition would be sought in $\underset{\sim}{P}$ (see Greenberg, 1965). The entries should increase, moving either to the right or to the left of the main diagonal. Thus, the property of large entries appearing above the main diagonal is not relevant and an anti-Robinson form is sought both above and below the main diagonal. These gradient conditions could be evaluated by a simple variation in the way $\overline{q}_{ijk}$ is defined in equation (5.3.2a), i.e.,

$$\overline{q}_{ijk} = \begin{cases} +1 & \text{if } i < j < k \text{ or } k < j < i \\ 0 & \text{otherwise} \end{cases}$$

Turning Points. As one use of a three-place function in a time-series context that has some historical interest, suppose $x_1, x_2, \ldots, x_n$ are n observations ordered through time consecutively with respect to the given subscripts. If we define

$$\overline{p}_{ijk} = \begin{cases} 1 & \text{for } (x_i - x_j)(x_j - x_k) < 0 \text{ for distinct } i, j, \text{ and } k \\ 0 & \text{otherwise} \end{cases}$$

$$\overline{q}_{ijk} = \begin{cases} 1 & \text{if } j = i + 1 \text{ and } k = i + 2 \\ 0 & \text{otherwise} \end{cases}$$

then $\mathcal{A}_3(\rho_I)$ is the number of "turning points in the sequence, i.e., one less than the number of runs up and down (Wallis and Moore, 1941). Our inference model based on $\mathcal{A}_3(\rho)$ provides the distribution for the number of turning points in a random sequence, and in turn, an obvious competitor to the temporal statistics considered in Section 4.3.2, particularly for trend. In a similar manner, k-place functions could be considered which would encompass the generalized serial correlation measures of Ghosh (1954).

Spatial Autocorrelation. Instead of the sequence $x_1, \ldots, x_n$ representing n given positions along a continuum or a temporal sequence, suppose these values denote n observations realized over the geographical locations O_1, $O_2, \ldots, O_n$. Assuming that the proximity matrix $\underset{\sim}{P}$ consists of symmetric measures of spatial separation, the raw index $\mathcal{A}_k(\rho_I)$ can be used to construct various indices of spatial autocorrelation. For instance, suppose we define

$$\bar{p}_{ijk} = \begin{cases} \text{sign}(p_{ik} - p_{ij}) & \text{for distinct i, j, and k} \\ 0 & \text{otherwise} \end{cases}$$

and

$$\bar{q}_{ijk} = \begin{cases} \text{sign}(|x_i - x_k| - |x_i - x_j|) & \text{for distinct i, j, and k} \\ 0 & \text{otherwise} \end{cases}$$

Here, $\mathcal{A}_3(\rho_I)$ is again the number of consistent minus the number of inconsistent triples. Explicitly, a consistent ordered triple, (O_i, O_j, O_k) is one for which the spatial separation between O_i and O_k is strictly greater (or strictly less) than between O_i and O_j and the difference between x_i and x_k is strictly greater (or strictly greater less) than the difference between x_i and x_j; an inconsistent triple is defined in the obvious way by reversing the roles of the terms "strictly greater" and "strictly less" in the condition on the differences between x_i and x_k and between x_i and x_j. In the spatial autocorrelation context, generalized indices of this type are discussed in more detail by Hubert, Golledge, and Costanzo (1981).

5.4 BOTTLENECK INDICES (C)

All of the raw measures of correspondence we have suggested have been defined in some way as a sum over pairs, triads, and so on. In the operations research literature, however, there also has been some discussion of what are called "bottleneck" objective functions. Specifically, suppose we wish to compare two n × n matrices $\underset{\sim}{P}$ and $\underset{\sim}{Q}$, but instead of using the QA measure $\mathcal{A}(\rho_I) = \Sigma_{i,j} p_{ij} q_{ij}$, we consider an index of the form

$$\sigma_{max}(\rho_I) = \max_{i,j}\{p_{ij}q_{ij}\}$$ (5.4a)

or

$$\sigma_{min}(\rho_I) = \min_{i,j}\{p_{ij}q_{ij}\}$$

depending on how the entries in $\underset{\sim}{P}$ and $\underset{\sim}{Q}$ are keyed. The corresponding optimization analogs (e.g., see Burkard, 1974) would be phrased either as

$$\min_{\rho} \max_{i,j}\{p_{\rho(i)\rho(j)}q_{ij}\}$$

or

$$\max_{\rho} \min_{i,j}\{p_{\rho(i)\rho(j)}q_{ij}\}$$

Obviously, extensions could also be developed for the nonmultiplicative measures of Section 5.2 or for the higher-order assignment indices of Section 5.3.

As one natural application of the measures in equation (5.4a), suppose $\underset{\sim}{P}$ is an $n \times n$ proximity matrix in which small proximities represent similar objects, and $\underset{\sim}{Q}$ is zero-one and represents a single subject (or a partition). The measure $\tilde{\sigma}_{max}(\rho_I)$ is the maximum proximity within the given subset (or within a class of the partition). Measures such as these have been very popular in hierarchical cluster analyses through what is called the complete-link method. Again, significance testing could proceed through a Monte Carlo strategy.

REFERENCES

Burkard, R. E. Quadratische bottleneck-probleme, Operat. Res. Verfahren 18 (1974), 26-41.

Coombs, C. H. Data and scaling theory, in G. Menges (Ed.), Handbook of Mathematical Economic Sciences, Vol. 2, West Deutscher Verlag, Dusseldorf, 1979.

Craytor, W. B., and L. Johnson, Jr. Refinements in Computerized Item Seriation, Museum of Natural History, University of Oregon, Bulletin No. 10, 1968.

Cunningham, J. P. Free trees and bidirectional trees as representations of psychological distance, J. Math. Psychol. 17 (1978), 165-188.

Dietz, E. J. Permutation tests for association between two distance matrices, Syst. Zool. 32 (1983), 21-26.

Frank, O. Sampling and estimation in large social networks, Social Networks 1 (1978), 91-101.

Ghosh, M. N. Asymptotic distribution of serial statistics and applications to problems of nonparametric tests of hypotheses, Ann. Math. Stat. 25 (1954), 218-251.

Greenberg, M. G. A method of successive cumulations for scaling of pair-comparison preference judgements, Psychometrika 30 (1965), 444-448.

Hartigan, J. A. Clustering Algorithms, Wiley, New York, 1975.

Hole, F., and M. Shaw. Computer analysis of chronological seriation, Rice Univ. Stud. 53 (1967), 1-166.

Holman, E. W. Completely nonmetric multidimensional scaling, J. Math. Psychol. 18 (1978), 39-51.

Hubert, L. J. Generalized proximity function comparisons, Br. J. Math. Stat. Psychol. 31 (1978), 179-192.

Hubert, L. J. Matching models in the analysis of cross-classifications, Psychometrika 44 (1979), 21-41.

Hubert, L. J. Analyzing proximity matrices: The assessment of internal variation in combinatorial structure, J. Math. Psychol. 21 (1980), 247-264.

Hubert, L. J., R. G. Golledge, and C. M. Costanzo. Generalized procedures for evaluating spatial autocorrelation, Geogr. Anal. 13 (1981), 224-233.

Hubert, L. J., R. G. Golledge, C. M. Costanzo, and N. Gale. Order-dependent measures of correspondence for comparing proximity matrices and related structures, in P. Nijkamp, H. Leitner, and N. Wrigley (Eds.), Measuring the Unmeasurable, Martinus Nijhoff, The Hague, 1985, pp. 399-423.

Hubert, L. J., and J. R. Levin. Measuring clustering in free recall, Psychol. Bull. 87 (1980), 59-62.

Jackson, D. M. Comparison of classifications, in A. J. Cole (Ed.), Numerical Taxonomy, Academic Press, New York, 1969, pp. 91-113.

Johnson, L., Jr. Introduction to imaginary models for archaelogical scaling and clustering, in L. L. Clarke (Ed.), Models in Archaeology, Methuen, London, 1982, pp. 309-379.

Johnson, R. M. Pairwise nonmetric multidimensional scaling analysis, Psychometrika 38 (1973), 11-18.

Kendall, M. G. Rank Correlation Methods, 4th ed., Hafner, New York, 1970.

Koopmans, T. C., and M. J. Beckman. Assignment problems and the location of economic activities, Econometrica 25 (1957), 53-76.

Kuzara, R. S., G. R. Mead, and K. A. Dixon. Seriation of anthropological data: A computer program for matrix ordering, Am. Anthropol. 68 (1966), 1442-1455.

Lerman, I. C. On two criteria of classification, in A. J. Cole (Ed.), Numerical Taxonomy, Academic Press, New York, 1969, pp. 114-128.

Lerman, I. C. Les bases de la classification automatique, Gauthier-Villars, Paris, 1970.

Pielou, E. C. Interpretation of paleoecological similarity matrices, Paleobiology 5 (1979), 435-443.

Sattath, S., and A. Tversky. Additive similarity trees, Psychometrika 42 (1977), 319-345.

Shepard, R. N. A taxonomy of some principal types of data and of multidimensional methods for their analysis, in R. N. Shepard et al. (Eds.), Multidimensional Scaling: Theory and Applications in the Behavioral Sciences I, Seminar Press, New York, 1972, pp. 21-47.

Shepard, R. N. The circumplex and related topological manifolds in the study of perception, in S. Shye (Ed.), Theory Construction and Data Analyses in the Behavioral Sciences, Jossey-Bass, San Francisco, 1978, pp. 29-80.

Sibson, R. Some thoughts on sequencing methods, in F. R. Hodson et al. (Eds.), Mathematics in the Archaeological and Historical Sciences, Edinburgh University Press, Edinburgh, 1971, pp. 263-266.

Wallis, W. A., and G. H. Moore. A significance test for time-series analysis, J. Am. Stat. Assoc. 36 (1941), 401-409.

6

Multiple Proximity Matrices

6.1 INTRODUCTION (A)[*]

A generalized notion of concordance was discussed in Chapter 3, using an LA index constructed from K-place assignment scores that could be additively decomposed into $\binom{K}{2}$ pairwise components. Each of the latter defined a separate LA index based on two-place assignment scores. In direct analogy to this discussion, a similar notion of concordance can be developed for the QA framework; in fact, most of the discussion in Chapter 3 extends immediately to this context merely by replacing each of the pairwise LA indices by a QA index. As usual, we are given a set S of n objects, $\{O_1,\ldots,O_n\}$, but there are now K proximity matrices denoted by $\underset{\sim}{P}_1,\ldots,$ $\underset{\sim}{P}_K$, each with zeros along its main diagonal, rather than K sequences as in Chapter 3. Typically, $\underset{\sim}{P} = \{p_{ij}^k\}$ is generated by a kth subject, and thus, the interpretation of K "independent" proximity functions will be tacitly assumed throughout. Also, it is implicitly assumed that all of the proximities in $\underset{\sim}{P}_1,\ldots,\underset{\sim}{P}_K$ are commensurable or have been made so by some initial transformation, e.g., to z scores for the n(n - 1) off-diagonal entries in each matrix.

Based on the collection of K commensurable matrices, $A = \{\underset{\sim}{P}_1,\ldots,\underset{\sim}{P}_K\}$, at least three different problem areas related to a general concept of concordance can be discussed separately (see Hubert, 1979):

1. Measures of concordance among the K elements of A
2. Procedures for relating the last K - 1 elements of A, $\underset{\sim}{P}_2,\ldots,\underset{\sim}{P}_K$, to a specific (fixed) matrix given a priori as $\underset{\sim}{P}_1$

---

[*]The reader is referred to the Preface for an explanation of the use of (A), (B), and (C) in text headings.

3. Techniques for evaluating the degree of concordance within and between
    the subsets defining a partition of A

In analogy with Chapter 3, each of these tasks will be reduced to a consideration of pairs of matrices chosen from A. Moreover, because most of
the literature that suggests procedures for comparing two matrices has
depended on a simple cross-product statistic, our discussion logically
begins here as well.

Given the two proximity matrices, $\underset{\sim}{P}_k$ and $\underset{\sim}{P}_{k'}$, the usual QA cross-
product measure of correspondence will now be denoted by

$$\mathcal{A}^{(k,k')}(\rho) = \sum_{i,j} p_{\rho(i)\rho(j)}^{k} p_{ij}^{k'}$$

where ρ is some permutation on the first n integers. As a convenience, if
ρ_I denotes the identity permutation, then $\mathcal{A}^{(k,k')}(\rho_I)$ is assumed to be the
raw index of agreement between the two matrices, $\underset{\sim}{P}_k$ and $\underset{\sim}{P}_{k'}$, when they
are considered by themselves. The direct analog in Chapter 3 to our reli-
ance on the QA cross-product measure would be the assumption of a sum of
squared differences defining the agreement between two sequences, which
is statistically equivalent to the use of a sum of cross-products between two
sequences.

6.1.1 A Generalized (QA) Measure of Concordance (B)

Using the notation of Section 6.1 and considering the first general problem
area indicated in (1), a raw measure of concordance, $\mathcal{C}$, among the K elements of A will be defined by the average[†] of the $\binom{K}{2}$ cross-product coefficients $\mathcal{A}^{(k,k')}(\rho_I)$

$$\mathcal{C} = \left[\frac{1}{\binom{K}{2}}\right] \sum_{k<k'} \mathcal{A}^{(k,k')}(\rho_I)$$

More precisely, if $\rho_1, \rho_2, \ldots, \rho_K$ denote K permutations on the first n
integers, let

$$\mathcal{C}(\rho_1,\ldots,\rho_K) = \left[\frac{1}{\binom{K}{2}}\right] \sum_{k<k'} \mathcal{A}^{(k,k')}(\rho_k,\rho_{k'})$$

[†]This is in contrast to Section 3.2.3 where the simple sum was used as a
starting point. Because of the product form for the constituent indices, the
use of an average from the beginning makes the construction of an appropriately normalized measure very easy.

where we extend our notation and define

$$\mathcal{A}^{(k,k')}(\rho_k,\rho_{k'}) = \sum_{i,j} p_{\rho_k(i)\rho_k(j)}^{k} \, p_{\rho_{k'}(i)\rho_{k'}(j)}^{k'}$$

Then, in terms of our previous notation

$$\mathcal{A}^{(k,k')}(\rho,\rho_I) = \mathcal{A}^{(k,k')}(\rho)$$

and

$$C = C(\rho_I,\rho_I,\dots,\rho_I)$$

Thus, to assess the hypothesis of no concordance, $C(\rho_1,\dots,\rho_K)$ is first evaluated over all possible permutations $\rho_1, \dots, \rho_K$, generating a distribution of $(n!)^K$ (possible nondistinct) equally likely realizations of the index. If C is sufficiently extreme with respect to this distribution, the hypothesis of no concordance is rejected. As in Section 3.2.3, the first permutation ρ_1 could, without loss of generality, be assumed equal to the identity ρ_I. In effect, the $(n!)^K$ equally likely permutations reduce to $(n!)^{K-1}$ under this assumption; i.e., the distribution of $C(\rho_I,\rho_2,\dots,\rho_K)$ is the same as that of $C(\rho_1,\rho_2,\dots,\rho_K)$.

Several comments regarding the index are in order. First of all, the first two moments of the permutation distribution for $C(\rho_1,\dots,\rho_K)$ are immediate, given the formulas for the first two moments of $\mathcal{A}^{(k,k')}(\rho)$ available in Chapter 4. In particular,

$$E(C(\rho_1,\dots,\rho_K)) = \left[\frac{1}{\binom{K}{2}}\right] \sum_{k<k'} E(\mathcal{A}^{(k,k')}(\rho_k,\rho_{k'}))$$

where

$$E(\mathcal{A}^{(k,k')}(\rho_k,\rho_{k'})) = E(\mathcal{A}^{(k,k')}(\rho))$$

Moreover, because the indices $\mathcal{A}^{(k,k')}(\rho_k,\rho_{k'})$ are uncorrelated in pairs,

$$V(C(\rho_1,\dots,\rho_K)) = \left[\frac{1}{\binom{K}{2}}\right]^2 \sum_{k<k'} V(\mathcal{A}^{(k,k')}(\rho_k,\rho_{k'}))$$

where

$$V(\mathcal{A}^{(k,k')}(\rho_k, \rho_{k'})) = V(\mathcal{A}^{(k,k')}(\rho))$$

In general, however, we rely on an approximate permutation test of $\mathcal{C}$ based on sampling from the complete permutation distribution, rather than on a possibly suspect normal approximation to the distribution of $\mathcal{C}(\rho_1, \ldots, \rho_K)$.

Secondly, and depending on how the matrices $\underset{\sim}{P}_1$, $\underset{\sim}{P}_2$, $\ldots$, $\underset{\sim}{P}_K$ are defined, various special cases of $\mathcal{C}$ can be obtained. The most well-known interpretation assumes that each of K subjects provides a rank ordering of the n objects as in Section 3.2.2. For example, if we denote the ranks assigned by subject k as r_{k1}, $\ldots$, r_{kn}, where r_{kj} is the value attached to object O_j, and let

$$p_{ij}^k = \frac{r_{ki} - r_{kj}}{\left[\sum_{i,j}(r_{ki} - r_{kj})^2\right]^{1/2}} \tag{6.1.1a}$$

then $\mathcal{C}$ is the average Spearman rank order correlation coefficient ρ_b between all pairs of rankings. In the case of untied ranks, a very simple transformation of $\mathcal{C}$ provides Kendall's well-known coefficient of concordance (Kendall, 1970). Alternatively, if proximity is redefined by

$$p_{ij}^k = \frac{\text{sign}(r_{ki} - r_{kj})}{\left[\sum_{i,j}\text{sign}(r_{ki} - r_{kj})^2\right]^{1/2}} \tag{6.1.1b}$$

then $\mathcal{C}$ is the average Kendall coefficient τ_b between all pairs of rankings (Ehrenberg, 1952; Hays, 1960); and finally, if the original observations are used (say, x_{k1}, $\ldots$, x_{kn} for subject k) and

$$p_{ij}^k = \frac{x_{ki} - x_{kj}}{\left[\sum_{i,j}(x_{ki} - x_{kj})^2\right]^{1/2}} \tag{6.1.1c}$$

then $\mathcal{C}$ is the average Pearson product-moment correlation between all pairs of the original sequences.

As a somewhat different application, suppose that each entry in a proximity matrix from A is a zero or a one. In this case, the index $\mathcal{C}$ is merely the average number of 1's that are placed in common positions over all pairs of matrices. Moreover, if the K elements in A represent paired-comparison matrices from each of K subjects, a simple transformation of $\mathcal{C}$ provides Kendall's coefficient of agreement. Alternatively, the K matrices could represent the results of a free-sort of the set of n objects by K subjects, where a matrix entry of 1 would now indicate that two objects were sorted together by the particular subject. In applications of this type in which partitions are provided by a subject rather than a matrix of paired-comparisons, the assumptions of $(n!)^K$ equally likely realizations of $\mathcal{C}(\rho_1,\ldots,\rho_K)$ appears to be a more justifiable "null" hypothesis than the traditional model used for paired comparisons discussed by Kendall (1970). This latter model considers the entries in a matrix to be independent, e.g., 0's and 1's occur with probability 1/2, and thus, the structure of a partition would be destroyed.

In general, some final normalization of $\mathcal{C}$ may be desirable having the $\mathcal{C}^*$ (or $\mathcal{C}^{**}$) form. Because of the product structure of each of the individual QA indices, each matrix entry could be initially normalized as in the rank correlation context, which in effect produces a normalized index directly. A transformation of this latter type is suggested for routine applications because it leads to a particularly simple interpretation for $\mathcal{C}$. Specifically, if each entry is defined as $[n(n-1)]^{-1/2}$ times its "z score," considering the $n(n-1)$ off-diagonal entries in a matrix as a sample, the index $\mathcal{C}$ is the average Pearson product-moment correlation between the $n(n-1)$ pairs of matrix entries taken over all $\binom{K}{2}$ pairs of matrices in A, i.e., an index of the $\mathcal{C}^*$ form. Obviously, for rank orders and if proximity is redefined as $p_{ij}^k = r_{ki} - r_{kj}$ or $p_{ij}^k = \mathrm{sign}(r_{ki} - r_{kj})$, respectively, this procedure generates average ρ_b or τ_b. More importantly, $\mathcal{C}^*$ provides a measure that is appropriate for general proximity matrices that do not have a particularly elementary structure, e.g., matrices containing entries that are not simple functions of the differences derived from rank orders.

6.1.2 A Priori Comparisons (B)

Instead of measuring concordance among all the K matrices, suppose the first matrix in A, $\underset{\sim}{P_1} = \{p_{ij}^1\}$, is unique in some particular way and we wish to relate the remaining K - 1 matrices in A to $\underset{\sim}{P_1}$. Generalizing from what was done in Section 6.1.1, the most obvious index would be the average of the K - 1 intercorrelations between $\underset{\sim}{P_1}$ and $\underset{\sim}{P_2}, \ldots, \underset{\sim}{P_K}$, or if no normalization is required, the average of the K - 1 raw cross-product statistics. More formally, let

$$\mathcal{D}(\rho_I, \rho_2, \ldots, \rho_K) = \left(\frac{1}{K-1}\right) \sum_{2 \leq k} \mathcal{A}^{(1,k)}(\rho_I, \rho_k)$$

where

$$\mathcal{A}^{(1,k)}(\rho_I, \rho_k) = \sum_{i,j} p_{ij}^1 p_{\rho_k(i)\rho_k(j)}^k$$

Again, if an observed index $\mathcal{D}$ is defined as $\mathcal{D}(\rho_I, \ldots, \rho_I)$, then $\mathcal{D}$ can be compared to an approximate permutation distribution based on the evaluation of $\mathcal{D}(\rho_1, \rho_2, \ldots, \rho_K)$ over all its $(n!)^{K-1}$ equally likely realizations (or an approximation based on sampling).

If we define P_1 and P_k through rank differences or the sign function, as in equations (6.1.1a) and (6.1.1b), then $\mathcal{D}$ provides a normalized measure of correspondence, having the $\mathcal{D}^*$ form, between K - 1 observed rank orders and a given rank order using average ρ_b or τ_b, respectively. These procedures are essentially the same as the suggestions of Page (1963), Lyerly (1952), Hays (1960), and Tate (1961) for testing homogeneity in a randomized-blocks context against an alternative of a particular treatment ordering (cf. Section 3.2.5). The use of differences between the original observations, as in equation (6.1.1c), would result in an average Pearson correlation. Obviously, this same measure $\mathcal{D}$ is appropriate more generally as well, e.g., P_1 could represent (i) a hypothesized rank order to be compared against K - 1 paired comparison matrices; (ii) a hypothesized partition as in the free-sort paradigm; (iii) an empirical norm generated from a second group of subjects; and so on. For these latter cases, in particular, a normalized index, $\mathcal{D}^*$, could use z scores for each proximity matrix; this would lead to an average Pearson correlation between P_1 and $P_2, \ldots, P_K$.

Even though Monte Carlo significance testing is the inference strategy suggested, we do note that asymptotic normality of $\mathcal{D}$ is very easy to establish for large K by a straightforward application of a central limit theorem because the index is composed of K - 1 independent random variables. Again, the first two moments of $\mathcal{D}$ are simple averages of the moments available for the comparison of two matrices.

Aggregation. Contrary to common practice, the use of the index $\mathcal{D}$ is generally preferable to the alternative strategy of comparing P_1 to a second (group) matrix $P_{SUM} = \Sigma_{2 \leqslant k} P_k$. The argument is more or less the same as given in Section 3.2.5. For example, in comparing P_1 to P_{SUM}, using the simpler procedure for comparing only two matrices, the corresponding z statistic will typically be smaller than the z statistic obtained by using the index $\mathcal{D}$. The numerators are the same in both instances, but the denominators for the two z statistics differ. More explicitly, $V(\mathcal{D}(\rho_1, \rho_2, \ldots, \rho_K))$ is the sum of the individual variances that would be obtained by separately comparing each of the matrices $P_2, \ldots, P_K$ to P_1 because all covariance terms are now excluded. Covariances of this latter type would have been present if $\underset{\sim}{P}_1$ were compared to $\underset{\sim}{P}_{SUM}$ and are generally expected to be positive if the matrices $\underset{\sim}{P}_2, \ldots, \underset{\sim}{P}_K$ have some of the hypothesized similarity to $\underset{\sim}{P}_1$. In short, a z statistic based on $\mathcal{D}$ will usually be larger than that obtained in the simpler comparison of P_1 to the group matrix P_{SUM}. (Although we rely on Monte Carlo distributions for significance testing rather than on comparing z statistics to a possibly suspect normal approximation, we would still expect that the use of a group matrix would lead to a larger significance level, other things being equal.)

6.1.3 Evaluating a Partition of the Set of Proximity Matrices (B)

Based on an a priori partition of A defined by different groups of subjects, there are several notions of "homogeneity" that could be pursued, i.e., the relationship of matrices within a class to those from different classes. One strategy, motivated by Hays (1960) uses the indices $\mathcal{A}^{(k,k')}(\rho_I)$ or their normalized variants as measures of proximity between pairs of elements chosen from A. If these values are placed within a K × K matrix with its main diagonal set equal to zero, our interest is in whether the matrix entries within classes are larger than the indices between classes. Because the given partition can also be represented by a second K × K matrix (e.g., an off-diagonal entry in the matrix could be set equal to 1 if the corresponding row and column object pair lies within the same class and 0 otherwise), the evaluation of a partition of A is reduced to the comparison of two K × K matrices. This is the type of QA comparison task discussed extensively in Section 4.3.4.

A second rather different problem of homogeneity, which is similar to our discussion in Section 3.2.6, relies on the $(n!)^K$ realizations of the K matrices under a hypothesis of independence. As an example, suppose there are two groups in the given partition and instead of an index of the form C or $\mathcal{D}$, a third measure $\mathcal{E}$ is defined through the following expression:

$$\mathcal{E}(\rho_1, \rho_2, \ldots, \rho_K) = \sum_{k,k' \in F} \sum_{i,j} p_{\rho_k(i)\rho_k(j)}^k p_{\rho_{k'}(i)\rho_{k'}(j)}^{k'}$$

where the set F contains pairs of indices that belong to separate subsets of the two-group partition of A. This proposal is developed in Section 3.2.6 as a way of assessing the concordance of rank orders provided by two separate groups of subjects. In general, a large value for $\mathcal{E}(\rho_I, \ldots, \rho_I)$ would denote that subjects between the classes of a partition are similar in terms of the matrix each provides as well as similar within classes; conversely, a small value for the index would denote a concordance within classes but a lack of concordance across classes. In an analogous manner, a J group partition of $\underset{\sim}{P}_1, \ldots, \underset{\sim}{P}_K$ could be approached as suggested in Section 3.2.6 for rank orders.

6.2 GENERALIZATIONS (C)

The previous discussion has been phrased completely in terms of indices
defined by simple cross-product measures and a pairwise comparison of
the elements in A. There are several obvious generalizations of this
approach that could be pursued. For example, instead of a simple cross-
product index of the form given by $\mathcal{A}^{(k,k')}(\rho)$, the multiplicative restriction
could be removed by defining an index using a four-place function of the
type discussed in Section 5.2. As a second possible extension, the cross-
product multiplicative condition could be maintained but a more complicated
function considered that relies on a comparison of entries within each
proximity matrix, e.g., the higher-order assignment measures of Section
5.3. In general, the principles behind the use of these alternative indices
would be exactly the same as laid down for the multiplicative QA measures.

REFERENCES

Ehrenberg, A. S. C. On sampling from a population of rankers, Bio-
    metrika 39 (1952), 82-87.
Hays, W. L. A note on average tau as a measure of concordance, J. Am.
    Stat. Assoc. 55 (1960), 331-341.
Hubert, L. J. Generalized concordance, Psychometrika 44 (1979), 135-142.
Kendall, M. G. Rank Correlation Methods, 4th ed., Hafner, New York,
    1970.
Lyerly, S. B. The average Spearman rank correlation coefficient, Psycho-
    metrika 17 (1952), 421-428.
Page, E. B. Ordered hypotheses for multiple treatments: A significance
    test for linear ranks, J. Am. Stat. Assoc. 58 (1963), 216-230.
Tate, R. F. On the use of partially ordered observations in measuring
    the support for a complete order, J. Am. Stat. Assoc. 56 (1961), 299-
    313.

7
Assignment Restrictions on the QA Model

7.1 INTRODUCTION (A)[†]

Throughout the first six chapters, we typically compared an observed statistic against a null model constructed from an assumption that all n! permutations of the first n integers were equally likely. Stated in another way, a uniform distribution imposed on the set of such permutations, Ω, led to a reference distribution for the given index. In this chapter, several alternatives to the use of a uniform distribution over Ω are suggested. Although the same descriptive (or normalized) statistics introduced earlier are still used, the null model from which the reference distribution is constructed is generalized in a number of ways.

To provide a context for the variations we discuss, suppose we start with the usual multiplicative QA framework based on the two n × n matrices $\underset{\sim}{P}$ and $\underset{\sim}{Q}$. The observed index is assumed, without loss of generality, to be defined by the identity permutation, ρ_I. The observed index, $\mathcal{A}(\rho_I)$, is compared to the distribution of $\mathcal{A}(\rho)$ over a possibly nonuniform structure defined on the set Ω.

As notation, suppose the integer set $\{1, 2, \ldots, n\}$ is partitioned into K subsets, $R_1, \ldots, R_K$, with sizes $n_1, \ldots, n_K$, where $\Sigma_k n_k = n$, $R_1 = \{1, 2, \ldots, n_1\}$, $R_2 = \{n_1 + 1, \ldots, n_2\}$, $\ldots$, $R_K = \{n_{K-1} + 1, \ldots, n_K\}$. The distributions over Ω we emphasize are constructed in the following way. First,

1. All permutations in Ω that map some element in R_k to an element in $R_{k'}$ where $1 \leq k \neq k' \leq K$ have probability 0. Thus, the only permutations in Ω that could have nonzero probabilities of occurrence are those that map the entries in R_k to R_k, $1 \leq k \leq K$. (Although $R_1, \ldots, R_K$ are not

---

[†]The reader is referred to the Preface for an explanation of the use of (A), (B), and (C) in text headings.

necessarily the same size, we are still "blocking" on these K subsets in the same way that is familiar to an analysis-of-variance context.)

2. Among all permutations defined in (1), we assume that each of the integer sets $R_1, \ldots, R_K$ is either fixed or random. If a set, say R_k, is to be considered fixed, then each permutation in Ω with nonzero probability must map all elements in R_k to exactly the same elements in R_k.

Once the fixed sets are identified and subject to condition (1), all permutations have the same probability of occurrence. In other words, after the permutations from Ω are eliminated that either map an element in R_k to $R_{k'}$ for $k \neq k'$ or violate a given restriction on fixed subsets, a uniform distribution is imposed over the remaining permutations. Thus, depending on what subsets are considered fixed, there are $2^K - 1$ nontrivial possible distributions over Ω that this structure defines (we eliminate the case in which all subsets are fixed because this trivial case produces only a single identity permutation ρ_I, which is given a probability of 1).

As might be expected, the moments of $\mathcal{A}(\rho)$ can be rather complicated. A way of finding the expectation is given below to help generate a normalized descriptive index when appropriate, but in general, Monte Carlo significance testing will be relied on as an inference strategy. To obtain $E(\mathcal{A}(\rho))$, we first note that

$$E(\mathcal{A}(\rho)) = E\left(\sum_k \sum_{i \in R_k} \sum_{k'} \sum_{j \in R_{k'}} p_{\rho(i)\rho(j)} q_{ij}\right)$$

$$= \sum_k \sum_{k'} E\left(\sum_{i \in R_k} \sum_{j \in R_{k'}} p_{\rho(i)\rho(j)} q_{ij}\right)$$

Depending on whether the sets R_k and $R_{k'}$ are (1) fixed or random, and (2) $k = k'$ or $k \neq k'$, the expectation,

$$E\left(\sum_{i \in R_k} \sum_{j \in R_{k'}} p_{\rho(i)\rho(j)} q_{ij}\right) \quad \text{differs:}$$

a. If $k = k'$, then $R_k = R_{k'}$. For fixed R_k, we obtain:

$$\sum_{i \in R_k} \sum_{j \in R_k} p_{ij} q_{ij}$$

and for random R_k:

$$\left[\frac{1}{n_k(n_k-1)}\right]\left(\sum_{j \in R_k}\sum_{j \in R_k}p_{ij}\right)\left(\sum_{i \in R_k}\sum_{j \in R_k}q_{ij}\right)$$

b. Suppose $k \neq k'$ with R_k fixed and $R_{k'}$ random, we obtain

$$\left(\frac{1}{n_k}\right)\sum_{i \in R_k}\left(\sum_{j \in R_{k'}}p_{ij}\right)\left(\sum_{j \in R_k}q_{ij}\right)$$

and

$$\left(\frac{1}{n_k}\right)\sum_{j \in R_{k'}}\left(\sum_{i \in R_k}p_{ij}\right)\left(\sum_{i \in R_k}q_{ij}\right)$$

when R_k is random and $R_{k'}$ is fixed. If R_k and $R_{k'}$ are both fixed, we have:

$$\sum_{i \in R_k}\sum_{j \in R_{k'}}p_{ij}q_{ij}$$

and when R_k and $R_{k'}$ are both random:

$$\left(\frac{1}{n_k n_{k'}}\right)\left(\sum_{i \in R_k}\sum_{j \in R_{k'}}p_{ij}\right)\left(\sum_{i \in R_k}\sum_{j \in R_{k'}}q_{ij}\right)$$

Certain of the $2^K - 1$ possible structures are of particular interest, and these are discussed in greater detail in the section that follows.

7.1.1 Some Special Cases (B)

The approach we have just introduced is very comprehensive and includes almost all of the work of the earlier chapters as special cases. Suppose we start with the LA context of Part 1 by letting $K = 2$ and assume that the two subsets R_1 and R_2 are of the same size, i.e., $R_1 = \{1, 2, \ldots, m\}$ and $R_2 = \{m+1, \ldots, 2m\}$, where $n = 2m$. The matrices $\underset{\sim}{P}$ and $\underset{\sim}{Q}$ are partitioned as follows:

$$
\underset{\sim}{P} =
\begin{array}{c}
\\
\begin{array}{c} R_1 \\ \left\{ \begin{array}{c} 1 \\ 2 \\ \vdots \\ m \end{array} \right. \end{array} \\
\begin{array}{c} R_2 \\ \left\{ \begin{array}{c} m+1 \\ \vdots \\ 2m \end{array} \right. \end{array}
\end{array}
\overbrace{\begin{array}{ccc} 1 & 2 & \cdots \; m \end{array}}^{R_1} \; \overbrace{\begin{array}{cc} m+1 & \cdots \; 2m \end{array}}^{R_2}
\left[
\begin{array}{c|c}
\underset{\sim}{0} & \underset{\sim}{C} \\
\hline
\underset{\sim}{0} & \underset{\sim}{0}
\end{array}
\right]
$$

$$
\underset{\sim}{Q} =
\begin{array}{c}
\begin{array}{c} R_1 \\ \left\{ \begin{array}{c} 1 \\ \vdots \\ m \end{array} \right. \end{array} \\
\begin{array}{c} R_2 \\ \left\{ \begin{array}{c} m+1 \\ \vdots \\ 2m \end{array} \right. \end{array}
\end{array}
\overbrace{\begin{array}{ccc} 1 & \cdots & m \end{array}}^{R_1} \; \overbrace{\begin{array}{cc} m+1 & \cdots \; 2m \end{array}}^{R_2}
\left[
\begin{array}{c|c}
\underset{\sim}{0} & \begin{array}{ccc} 1 & & \underset{\sim}{0} \\ & \cdot & \\ & \cdot & \\ \underset{\sim}{0} & & 1 \end{array} \\
\hline
\underset{\sim}{0} & \underset{\sim}{0}
\end{array}
\right]
$$

where $\underset{\sim}{C}$ is of size $m \times m$ and corresponds to the assignment score matrix used to obtain the LA index of Chapter 1. In other words, if R_1 is considered fixed and R_2 random, then the QA index $\mathcal{A}(\rho)$ is actually an LA index of the form $\mathcal{A}(\rho)$ based on $\underset{\sim}{C}$ as the matrix of assignment scores. Incomplete selection as discussed in Section 2.3.5 could be handled by merely restricting the number of 1's in Q.

Continuing in this manner, we can also encompass the material in Chapter 2 on comparing two rectangular matrices of the same size. Here,

K is again 2, but the sets R_1 and R_2 now have sizes n_1 and n_2: $R_1 = \{1, \ldots, n_1\}$ and $R_2 = \{n_1 + 1, \ldots, n_1 + n_2\}$, where $n_1 + n_2 = n$. The matrices $\underset{\sim}{P}$ and $\underset{\sim}{Q}$ are partitioned as

$$
\underset{\sim}{P} =
\begin{array}{c}
\\
R_1 \left\{ \begin{array}{c} 1 \\ \vdots \\ n_1 \end{array} \right. \\
R_2 \left\{ \begin{array}{c} n_1 + 1 \\ \vdots \\ n_1 + n_2 \end{array} \right.
\end{array}
\left[
\begin{array}{c|c}
\underset{\sim}{0} & \underset{\sim}{C} \\
\hline
\underset{\sim}{0} & \underset{\sim}{0}
\end{array}
\right]
$$

$$
\underset{\sim}{Q} =
\begin{array}{c}
\\
R_1 \left\{ \begin{array}{c} 1 \\ \vdots \\ n_1 \end{array} \right. \\
R_2 \left\{ \begin{array}{c} n_1 + 1 \\ \vdots \\ n_1 + n_2 \end{array} \right.
\end{array}
\left[
\begin{array}{c|c}
\underset{\sim}{0} & \underset{\sim}{B} \\
\hline
\underset{\sim}{0} & \underset{\sim}{0}
\end{array}
\right]
$$

If R_1 is considered fixed and R_2 random, then a comparison of $\underset{\sim}{C}$ and $\underset{\sim}{B}$ is obtained through the index $\Lambda(\rho)$ that matches across columns. Conversely, if R_1 is random and R_2 fixed, then a matching across rows is effected through an index of the form $\Lambda(\phi)$. When R_1 and R_2 are both random, the inference

model of Section 2.4 based on the index $\Lambda(\phi,\rho)$ and row and column matching is defined.

The extensions of the LA model discussed in Chapter 3 that involve
the pairwise decomposition of K-place assignment scores can be included
within our more general structure as well. For instance, the notion of
concordance would involve K subsets, $R_1, \ldots, R_K$, where each is of the
same size and considered random (or equivalently, R_1 could be assumed
fixed and only $R_2, \ldots, R_K$ random):

$$R_1 = \{1,\ldots,m\}, \quad R_2 = \{m+1,\ldots,2m\}, \quad R_3 = \{2m+1,\ldots,3m\}, \quad \ldots,$$

$$R_K = \{(K-1)m+1,\ldots,Km\}, \quad \text{for } Km = n$$

The matrices $\underset{\sim}{P}$ and $\underset{\sim}{Q}$ would have the form

$$\tag{7.1.1a}$$

$$
\underset{\sim}{Q} =
\begin{array}{c}
\left.\begin{array}{c} \\ R_1 \\ \\ \end{array}\right\{
\left.\begin{array}{c} \\ R_2 \\ \\ \end{array}\right\{
\\
\left(\begin{array}{c} \\ R_K \\ \\ \end{array}\right.
\end{array}
\begin{array}{c}
\overbrace{}^{R_1}\ \overbrace{}^{R_2}\ \cdots\ \overbrace{}^{R_K}
\end{array}
$$

$$
\begin{array}{ccc}
\begin{pmatrix} 1 & & \underset{\sim}{0} \\ & \ddots & \\ \underset{\sim}{0} & & 1 \end{pmatrix} &
\cdots &
\begin{pmatrix} 1 & & \underset{\sim}{0} \\ & \ddots & \\ \underset{\sim}{0} & & 1 \end{pmatrix} \\
\underset{\sim}{0} & \cdots & \begin{pmatrix} 1 & & \underset{\sim}{0} \\ & \ddots & \\ \underset{\sim}{0} & & 1 \end{pmatrix} \\
\vdots & & \vdots \\
\underset{\sim}{0} & \cdots & \underset{\sim}{0}
\end{array}
$$

(Figure: matrix $\underset{\sim}{Q}$ partitioned into blocks indexed by columns R_1 ($1\cdots m$), R_2 ($m+1\cdots 2m$), ..., R_K ($(K-1)m+1\cdots Km$) and rows R_1 ($1\cdots m$), R_2 ($m+1\cdots 2m$), ..., R_K ($(K-1)m+1\cdots Km$).)

For example, in measuring concordance among K sequences, the submatrix $\underset{\sim}{C}_{kk'}$ in $\underset{\sim}{P}$ would be defined as the m × m (symmetric) assignment matrix that relates the m entries of the kth and the k'th sequences.

The inference model based on $\underset{\sim}{P}$ and $\underset{\sim}{Q}$ and the explicit structure given in (7.1.1a) are discussed in some detail by Klauber (1975). When the proximities in $\underset{\sim}{P}$ are symmetric, this latter reference gives a general formula for the variance of $\mathcal{A}(\rho)$ under any of the $2^K - 1$ different variations that could be constructed from the fixed/random option for each of the K subsets $R_1, \ldots, R_K$.

Modifying the pattern for $\underset{\sim}{Q}$ in (7.1.1a) to test for an a priori order in K sequences is fairly easy. Assuming that the first sequence (corresponding to R_1) is fixed, the matrix $\underset{\sim}{Q}$ would take the form

$$
\underset{\sim}{Q} =
\begin{array}{c}
\\
R_1\\
\\
R_2\\
\\
R_K
\end{array}
\left[
\begin{array}{c|c|c|c}
& 1 \cdots m & m+1 \cdots 2m & \cdots & (K-1)m+1 \cdots K_m \\
\end{array}
\right]
$$

The matrix $\underset{\sim}{Q}$ is partitioned with column groups R_1 (columns $1 \cdots m$), R_2 (columns $m+1 \cdots 2m$), ..., R_K (columns $(K-1)m+1 \cdots K_m$), and row groups R_1 (rows $1 \cdots m$), R_2 (rows $m+1 \cdots 2m$), ..., R_K (rows $(K-1)m+1 \cdots Km$).

The blocks contain: the R_1–R_1 block is $\underset{\sim}{0}$; the upper-triangular off-diagonal blocks (R_1 row group with $R_2, \dots, R_K$ column groups) are diagonal-strip identity matrices with 1's on the diagonal and $\underset{\sim}{0}$ off-diagonal; all lower blocks are $\underset{\sim}{0}$.

In the multigroup concordance framework, defined by some partition of the
K sequences, the appropriate definition of $\underset{\sim}{Q}$ would have a "diagonal strip"
of 1's in the upper-triangular portion of the matrix whenever that particular
assignment matrix is defined for two sequences in different subsets of the
partition. Finally, in the two-dependent sample context based on an additive
index, each matrix $\underset{\sim}{C}_{kk'}$, $k \neq k'$ would be of size 4×4; a single one would
be placed in the first row and column in each of the corresponding sub-
matrices in the upper-triangular portion of $\underset{\sim}{Q}$.

It should be obvious that the multiplicative QA model of Chapter 4 is
itself included trivially in this more general inference structure. Here,
only one set $R_1 = \{1, \dots, n\}$ is defined, which is assumed random. In fact,
it should also be apparent that we could move away from the simple multiplicative structure for defining a measure of correspondence between $\underset{\sim}{P}$ and
$\underset{\sim}{Q}$, e.g., to higher-order assignment measures (Section 5.3) to four-place
indices (Section 5.2), or to bottleneck measures (Section 5.4). Even though
the index we use may change, the inference structure based on the decomposition $R_1, \dots, R_K$ and the need for some particular specification of which
sets are assumed fixed or random would remain exactly the same as before.

Finally, we might note that the approach to multiple proximity matrices in Chapter 6 could also be included in our general framework. Unfortunately, for assessing concordance or a partition of the set of proximity matrices $A = \{P_1, \ldots, P_K\}$, we would need very complicated indices of correspondence between our two matrices P and Q that would take the same form

$$
\begin{array}{c}
\begin{array}{cccc} R_1 & R_2 & \cdots & R_K \end{array} \\
\begin{array}{c} R_1 \\ R_2 \\ \vdots \\ R_K \end{array}
\left[
\begin{array}{c|c|c|c}
P_1 & 0 & \cdots & 0 \\
\hline
0 & P_2 & \cdots & 0 \\
\hline
\vdots & \vdots & & \vdots \\
\hline
0 & 0 & \cdots & P_K
\end{array}
\right]
\end{array}
$$

e.g., we would need to relate the submatrices $P_1, \ldots, P_K$ in P to these same submatrices in Q. Because of these difficulties, we will not pursue the topic any further. However, for relating a fixed matrix, P_1, to the remainder, $P_2, \ldots, P_K$, the appropriate structure for P and Q would be straightforward:

$$
P =
\begin{array}{c}
\begin{array}{cccc} R_1 & R_2 & \cdots & R_K \end{array} \\
\begin{array}{c} R_1 \\ R_2 \\ \vdots \\ R_K \end{array}
\left[
\begin{array}{c|c|c|c}
P_1 & 0 & \cdots & 0 \\
\hline
0 & P_1 & \cdots & 0 \\
\hline
\vdots & \vdots & & \vdots \\
\hline
0 & 0 & \cdots & P_1
\end{array}
\right]
\end{array}
$$

$$
\underset{\sim}{Q} =
\begin{array}{c}
R_1 \left\{ \right. \\
R_2 \left\{ \right. \\
\vdots \\
R_K \left\{ \right.
\end{array}
\left[
\begin{array}{c c c c}
\underset{\sim}{0} & \underset{\sim}{0} & \cdots & \underset{\sim}{0} \\
\underset{\sim}{0} & \underset{\sim}{P}_2 & \cdots & \underset{\sim}{0} \\
\vdots & \vdots & \vdots & \vdots \\
\underset{\sim}{0} & \underset{\sim}{0} & \cdots & \underset{\sim}{P}_K
\end{array}
\right]
$$

Thus, at least at a theoretical level this latter comparison can be handled fairly easily.

Some Particular Applications. The special cases discussed above all illustrate ways in which the restriction of assignment can be used to incorporate several of the general analysis paradigms discussed in the first six chapters. There are also a few specific applications that may be of interest to mention, if only in passing, which indicate some of the great flexibility inherent in these generalizations. For instance:

1. Using the cluster statistic notion of Section 4.3.3, we can evaluate the effect of adding a certain number of objects to a fixed subset.
2. Confirmatory comparisons of a posited seriation can be made when certain "anchor" objects are already placed in fixed positions in the sequence.
3. Partitions can be evaluated in which some or all of the classes contain specific objects.
4. If the basic object set $\{O_1, \ldots, O_n\}$ consists of two distinct types of entries (e.g., people and stimuli), we can fix one set and evaluate the pattern of proximities that refer to the second. This latter application relates directly to the Coombsian ideal-point model in which people and stimuli both share the same "space."

There is one application, discussed in detail by Pike and Smith (1974), that should be mentioned in a little more detail. Here, the matrix $\underset{\sim}{P}$ takes the general form

$$
\underset{\sim}{P} =
\begin{array}{c}
\begin{array}{cccc}
\quad R_1 \quad & \quad R_2 \quad & \cdots & \quad R_K \quad
\end{array} \\[4pt]
\begin{array}{c}
R_1 \\[14pt]
R_2 \\[14pt]
\vdots \\[14pt]
R_K
\end{array}
\left[
\begin{array}{c|c|c|c}
\underset{\sim}{P}_{11} & \underset{\sim}{P}_{12} & \cdots & \underset{\sim}{P}_{1K} \\[6pt]
\hline
\underset{\sim}{P}_{21} & \underset{\sim}{P}_{22} & \cdots & \underset{\sim}{P}_{2K} \\[6pt]
\hline
\vdots & \vdots & & \vdots \\[6pt]
\hline
\underset{\sim}{P}_{K1} & \underset{\sim}{P}_{K2} & \cdots & \underset{\sim}{P}_{KK}
\end{array}
\right]
\end{array}
$$

and

$$
\underset{\sim}{Q} =
\begin{array}{c}
\begin{array}{cccc}
\quad R_1 \quad & \quad R_2 \quad & \cdots & \quad R_K \quad
\end{array} \\[4pt]
\begin{array}{c}
R_1 \\[14pt]
R_2 \\[14pt]
\vdots \\[14pt]
R_K
\end{array}
\left[
\begin{array}{c|c|c|c}
\underset{\sim}{Q}_{11} & \underset{\sim}{Q}_{12} & \cdots & \underset{\sim}{Q}_{1K} \\[6pt]
\hline
\underset{\sim}{Q}_{21} & \underset{\sim}{Q}_{22} & \cdots & \underset{\sim}{Q}_{2K} \\[6pt]
\hline
\vdots & \vdots & & \vdots \\[6pt]
\hline
\underset{\sim}{Q}_{K1} & \underset{\sim}{Q}_{K2} & \cdots & \underset{\sim}{Q}_{KK}
\end{array}
\right]
\end{array}
$$

where

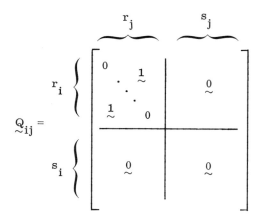

and $\underset{\sim}{1}$ denotes an upper or lower triangular matrix of all 1's. Each subset R_k consists of n_k subjects of which r_k are "patients" and s_k are "controls," $n_k = r_k + s_k$. The proximity measures in $\underset{\sim}{P}$ are indicators of subject contact; thus, the subset structure of each submatrix $\underset{\sim}{Q}_{ij}$ merely sums the patient-to-patient contact measures between stratum i and stratum j. The inference model based on all sets $R_1, \ldots, R_K$ being random induces a random selection of r_k subjects from the pool of n_k to serve as patients. The index $\mathcal{A}(\rho)$ is then the sum of all patient-to-patient contacts between the randomly selected subsets. For this special structure, Pike and Smith (1974) have given explicit formulas for the mean and variance of $\mathcal{A}(\rho)$, along with several numerical examples.

REFERENCES

Klauber, M. R. Space-time clustering for more than two samples, Biometrics 31 (1975), 719-726.

Pike, M. C., and P. G. Smith. A case-control approach to examine diseases for evidence of contagion, including diseases with long latent periods, Biometrics 30 (1974), 263-279.

Index

DATE DUE

	261-2500		Printed in USA